Molecular Genetic Information Systems
Modelling and Simulation

Molecular Genetic Information Systems

Modelling and Simulation

edited by

Klaus Bellmann

with Contributions by

K. Bellmann, R. Böttner, J. Born,
T. Cierzynski, A. Knijnenburg, U. Kreischer,
R. Lindigkeit, H. Neumann, V. A. Ratner,
R. Rosen, R. Schulz, R. N. Tschuraev

115 Figures and 23 Tables

AKADEMIE-VERLAG · BERLIN

1983

Erschienen im Akademie-Verlag, DDR-1086 Berlin, Leipziger Straße 3—4
Lektor: Christiane Grunow
Einband und Schutzumschlag: Karl Salzbrunn

Lizenznummer: 202 · 100/497/82
Gesamtherstellung: VEB Druckhaus „Maxim Gorki“, 7400 Altenburg
Bestellnummer: 763 010 6 (6647) · LSV 1325
Printed in GDR
DDR 50,— M

Preface

In this volume the reader will find contributions elaborated by our interdisciplinary team in the seventies at a border line between molecular biology, systems theory and informatics, and that are inspired by the new cybernetic wag of thinking.

From the point of view of both cybernetics/informatics and molecular biology our interest was pointed towards the analysis of self-organizing systems. These open systems which take up energy and negentropy control themselves in the process of becoming organized. Any organism needs a sufficient amount of internal controling information to develop such an organized structure. This information, available in the developmental process, is mainly stored in the genetic memory.

Therefore, all of us were interested in processes and mechanisms of control of genetic information release and processing and especially in cases which at least to some extend are accessible experimentally. Microorganisms are one example. They serve as model objects which become suitable structured and organized during evolution. These suitable structures are the basis of control of genetic information release and processing for a broad clase of systems. We all considered the task as a cybernetical problem envisaging the aim as a search for facts which characterize these suitable structures and functions using the technique of modelling and simulation. Furthermore, the aim was to derive some hypotheses that might contribute to developing a theory of genetic control systems. First of all, in our collaboration, we systematized the knowledge previously existing, and in doing this we found a common language that served in setting up problems and aims. We recall with pleasure the creative discussions, rich in ideas, with Dr. LINDIGKEIT, Akad. Dr. SINAIDA and Dr. HANS-ALFRED ROSENTHAL, Dr. G. RICHTER (Berlin), and the deep and impressive talks with Akad. A. A. LJAPUNOV and Dr. V. A. RATNER (Novosibirsk). Moreover, we would like to point out that our common modelling and simulation efforts were supported by the work of Dr. ROSEN (Halifax) and Dr. R. TSANEV (Sofia) and by helpful discussions with them.

Now, as a result of our work some algorithms and simulation systems are available serving as tools for simulation experiments which should provide insights suitable for deductions of more general properties of the systems considered. First results in this direction may be found in the contributions of A. KNIJNENBURG/U. KREISCHER, H. NEUMANN/U. KREISCHER, R. BÖTTNER/K. BELLMANN/R. N. TSCHURAEV/V. A. RATNER (structure and function of DNA, RNA, genetic networks).

The reader should become aware of these tools, now available for treating corresponding research problems. All algorithms are implemented on BSM6-FORTRAN.

It should be mentioned that six of us, A. KNIJNENBURG, H. NEUMANN (systems molecular geneticists), R. BÖTTNER, R. N. TCHURAEV (physicists), U. KREISCHER (computer scientist), J. BORN (mathematician), received their Ph. D. for research work connected with their contribution to this volume.

KLAUS BELLMANN

List of Contributors

BELLMANN, K., Dr. habil.

Central Institute of Cybernetics and Information Processes, Acad. Sci. GDR., DDR-1086 Berlin, Kurstraße 33

BÖTTNER, R., Dr.

Central Institute of Cybernetics and Information Processes, Acad. Sci. GDR., DDR-1086 Berlin, Kurstraße 33

BORN, J., Dr.

Central Institute of Cybernetics and Information Processes, Acad. Sci. GDR., DDR-1086 Berlin, Kurstraße 33

CIERZYNSKI, T., Dipl.-Ing.

Institute of Mathematics, Acad. Sci. GDR., DDR-1080 Berlin, Mohrenstrasse 39

KNIJNENBURG, A., Dr.

Central Institute of Cybernetics and Information Processes, Acad. Sci. GDR., DDR-1086 Berlin, Kurstraße 33

KREISCHER, U., Dr.

Centre of Computing Technique, Acad. Sci. GDR DDR-1199 Berlin, Rudower Chaussee 5

LINDIGKEIT, R., Prof. Dr. habil.

Central Institute of Molecular Biology, Acad. Sci. GDR. DDR-1115 Berlin, Lindenberger Weg 70

NEUMANN, H., Dr.[1]

Central Institute of Molecular Biology, Acad. Sci. GDR. DDR-1115 Berlin, Lindenberger Weg 70

RATNER, V. A., Prof. Dr. sc.

Institute of Cytology and Genetics, Siberian Branch of Acad. Sci. USSR, 630090 Novosibirsk USSR.

---

[1] present address: Institute of Plant Breeding, Acad. Agric. Sci. GDR., DDR-2601 Gülzow

Rosen, Robert, Prof. Dr. sc.

Department of Physiology and Biophysics, Dalhousie Univ. Halifax, Nova Scotia, Canada B3H 447

Schulz, R., Dr.

Central Institute of Mathematics and Mechanics, Acad. Sci. GDR., DDR-1199 Berlin, Rudower Chaussee 5

Tchuraev, R. N., Dr.

Institute of Cytology and Genetics, Siberian Branch of Acad. Sci. USSR, 630090 Novosibirsk, USSR.

Contents

Introduction

K. Bellmann

Understanding genetic information processing in biological selforganisation needs detailed knowledge about subprocesses and laws of their composition to functional units.

A comprehensive representation of the experimental knowledge about construction and behaviour of molecular genetic control systems and of imaginable general functional properties of them was given by Ratner (1977). In this volume results obtained mainly in procaryotic objects are collected, analyzed, synthetized and often formalized focusing on aspects of systems theory, control, and information.

An analysis of the complex control mechanisms that govern the development is only possible by decomposition of the whole system into simpler interacting subsystems, and subsequent analysis of their interactions and dynamics.

R. Rosen has devoted a special volume (1972) to the specification of such simpler units, to the approaches by which such units can be studied as systems in their own sight and to a determination of how the properties of the units are reflected in higher-level organization constructed by them.

A general representation of this problem outlined on a high abstraction level is given by R. Rosen (1972) in this volume.

Global models and concepts of epigenetic control were elaborated from a macroscopical point of view by Waddington, (1970, 1957), Thom (1970), Martinez (1972), and Holtzer (1963). The work of Harris (1968), Buttough (1967), Rosenthal/Grob (1973), Strehler et al. (1973), Jacob/Monod (1963), Britten/Davidson (1969, 1971), Nover (1973), Luckner/Nover (1973), Wolpert (1970) and Molik (1978) has contributed to an elucidation of what differentiation is and how it is to formalize in an appropriate comprehensive framework. The development (self organisation differentiation) of the growing biological system in a given environment is based (a) on the processes of generating the sets of genetic signals from the genetic storage to metabolites carrying information which is needed in different stages of development (b) on modifications of these signals, and (c) on their effects on metabolic control.

The basic structure of gene expression consists in the following steps:

I. Selection of information to be released from the storage at a given time interval, depending on the concentration of different molecule species.

II. Generating a signal sequence by coding the information of a given storage unit (template site) on a primary physical signal carrier (transcription, RNA carrier),

III. RNA-signal processing (as post-transcriptional channel processes): inactivation, reactivation, decay of signals, cutting of signal sequences in single signals.

IV. Recoding of the information of the single RNA-signals on a secondary carrier (translation, protein carrier, i.e. enzyme) addressed to a specific metabolite species (key-lock-principle).

V. Protein signal processing (as posttranslational processes): inactivation reactivation, decay.

Theoretical analysis of control of genetic information processes was done up to now by methods of continuous and methods of discrete mathematics. In both cases the time depending state of the system is described by a vector of concentrations of different molecule species; a special subset are different genetic signal molecules (activated and inactivated RNA and enzymes). The very different approaches for various aspects and abstraction levels are represented as continuous approaches, as ordinary differential equations (RASHEWSKY (1940, 1961), TURING (1952), KASCER (1957, 1963), GOODWIN (1963), JACOB, MONOD (1961), TSANEV, SENDOV (1971), KNORRE (1969, 1968), KNORRE/BÖTTER (1971)) or as discrete approaches, as a single abstract automaton or a net of them (SUGITA (1961, 1963), EDEN (1958), ULAM (1962), KAUFFMAN (1969), TCHURAEV, RATNER (1972), STAHL (1965, 1967), VON NEUMANN (1966), APTER (1966), ARBIB (1966, 1972), LINDENMEYER (1971), MERZENICH (1974), MOLIK (1978)).

In these contributions different abstraction levels and some of the processes I.—V. were considered. An intermediate model type (differential equations with discrete variable parameters, GILDEMAN, KUDRINA, POLETAEV (1970), was utilized in modelling genetic information processing by TCHURAEV, RATNER (1972) (contribution in this volume).

As an illuminating example for modelling cellular differentiation taking into account feedbacks from V. to I. in connection with somatic growth processes, the work of SENDOV, TSANEV (1974) may be mentioned. This model was elaborated in order to demonstrate the effect of control of genetic information release from the genetic storage on cell differentiation and is represented as an algorithmic simulation model of an eucaryotic as an algorithmic simulation model of an eucaryotic cell including a net of interconnected control circuits. Simulation experiments led to the insight that the known mechanisms of genetic information processing can indeed account for the basic phenomena characterizing cell differentiation; namely, the emergence of new cellular types through successive cellular divisions. Since in the research activities mentioned till now the full set of processes outlined in I.—V. was not completely investigated, we tried to do this. In the contribution of BÖTTNER, BELLMANN, CIERZYNSKI in this volume an attempt is made to elaborate a non linear discrete simulation model mapping the whole spectrum of the process of genetic information processing. In this model stochastic features of the processes signal generation (transcription), RNA-signal decay, signal receding (translation), protein signal decay are included in order to obtain information about global reliability of the whole system in the presence of noisy subprocesses.

In order to investigate dynamical properties of the developemental process of a well investigated object with sufficient complexity the models of BÖTTNER, BELLMANN, CIERZYNSKI and RATNER, TCHURAEW were applied to the DNA λ plage system (see contribution of BÖTTNER, BELLMANN, TCHURAEW, RATNER in this volume) with special emphasis to stability (reliability) of the system against external and internal stochastic influences.

Undisputedly, the mathematical character of the model to be constructed depends on the objectives for the application of the model, i.e. the nature of the model should suit the nature of the problem. In applying well-known continuous (differential equations) and discrete methods (abstract automata) to epigenetic control nets (mostly in a very abstract form), the aim was to analyse the dynamical properties of the system considered, usually in dynamical models in biochemistry (kinetic models). But, instead of considering phenomena from the exterior as before (e.g. by rate equations), an attempt was made to reconstruct them from within (bottom up technique).

Consequently, if a more detailed analysis of the genetic signal processes is envisaged, another type of models is needed because necessarily restrictions arising from the mainly utilized mathematical methods are to be taken into account, so that the detailed mechanisms of information release from the storage and its control by special features of the genetic storage had to be omitted in modelling. Now, in order to understand the dependency of the kinetics of the different signal species and the intergrative behaviour of linked genetic information subprocesses on the construction and organisation of information storage and information release from this storage (template) it is necessary to include into the model structural properties of the template. These properties may be nbr. and arrangement of promotor, operator and terminator sites when investigating possible (on principle) organisation forms of procaryotic transcription units, (see contributions of NEUMANN, KREISCHER in this volume). In special cases where eucaryotic transcription processes influenced by the state of the cell milieu (e.g. by application of ammonium sulfate) are investigated, time depending conformation of the template caused by salt ions and its consequences in transcription regulation is to be mapped in a suitable manner (see contribution of BELLMANN, BORN, LINDIGKEIT, SCHULZ in this volume). To recognize the importance of the organisational and construction properties of the genetic storage for the qualitative properties of the developmental and propagating process of an elementary genetic object, the genetic storage of a $Q\beta$ like RNA phage and the linked replication and translation processes controlled by it were described on a very molecular level (see KNIJNENBURG, KREISCHER in this volume).

The general line of this work of our team (and some results) was sketched some time ago (BELLMANN, BÖTTNER, KNIJNENBURG, NEUMANN 1974; BELLMANN, 1977).

In contrast to the previous investigations where rates of the processes are utilized, in the models of NEUMANN, KREISCHER, KNIJNENBURG, BELLMANN, LINDIGKEIT signal generation (transcription) and signal recording (translation) was mapped by simulating polymerase, ribosome, and replicase molecules moving along the DNA and RNA template resp., they move unrestrictedly along "open" regions and are temporarily or permanently stopped at special template sites that serve in regulation of transcription. A similar approach was published by SINGH (1969).

Obviously, models with such properties can not be elaborated classically. We utilized systems of non-linear (sometimes stochastic), liked difference equations with time lags mapping the constructional and organisational features of the genetic information system. These linked algebraic and difference equations are to be solved recursively on event orientated discrete time parameter. These equation systems were elaborated as modular models. They proved to be a suitable basis for appropriate simulation systems with high level service for molecular geneticists. An example for this is the simulation system SIMTRA of KREISCHER, NEUMANN (in this volume).

By means of such models regularities in genetic information processing in nets of transcription systems, as e.g. qualitative behaviour of interacting storage units (transcription), reliability of the whole process under noisy subprocesses, developemental patterns depending on the organization of information storage, may be recognized.

Because dynamical phenomena of procaryotic transcription processes are far better known then such relatively "dark" areas as the eucaryotic transcription process, we were mainly concerned with procaryotic systems. When bacteriophages are involved, the model structure is almost completely determined by experimental measurements and consequently the model is an ideal tool for gaining insight and for instruction purposes.

In treating such problems one has to go through the wellknown stages of applying mathematical methods in modelling:

1. Definition of the real object and its mechanisms as a system and determination (as strict as possible) of the (measurable) parameters to be taken into account.
2. Formulation of the model by carefully specifying the hypotheses assumed and experimental data available. The architecture of the model should be the best reflection of all known facts.
3. Mathematical formulation of the subprocesses and their interactions leading to a simulation algorithm that may include non-linearities, time lags, binary decisions and other unfavourable mathematical features.
4. Validation of the simulation model:
   - qualitative approach to the experimental data, falsification of basic hypotheses, of function types of subprocesses and of feedback and feedforeward relations
   - solution of the inverse problem (parameter adaptation), which is defined here as search for the best numerical values of the unknown (or badly estimable) parameters of the model that account best for a given set of experimental data by permanently comparing the output of the simulation system with the output data of the real object.

In doing this, it is often necessary to adapt already known search methods, and even sometimes to develop new methods suitable to the problem to be solved. In the case of parameter adaptation of high dimensional complex simulation models without favourable mathematical properties (as e.g. linearity, unimodality) stochastic search algorithms have been proved very useful. A globally effective method is desribed in this volume by Born, Bellmann, applied to an adaptation problem outlined in the contribution of Bellmann, Born, Lindigkeit, Schulz (in this volume).

When modelling and simulation of such processes is carried out, this work will be found to provide an irreplaceable tool in research:

it demonstrates the incompatibility of certain hypotheses and it reveals our ignorance.
it enables the interpretation of experimental data to be based no longer on few morphological accidents but on their underlying causes and consequently it opens the vista to human control of genetic information processes.

Further on, it should be underlined, that in the course of modelling and simulation studies with the model (sensitivity analysis, recognizing of key processes, parameter adaptation, investigation of qualitative behaviour of linked subsystems) we learned that such a type of models supports experimental research because these models may be utilized to help classifying the experimental results and rearranging hypotheses previously existing and in proposing special crucial experiments.

Going this way in molecular genetic research means being on the road to an important area of quantitative biology.

References

Arbib, M. A. (1966): A simple self-reproducing universal automaton. Inform. Control **9**, 177–189

Arbib, M. A. (1972): Automate theory in the context of theoretical embryology. In: Biological Organization at the Supercellular Level (ed. R. J. C. Harris), Academic Press, New York

Apter, M. J. (1966): Cybernetics and Development. Pergamon Press, Oxford

Bellmann, K. (1977): Mathematisch-kybernetische Simulationsmodelle der Regulation der Genexpression. In: V. A. Ratner, Molekulargenetische Steuerungssysteme (ed. K. Bellmann), VEB Fischer Verlag Jena and Gustav Fischer Verlag: Stuttgart, New York, Anhang 287–300

Bellmann, K., R. Böttner, A. Knijnenburg and H. Neumann (1976): Computer Simulation models of gene expression. In: Mathematical Models of Metabolic Regulation (T. Keteli, S. Lakatos, eds.), Akademiai Kiado, Budapest, 227–256

Britten, R. J. and E. H. Davidson (1969): Gene regulation for higher cells. A theory. Science **165**, 349.

Bullough, W. S. (1967): The Evolution of Differentiation. Academic Press, New York, London

Davidson, E. H. and R. J. Britten (1971): Note on the control of gene expression during development. J. Theor. Biol. **32**, 123–130

Eden, M. (1958): Symp. Inform. Theory Biol. Gatlinburg, Tenn.

Gilderman, Ya, I., K. N. Kudrina and I. A. Poletaev (1970): Models of L-Systems (Systems with limiting factors). In: Studies on Cybernetics, Sov. Radio Moskau, 165–210

Goodwin, B. (1963): Temporal organization in cells. Academy Press, New York

Harris, H. (1968): Nucleus and cytoplasma. Clarendon Press, Oxford University Press, Oxford

Heinmets, F. (1966): Analysis of normal and abnormal cell growth. Plenum Press, New York

Heinmets, F. (1970): Quantitative cellular biology — an approach to the quantitative analysis of life processes. Marcel Dekker, Inc. New York

Holtzer, H. (1963): In: General physiology of cell spezialization. McGraw Hill, New York, pp. 80

Jacob, F., and J. Monod (1961): On the regulation of gene activity. Cold Spring Harbor Symp. Quant. Biol. Cellular Regulatory Mechanisms. **26**, 193–211

Jacob, F., and J. Monod (1963): In: Cytodifferentiation and Macromolecular Synthesis. (ed. M. Locke). Academic Press, New York, London

Kascer, H. (1957): Appendix. In: The strategy of genes. (ed. C. H. Waddington). Edinburgh Univ. Press, Edinburgh.

Kascer, H. (1963): The kinetic structure of organisms. In: Biological Organization at the Supercellular Level. (ed. R. J. C. Harris). Academy Press, New York.

Kauffmann, S. (1969): Metabolic stability and epigenesis in randomly constructed genetic nets. J. Ther. Biol. **22**, 437–467.

Knorre, W. A. (1968): Oscillations of the rate of synthesis of β-galactosidase in *Escherichia coli* ML 30 and ML 308. Biochem. Biophys. Res. Comm. **31**, 812–817.

Knorre, W. A. (1969): Mathematische Modelle der Enzymsynthese in Bakterien. studia biophysica **14**, 29–42.

Knorre, W. A., and R. Böttner (1971): Some considerations on dynamics of biological systems. Vortrag auf dem 1. Europäischen Biophysik-Kongreß 1971, Baden bei Wien.

Lindenmayer, A. (1971): Developmental systems without cellular interactions, their languages and grammars. J. Theoret. Biol. **30**, 455–484.

Luckner, M., and L. Nover (1973): Thesen zur Diskussionsveranstaltung „Differenzierungsprozesse bei Mikroorganismen, Tieren und Pflanzen", Halle.

Martinez, Hugo, M. (1972): Morphogenesis and Chemical dissipative structures. A computer simulated case study. J. Theor. Biology, **36**, 479–501.

Merzenich, W. (1974): Cellular automata. In: Lecture Notes in Biomathematics, Physics and Mathematics of the Nervous System. (eds. M. Conrad, W. Güttinger, M. DalCin). Springer-Verlag, Berlin, Heidelberg, New York.

Molik, B. (1978): Die Modellkonzeption des epigenetischen Systems zur theoretischen und mathematischen Behandlung der organismischen Selbstorganisation. Forschungsbericht, ZKI Berlin.

von Neumann, J. (1966): Theory of Self-Reproducing Automata. Univ. of Illinois Press, Urbana.

Nover, L. (ed.) (1973): Bericht über Diskussionsveranstaltung „Differenzierungsvorgänge bei Mikroorganismen, Tieren und Pflanzen", Halle.

Nover, L., and M. Luckner (1973). Differenzierungsvorgänge bei Mikroorganismen, Tieren und Pflanzen — ein allgemeines Konzept. Bericht über Diskussionsveranstaltung, Halle.

RASHEVSKY, N. (1940): Bull. Math. Biophys. 2, 15.
RASHEVSKY, N. (1961): Mathematical Biophysics (3rd. ed.). Dover, New York.
RATNER, V. A. (1977): Molekulargenetische Steuerungssysteme. (ed. K. BELLMANN). Akademie-Verlag, Berlin and Gustav-Fischer-Verlag, Stuttgart, New York.
ROSEN, R. (1972): Foundations of Mathematical Biology, Vol. 2, Cellular Systems. Acad. Press, New York, London.
ROSENTHAL, S., and J. GROSS (1973): Molekularbiologische Probleme der Zellreifung am Modell der roten Blutzelle. Thesen zur Diskussionsveranstaltung „Differenzierungsprozesse bei Mikroorganismen, Tieren und Pflanzen", Halle
SENDOW, Bl. and R. TSANEV (1974): A mathematical model of cellular differentiation. In: Mathematical Models in Biology and Medicine. (N. T. J. BAILEY, Bl. SENDOV. R. TSANEV eds.) North-Holland/American Elsevier, Amsterdam—London—New York.
SIMON, Z., and E. RUCKENSTEIN (1966): Regulation and synthesis in the living cell. J. Theoret. Biol. 11, 282—333.
SINGH, U. N. (1969): Polyribosomes and unstable messenger RNA. A stochastic model of protein synthesis. J. Theor. Biology, 25, 444—460.
STAHL, W. R. (1965): Self-reproducing automata. Perspect. Biol. Med. 8, 373—393.
STAHL, W. R. (1967): J. Theor. Biol. 14, 187—202.
STEHLER et al. — c.f. PRITZ, F. (1973): Currents in modern biology, 5, 66.
SUGITA, M. (1961): Functional analysis of chemical systems in vivo using a logical circuit equivalent. J. Theoret. Biol. 1, 415—430.
SUGITA, M. (1963): Functional analysis of chemical systems in vivo using a logical circuit equivalent. II. The Idea of a molecular automaton. J. Theor. Biol. 4, 179—192.
TCHURAEV, R. N., and V. A. RATNER (1972): Modelling of molecular genetic control systems in the language of abstract automata theory. In: Research in theoretical genetics. Novosibirsk (in russ.).
THOM, R. (1970): Topological models in biology. In: Towards a theoretical biology Vol. 3. (ed. C. H. WADDINGTON). Edinburgh Univ. Press, Edinburgh.
TSANEV, R., and B. SENDOV (1971): Possible molecular mechanism for cell differentiation in multicellular organisms. J. Theor. Biol. 30, 337—393.
TURING, A. M. (1952): The chemical basis of morphogenesis. Phil. Trans. Roy Soc. B. 237, 5—72.
ULAM, S. (1962): Proc. Symp. Appl. Math. 14, 215.
WADDINGTON, C. H. (1957): The strategy of genes. Allen and Unwin, London.
WADDINGTON, C. H. (ed.) (1970): Towards a theoretical biology. Vol. 3, Edinburgh Univ. Press.
WOLPERT, L. (1970): Positional information and pattern formation. In: Towards a Theoretical Biology Vol. 3. (ed. C. H. WADDINGTON). Edinburgh Univ. Press, Edinburgh.

1. Dynamic modelling of genetic and epigenetic control

R. ROSEN

Contents

1. Introduction

Biology is concerned with the understanding of the properties of living systems. It is in many ways an exceedingly difficult science, because (a) individual organisms, considered as physical systems, are of the utmost complexity, and (b) distinct organisms, again considered as physical systems, can be of the utmost diversity. These two features of biology have made it very difficult to formulate universal "laws", valid for all organisms, which would codify organic behavior in the same way as, for example, NEWTON's Laws codify the behaviour of mechanical systems. Indeed, the currently dominant view in theoretical biology is that we can only make progress in this direction by reducing biological behaviour to underlying physical (i.e. molecular) behaviour, thereby bringing known physical principles directly to bear upon biology.

Almost the only useful generalities regarding biological behaviour which are of a purely biological origin are those concerned with evolution and its genetic basis. We do not need to review the details of these generalities here, except to note the following: in order for evolution to be possible through natural selection all characters on which selection acts must ultimately be under genetic control. That is, all characters of organisms which have evolved, or can evolve, ultimately must rest on some property of the genome. This generalization is supported by many decades of experience with heredity in biological systems.

The science of genetics originally arose through a study of the inheritance of gross morphological characteristics in multicellular organisms. For instance, MENDEL's formulated his hypothesis of the atomic gene, and the laws bearing his name which govern the assortment of genes in subsequent generations, by studying characters such as plant size, and the color and surface of the cotyledon, in peas. The role of mutation was studied in many organisms, but particularly in the fruit fly *Drosophila,* through the appearance and heritability of gross morphological abnormalities; thousands of examples of such single-gene mutations are known. The point here is that gross morphological properties of multicellular organisms are ultimately controlled by single genes, or by groups of genes, obeying simple rules of assortment in subsequent generations.

On the other hand, with the growth of molecular biology over the past thirty years, the emphasis in genetics has shifted from morphology to biochemistry. The circle of ideas involved here is, roughly, the following: (a) according to the cell theory, all properties of multicellular organisms devolve onto properties of their constituent cells; (b) the properties of a cell are determined by the kinds of specific catalysts (enzymes) possessed by the cell, and by the concentrations of these catalysts; (c) the role of the genome in a cell is to d termine the character and concentration of these catalysts. With the identification of enzymes with protein, the elucidation of protein structure as linear copolymers of amino acids, the identification of genetic material with nucleic acids, and the WATSON-CRICK model of DNA as a helical copolymer of nucleotides, the role of the gene has become firmly tied to the intracellular synthesis of specific proteins. Once again, the details of this picture are too well-known to require detailed discussion here.

However, it must be noted that the biochemical picture of primary genetic activity as the determinant of primary structure of protein is almost entirely an **intracellular** picture. On the other hand, the original concept of the functional role of the gene was in terms of gross morphological properties of multicellular organisms. The morphological features of multicellular organisms involve properties of **populations** (clones) of cells, and therefore must involve **intercellular** phenomena. Thus the fundamental question arises: how can we understand the intercellular properties through which genes control morphology, in terms of the intracellular picture of primary genetic activity which has arisen from molecular biology? In particular, how can we build an integrated theory of the behavior of cell populations, and its genetic mediation, from the intracellular picture we have of the biochemical properties of the genome? It is with this kind of question that we shall be concerned in the present paper.

It is fairly clear how we would want to proceed in considering such a question. Namely, if the properties of an individual cell are determined by its genome, then the cell itself can be regarded as a collection of molecules which are determined by the specificities of the enzymes "coded" in the genome. Therefore, the interaction between cells, regarded as arising from the interactions of their specific molecules, is itself ultimately determined by the genomes of the cells. The genomes are not **directly** involved in the interaction, in general, but the interaction itself is specified **indirectly.** Insofar as such interactions determine the behavior of a cell population as a whole, that behavior is under genetic control. But since such interactions arise at the level of **products** of primary intracellular genetic activity, and not directly at the genetic level itself, such interactions may be regarded as **epigenetic.** Thus the problem we are addressing involves the nature of the genetic control of epigenetic processes.

This kind of problem is complicated by the fact that interactions between gene products, arising at an epigenetic level, can have a direct effect on the genome itself. In particular, new substances arising through epigenetic interactions can modulate (control) the way in which specific genes in the genome are expressed. We may formulate this by saying that the functional properties of a cell genome are strongly coupled to the epigenetic interactions in which the cell is involved. Thus the qualitative picture we obtain of the way in which the properties of cell populations depend on the genome is a very complex one, which has only recently begun to be attacked. The prime method of gaining insight into such systems has been through the study of mathematical model systems, in which the specific consequences and implications of postulated interaction mechanisms can be determined. It is not surprising that the mathematical study of such complex, strongly interacting systems have already led to many insights, not expected on simple reductionistic grounds. In the subsequent sections, we shall explore some of these model systems, examine their properties, and consider what we can learn from them which can be incorporated into a general genome-directed theory of the properties of cellular populations.

2. Morphogenesis

All of the questions raised above, regarding the genetic control of properties of sell populations, can be brought to a focus in the consideration of developmental phenomena in multicellular organisms. Development, of course, is the process by which a single cell (zygote) can give rise to a fully functional multicellular organism of a particular type. Closely allied to development are phenomena of regeneration, since we may suppose that the processes which give rise to a particular morphology also **stabilize** that morphology. Thus, developmental phenomena in their broadest sense concern the generation of pattern and form in populations of cells; i.e. with **morphogenesis.**

Morphogenetic phenomena in development bring into sharp relief the interplay between genetic and epigenetic processes. For a developing multicellular organism constitutes a clone of cells, derived through a genetically conservative process of cell division from an initial zygote. Thus each cell in this clone possesses the same genome. Nevertheless, the cells in such a clone become progressively differentiated from one another, giving rise to cellular subpopulations as different from one another, in molecular terms, as *Paramecium* is different from *Amoeba*. This elaboration of entirely new cell types, different from each other and from the zygote from which they were derived, can only be explained in terms of differential gene expression in the various cell types; in its turn, differential gene expression must arise from interactions between the genome and the successive products of its own activity.

In the study of morphogenetic phenomena, we can broadly recognize a small number of **fundamental morphogenetic mechanisms**, which allow specific kinds of patterns to be generated and maintained in cellular populations. These mechanisms can be specified as follows:

1. **Morphogenetic Movement:** differential changes in the relative spatial positions of the cells within a population:
2. **Differentiation:** changes in the internal chemical constitution of the cells in a population;
3. **Differential Birth and Death:** differential changes in the relative numbers of cell types within a population.

(See the Bibliographical Notes in Section VI below.)

Any real morphogenetic process involves a combination of these fundamental mechanisms. Nevertheless, for many purposes it is important to study these mechanisms in a "pure" form, so that their morphogenetic capabilities may be assessed, and the interactions between them more fully understood. In the present article, we shall study in particular the mechanisms of morphogenetic movement and of differentiation, and attempt to characterize the genetic and epigenetic bases of each of them. If we can succeed in this more limited task, we shall have at our disposal a suggestive basis from which to proceed to the general question of how the genome exerts control over arbitrary properties in populations of cells.

Before proceeding to a consideration of how these basic morphogenetic mechanisms can be represented in mathematical model systems, let us consider a number of specific biological examples of each of them. This will serve to place the model systems into an empirical context, and this context will in turn determine the character of the mathematical analysis.

2.1. Morphogenetic movement

As noted above, morphogenetic movement involves relative changes in spatial position occurring between the units or elements of a population. To study this phenomenon in its pure form, we suppose that (a) the properties of the elements of the population are not changing in time (i.e. that there is no differentiation), and (b) that the relative numbers of the different kinds of elements present in the population are fixed (i.e. that there are no sources or sinks for these elements in the system). There are many examples of important morphogenetic systems which satisfy these conditions.

2.1.1. The folding of protein

Our first example of morphogenetic movement occurs already at the molecular level. The primary structure of a polypeptide chain (i.e. the linear sequence of amino acids along the chain) is what is specified directly by the genome. In order to become biologically active, such a chain must acquire a characteristic geometric conformation; i.e. it must **fold** into a specific three-dimensional structure, the tertiary structure of the protein. It has long been conjectured that the primary structure uniquely determines the tertiary structure, in the following sense: under specific ambient conditions, a polypeptide chain will spontaneously assume a configuration of minimal free energy. The folding process is precisely the movement to such a minimal free-energy state, which is therefore also the state of biological activity. Now it should be noted that the dynamics which governs the folding process itself is not directly coded in the genome; this dynamics depends upon the specific chemical properties of the constituent amino acids, and the constraints upon their interaction arising from the primary structure. The amino acid interactions which generate the folded structure are **epigenetic** interactions. The genome controls the folded structure only indirectly by its specification of primary structure. As far as the dynamics of folding itself is concerned, the genome acts by specifying **initial conditions** for that dynamics; once the initial conditions are specified, the genome effectively disappears from view, and the epigenetic process of folding proceeds spontaneously. Protein folding is a classical example of **self-organization**, or **self-assembly**, in that the information required to specify the final structure is in some sense inherent in the system itself.

Let us also note that the structure so generated is **stable** to various kinds of perturbations. It is stable to **randomizations**, in the sense that if such a tertiary structure is destroyed (e.g. through gentle heating or other mechanisms of denaturation) it will be spontaneously re-established when the original ambient conditions are restored. It is stable to at least certain kinds of **amputations**; large parts of the molecule can be removed with proteolytic enzymes without appreciable loss of biological activity (and hence without destroying the existing tertiary structure). Finally, it is stable to various kinds of **hybridization**; replacement of particular amino acids along the chain by others. Any mechanism proposed to explain the generation of tertiary structure must also account for these remarkable stability properties.

2.1.2. Viral assembly

Simple spherical and cylindrical viruses are also produced by a mechanism of morphogenetic movement. Such viruses consist of a closed shell consisting of a definite number of protein coat molecules, surrounding a core of genetic material (nucleic acid). This characteristic morphology arises through a process of self-assembly very similar to that involved in the folding protein, especially in that the biologically active form can be regarded as minimal free-energy configuration. The role of the genome here lies in determining the primary structure of the protein coat molecules. These molecules then fold, as described above; it is the propertied of the **folded** molecules which then determine the morphology of the active virus, and the dynamical process by which it assembles. The genome thus appears here at two removes from the epigenetic interactions which generate the form of the active virus.

2.1.3. Cell sorting

Populations of cells taken from sufficiently early embryonic stages, or from sufficiently simple multicellular organisms (e.g. sponges) exhibit remarkable stability properties to perturbations such as randomization and hybridization. For instance, if cells taken from embryonic lung or kidney are spatially randomized, and then returned to physiological conditions, the cells will move relative to one another in such a way as to re-establish at least an approximation to the histology of the organ from which they were taken. Such phenomena are collectively called **sorting-out**, or **cell sorting**. The study of cell sorting is particularly interesting because the original structure was not generated through a process of morphogenetic movement alone. Therefore we have a separation between the mechanisms by which developmental patterns are generated and the mechanisms through which they are stabilized. As we shall see, the interplay between generation and stabilization processes, and the manner in which they must be integrated, throws considerable light on the genetic and epigenetic factors governing morphogenesis.

2.2. Differentiation

In its simplest form, differentiation phenomena involve populations of identical cells which are assumed to be (a) initially identical, (b) fixed in space relative to one another, and (c) neither dividing nor dying. The question is: under what circumstances can such a population of initially identical cells become differentiated from one another?

The prototypic problem in the study of differentiation is the establishment of a **gradient** in the initially homogeneous aggregate of cells. The establishment of such gradients are almost the initial phenomena of development, and underlie all subsequent developmental phenomena. Gradient generation has always appeared mysterious from a purely physical point of view; in most systems with which physicists deal, systems tend to pass spontaneously from asymmetric to symmetric; from organized to disorganized; from inhomogeneous to homogeneous. Gradient generation involves systems which spontaneously proceed in the other direction. Therefore it is not surprising that an understanding of spontaneous generation of gradients in initially homogeneous systems is the crucial step in understanding differentiation in general, and the relation between biological systems and purely physical ones.

With these particular examples in hand, let us now turn to a consideration of the mathematical representation of the phenomena of morphogenetic movement, and of gradient generation.

3. Mathematical models of morphogenetic movement

Recalling our examples of morphogenetic movement in the preceding section, we consider a population consisting of a finite number of "cell types", and a fixed number N_i of cells of each type. In the pure situation we assume that, neither the cell types nor their numbers are changing. That is, we assume there is no differentiation in the system, and no cell division or cell death. The total number of cells in our population is thus $N = N_1 + N_2 + \cdots + N_n$.

Let us imagine a region of space, partitioned into N subregions in some regular way (e.g. into squares or cubes). Each way of assigning a cell of our population to a subregion of the partition defines a **pattern** of our population in space. Thus, a pattern is simply a mapping from our population of cells onto the subregions of our fixed patrition.

For purposes of illustration, let us assume that we have two cell types A, B, and a fixed number of cells N_A, N_B respectively. Consider a two-dimensional region, partitioned into squares. A pattern is then an assignment of cells to the squares in the partition.

Intuitively, the change of pattern over time (i.e. the change in the manner in which cells are associated with fixed squares in the partition) must proceed in such a way as to minimize some measure of the "free energy" of the population. To represent this, we must (a) define a general notion of the "free energy" of a pattern, and (b) specify how the assignment of cells to squares changes in time.

In the special case we are considering, we may have the following situations: (1) a cell of type A may abut a cell of type B; (2) a cell of type A may abut a cell of type A; (3) a cell of type B may abut a cell of type B. Let us associate with each of these situations the non-negative numbers λ_{AB}, λ_{AA}, λ_{BB}, which represent the respective affinities of an A-cell for a B-cell, an A-cell for another A-cell, and a B-cell for another B-cell. Given any specific pattern P, the assignment of cells to squares determines how many abutments of each type are present; suppose that there are N_{AB} abutments of A-cells with B-cells, N_{AA} abutments of A-cells with A-cells, and N_{BB} abutments of B-cells with B-cells. Then the total affinity of the patterns is just

$$E = N_{AA}\lambda_{AA} + N_{BB}\lambda_{BB} + N_{AB}\lambda_{AB}.$$

This total affinity is inversely related to the "total free energy" of the pattern in question; a maximization of this function $E(P)$ amounts to a minimization of the total free energy of the pattern.

Now let us consider how the given pattern may change in time (which we suppose for convenience to be discrete). A pattern P' will be said to be a **neighbor** of P if P' can be obtained from P by one or more non-overlapping exchanges of cells between adjacent squares in the partition. We allow each cell to be moved at most once. Thus, to each pattern P, we can associate the set $U(P)$ of all neighboring patterns. To each pattern P' in $U(P)$, we can associate a total affinity $E(P')$. We choose that pattern P' in $U(P)$ for which $E(P')$ is maximal; such a choice can always be made in a consistent way. We say that in one time unit the system will undergo a transition from the initial pattern P to the new pattern P'.

Likewise, we can form the set of neighboring patterns $U(P')$ to P'. In this set, we can determine that pattern P'' such that $E(P'')$ is maximal in $U(P')$. We can continue this process, thereby generating a sequence of patterns

$$P \to P' \to P'' \to \cdots \to P^{(k)}.$$

This sequence will terminate when we reach a pattern $P^{(k)}$ such that $E(P^{(k)}) > E(P)$ for all patterns P in $U(P^{(k)})$.

The above procedure can be generalized to an arbitrary number of cell types, and to arbitrary ways of partitioning space. It provides a framework for discussing morphogenetic movement as a transition to a minimal free-energy state. In such a situation, we can characterize not only the final stable configuration arising from an arbitrary initial pattern P, but also the temporal sequence (trajectory) of intermediate patterns, as a function of the system parameters (i.e. of the numbers of cells of each type, and of the affinities λ).

The technical exploration of models of this type completely solves the problem of morphogenetic movement, in the sense that it is possible to characterize the terminal patterns which can arise in any population of cells, once the system parameters are specified. This is true even in more general situations, in which the cells are assumed to be anisotropic (i.e. even when the affinities vary in an arbitrary way over the total cell surface). It can be said that all experimental results on cell sorting can be understood on the basis of assuming only a differential affinity between cells, and a general motility of cells, which allows them to "explore" a variety of local environments.

The sorting models not only allow us to deal with the phenomena of cell sorting, but also with the other examples of morphogenetic movement which were described in the preceding section. For instance, the protein folding problem differs from a sorting problem only in the presence of an additional constraint, corresponding to the imposition of a primary structure. In the sorting model consisting of two cell types, for example, we may imagine the B-cells to be tied together in a linear array. Then we ask what is the configuration of this array which minimizes the total free energy (i.e. maximizes the affinity E). Such an approach to the protein folding problem is of quite a different character from those presently being adopted, and in some ways is far superior to them.

The importance of the sorting models, for our present purposes, is not so much that it allows us to understand the capabilities of morphogenetic movement, but that it relates such movement to a wide class of biological and physical phenomena. For instance, the models we have described are equally applicable to physical phenomena such as phase separation (i.e. the breaking of emulsions). Such processes are also minimal free-energy situations occurring in populations of differentially adhesive and motile units. A standard way of treating such problems in physics is through the use of models like the Ising Model; indeed, it will be recognized that our approach has many points of contact with the Ising Model.

Thus, the cell sorting models belong to a wider context than that arising in morphogenesis; they pertain to general populations of units satisfying the conditions described above. Such systems encompass a spectrum ranging from the purely inorganic (where the units are atoms or molecules) through the cellular up to ecological and social situations (where the units are cells or entire organisms). As we shall see subsequently, it is this wider context which will allow us to give a deeper and more general meaning to the concept of the genome, and to the functional role of the genome in determining the behaviours of sorting populations.

4. Mathematical models of differentiation

In the present section, we shall consider the mathematical representation of populations of cells which are exhibiting pure differentiation. That is, we shall envisage a population of initially identical cells, distributed in space in some way as in the preceding section; as before, we assume that the number of cells in the population is not changing, and that the pattern of spatial distribution of the cells is fixed as well. The fixing of such a spatial pattern establishes a corresponding pattern of neighbourhood relationships between the constituent cells; these neighbourhood relationships in turn determine the pattern of interaction which is established between them.

Intuitively, differentiation means that the various cells in the population, assumed initially identical, are becoming progressively different from one another in time. Therefore, we must now impose a certain amount of internal structure to the cells themselves. We may do so by supposing that each cell in the population can be represented as a point in some "state space" spanned by a set of state variables $x_1, \ldots, x_n$. These state variables may be thought of as representing concentrations of specific chemical species.

Let us consider a particular cell (say the ith) in the population. We suppose that it is capable of maintaining an **autonomous** dynamics, which would characterize its behavior if it were completely isolated. Let us take this dynamics to be of the general form

$$dx_{ik}/dt = f_{ik}(x_{i1}, \ldots, x_{in}) \tag{1}$$

where the index k, which characterizes the state variables of the cell, ranges from 1 to n. Since the cells in the population are assumed identical, they all exhibit the same autonomous dynamics.

Now let us suppose that the cells can interact; e.g. by diffusion. As far as the ith cell is concerned, its autonomous dynamics (1) must now be supplemented by additional terms, representing diffusive exchange with the cells with which it is in contact in the population. Thus the equations (1) become, in general terms,

$$dx_{ik}/dt = f_{ik}(x_{i1}, \ldots, x_{ik}) + \Sigma_j D_{jk}(x_{jk} - x_{ik}) \tag{2}$$

where we have taken the simplest form of diffusive interaction, i.e., proportional to the difference in reactant concentration between adjoining cells.

Following the ideas of RASHEVKSY and of TURING, let us suppose that the coupled system (2) is in a steady state, which is **homogeneous** throughout the population;

i.e. such that

$$x_{ik} = x_{jk}$$

for all i, j, k. We can inquire now into the **stability** of steady state; i.e. for the response of the system to a small perturbation from that steady state. In general terms, this kind of question can be handled in a standard kind of way. Around any steady state, the dynamical equations (2) can be linearized. The state variables in the linearized system are the deviations of the original state variables from the steady state values. Under quite general conditions, the stability of the original steady state of (2) will be the same as the stability of the system linearized about that steady state. Moreover, the stability properties of the linearized system can be completely determined in terms of the eigenvalues of its matrix.

Now it can obviously happen that the steady state of (2), which we have supposed to correspond to a homogeneous situation in which all the cells were identical, can be unstable. If this should be the case, then a random perturbation away from the steady state will be progressively amplified by the system. Intuitively, this means that the constituent cells are becoming more and more different from one another, and therefore that the system is differentiating. Eventually, a new stable pattern will be established, corresponding to the nearest stable attractor of the system (2).

The above approach has provided one of the basic schemes for the investigation of differentiation in biological systems of all kinds. In addition to its obvious role in seeking to understand differentiation phenomena in development, it has been applied to such situations as the aggregation of slime molds. By generalizing the form of the interactive terms in (2), the formalism becomes applicable to such problems as the propagation of excitation in networks of excitable cells (a problem which is not usually thought of as a morphogenetic problem, but which is revealed in this fashion to be exactly analogous to problems of differentiation). Also, insofar as the transition of (2) to a new attractor involves accumulations of reactants against concentration gradients, the formalism bears on such purely physiological problems as active transport.

Furthermore, the above scheme has been applied to other kinds of reaction-diffusion systems which are of a purely inorganic character. It has played a significant role in the analysis of such systems as the ZHABOTINSKII reaction. More generally, PRIGOGINE and his co-workers have considered such situations from the standpoint of irreversible thermodynamics, under the rubric of "symmetry breaking" (i.e. the transition from a homogeneous to an inhomogeneous state) in general open systems far from thermodynamic equilibrium. Thus, just as with the cell sorting models, the formal approach to differentiation during development becomes embedded into a wider context, encompassing inorganic systems as well as a wealth of biological phenomena from all levels of biological organization. This context will also be helpful to us in assessing the functional role of the genome in controlling the dynamics which gives rise to developmental phenomena.

One final aspect of the mathematical structure of systems like (2) can be mentioned. We have seen that according to our scheme, differentiation is associated with the stability properties of a homogeneous steady state for a population of interacting cells. These stability properties in turn depend upon the specific system parameters; namely, the rate constants for the autonomous reactions (1), and the diffusion constants which measure the magnitudes of the cellular interactions. We may ask how the stability properties and the subsequent dynamical behaviour of the system change when these system parameters are varied. This kind of stability problem belongs in general to

bifurcation theory. It is at this point that the analysis of systems like (2) makes contact with notions of structural stability, and thence with the catastrophe theory of Thom. We cannot go deeply into these matters here, but we shall touch on some of them subsequently.

5. The role of the genome

Insofar as the morphogenetic mechanisms we have discussed in the preceding sections refers to biologically meaningful situations, its character is entirely **epigenetic**. For instance, in the cell-sorting models, the nature of the pattern generated in a particular set of circumstances is entirely determined by the differential affinities for one another possessed by the cells in a sorting system. These affinities in turn must devolve upon the properties of the cell surfaces, and hence arise from interactions between remote products of primary genetic activity. Indeed, the genome is never mentioned in discussing the properties of a sorting model; this is in fact why the same sorting models which account for cell sorting are also applicable to purely inorganic situations, such as phase separations, in which the concept of a genome is apparently meaningless. Nevertheless, we can see that the genome can be regarded as **tacitly** involved in the sorting models; it is that which, through a dynamics distinct from the one which governs the sorting itself, establishes the initial conditions and parameter values which determine the properties of the sorting model.

The same is true of the differentiation models described in the preceding section. Once again, their character is entirely epigenetic; the concept of the genome is never mentioned. As in the sorting models, it is this feature which allows the formalism to be meaningfully applied to arbitrary open systems far from thermodynamic equlibrium, as in the Zhabotinskii reaction. The dynamics of such a system is completely determined by the specific system parameters, such as rate constants and diffusion constants, which characterize it; these parameters specify dynamical interactions between reactants which are only remotely related to primary genetic activity. Once again, then, the role of the genome is at best tacit; it is to establish the parameter values and initial conditions which govern a particular differentiating system.

Thus, to a certain extent the genome always sits **outside** any particular dynamical model of a biological process. As far as such a model is concerned, its role disappears once the model is specified. What we wish to do in the present section is to see whether we can embody the genome **explicitly** in such dynamical models, without generating an infinite regress.

Ironically, the clue as to how this may be done comes from the purely inorganic systems governed by the formalisms we have described above, for which the concept of a genome seems meaningless. In fact, as we shall see, the concept is far from meaningless in these situations; once it is formalized, it allows us to see more clearly the unique character of the genetic aspect of biological behavior, and to better assess the role of inorganic metaphors (such as the Zhabotinskii reaction) for pattern generation in biological systems.

For simplicity, let us begin by considering a very simple inorganic system, such as a real (i.e. non-ideal) gas. Such a gas is described in terms of the classical thermodynamic state variables (p, v, T) (pressure, volume and temperature, respectively). The equilibrium behaviour of such a gas satisfies some kind of **equation of state**, such as the

VAN DER WAALS

$$(p + a/v^2)\,(v - b) = rT. \tag{3}$$

This relation has the following significance: if we impose a pressure p on a quantity of gas, maintained at a fixed temperature T, and allow the gas to come to equlibrium under these conditions, then the equilibrium volume v will be related to the imposed temperature and pressure through the relation (3).

Let us notice that the relation (3) contains a number of **parameters**; namely (a, b, r). In physical terms, these parameters pertain to the molecular structure of the gas; they are determined by such things as the molecular volume, and the forces of attraction manifested between the molecules. Specifying these parameters, then, is the same as specifying the "species" of gas under consideration; therefore it is not unreasonable to identify (a, b, r) with the **genome** of the gas.

Once a particular species of gas is specified, the equation of state (3) can be looked at in the following way: for each pair (p, T) of imposed pressure and temperature, the volume of the gas at equilibrium is determined by (3). We may look upon the choice of a pair (p, T) as the establishment of a particular set of environmental conditions; (3) then asserts that the resultant volume v is uniquely determined by the genome (a, b, r) of the gas, and by the environmental conditions (p, T). Thus, it is further not unreasonable to look upon the equilibrium volume v as the **phenotype** of the gas; and upon the equation of state (3) as describing a relation between a genome, a set of environmental conditions, and the resultant phenotype.

Let us note that the specific dynamical process whereby an imposed set of environmental conditions are transduced by the genome into a corresponding phenotype is not at all embodied in the equation of state. What that equation says is simply that to each genome (a, b, r), there corresponds a **mapping** from the set of environment states to the set of phenotype states, of the form

$$\Phi_{(a,b,r)} : E \to P$$

where in this case, E is the set of all (p, T)-pairs, and P is the set of resultant v. Thus an equation of state can be regarded as establishing a parameterized family of mappings between environmental conditions and phenotypes, parameterized by the set of possible genomes.

This kind of picture has many interesting implications. For example, we can ask what happens when the genome is changed for some initial value (a, b, r) to some new value (a', b', r'). In the case of the VAN DER WAALS equation considered above, it happens that **any** such change of genome can be "annihilated" by isomorphisms (i.e. co-ordinate transformations) $\alpha: E \to E$, $\beta: P \to P$, which in the present case are simply changes of **scale**. This assertion is the essence of the "Law of Corresponding States" for such gases. More precisely, we can find isomorphisms α, β such that the diagram

$$\begin{array}{ccc} E & \xrightarrow{\Phi_{a,b,r}} & P \\ {\scriptstyle\alpha}\downarrow & & \downarrow{\scriptstyle\beta} \\ E & \xrightarrow[\Phi_{a',b',r'}]{} & P \end{array}$$

always commutes; hence all the mappings $\Phi_{a,b,r}$ are **conjugate** to each other.

Conjugacy is an equivalence relation between mappings from E to P. In general, a parameterized family of such mappings will decompose into a number of conjugacy classes. If by a small change in parameter values (i.e. in genome) we cause the corresponding maps to lie in different conjugacy classes, we say that a **bifurcation** has occured; the systematic study of such bifurcations in parameterized families of mappings is one of the main objects of the catastrophe theory of Thom. It is also at the root of the employment of dimensional-analytic arguments to the theory of analog (scale)-modelling in engineering, and the D'Arcy Thompson "Theory of Transformations". We have explored the relationships between these ideas elsewhere, and we will not repeat the analysis here.

In general, the above arguments are applicable to any class of phenomena governed by an equation of state. Insofar as we believe that phenotype is uniquely determined by genome and environment, and insofar as we identify phenotype with the properties of clones (i.e. with populations of genetically identical cells), we must suppose that there is an equation of state relating them. As we have seen, such an equation of state can generally be expressed as a family of mappings from environments to phenotypes, parameterized by possible genomes. The picture we then arrive at, through this kind of heuristic argument, is that the properties of a particular mapping in such a family are manifestations of epigenetic properties, while the **choice** of such a map (i.e. the association of a map with a corresponding genome) represents the genetic aspects.

Let us briefly consider the morphogenetic models of cell sorting and differentiation developed previously in the light of the above picture.

In the cell sorting models, we can consider that there is an equation of state relating an initial pattern P_0, a final pattern $P^{(n)}$, and the affinities λ_{ij} between pairs of cell types. Here it is natural to consider the parameters λ_{ij} as representing genome, the initial pattern P_0 as representing the environment on which the genomic parameters act, and the equation of state itself as a family of mappings of the form

$$\Phi_{\lambda ij} : \Pi \to \Pi$$

where Π is a space of patterns. Without going into full details, we can see: (a) that the separation of variables occurring in a sorting problem into genome, environment and phenotype is meaningful even in inorganic situations such as phase separations; (b) that the set of possible genomes is divided into conjugacy classes, which characterize all the possible ways in which a genome can determine sorting patterns, and (c) at the level of morphogenesis through sorting mechanisms, the biological and nonbiological systems are completely analogous.

Likewise, in the differentiation models we have considered, the equation of state relates (a) the geometric configuration of cells; (b) the reaction parameters (rate constants) which govern the autonomous reaction within the cells, and the diffusion constants between them; (c) the initial perturbation away from the homogeneous steady state, and (d) the final pattern of reactants so generated. Here we can take (a) and (c) to be the environmental quantities, (b) to represent the effect of the genome, and (d) to be the resultant phenotype. In the local linearized analysis we have described above, the genomic quantities determine the eigenvalues of the linearized system, and therefore the stability properties of the homogeneous steady state. The actual phenotype to be observed in such a situation is determined in part by these stability properties, and in part by the kind of perturbation imposed on the system.

Let us make two concluding observations regarding this kind of picture. The first is that the effect of the genome on epigenetic interactions is always representable through parameters governing rates. This is in accord with our intuitive understanding that the genome acts through rate determination and rate modulation. Furthermore, we can concatenate equations of state in a hierarchical fashion, in the following way: we can suppose that the phenotype arising at a particular epigenetic level serves to determine the rates at which epigenetic interactions occur at other levels. Thus, the primary effect of the genome can be regarded as cascading through a sequence of epigenetic levels, so that at any particular level, the genome appears as a set of rate constants appropriate to determine the relation between environment and phenotype at that level. A corollary of this picture is that the distinction between genome, environment and phenotype is a **relative** one, depending on the level of analysis at which we are working; what appears as phenotype at one level can appear as genome at higher levels, and conversely. According to this picture, then, the distinction between organic and inorganic systems is precisely the following: whereas in inorganic systems the parameters which we functionally identify as genome are **invariant**, so that their effects propagate only upward through an epigenetic hierarchy, biological systems exhibit the possibility that phenotype at one level of such a hierarchy may modify genome at a **lower** level. This characteristic feature of organisms is not seen when we consider only a single epigenetic level, but it must appear when we consider the interactions between several such levels.

Finally, we may observe that the relation between genome, environment and phenotype which we have developed above may be regarded as a species of **pattern recognition**. More specifically, if we fix an environment e in E, then we may regard the **generation** of a particular phenotype p as the **recognition** of the genome which determines that phenotype in the environment e. In this way, we may regard the generation of morphological patterns in cell populations to be the recognition of the genomic patterns which underlie them. This turns out to be a useful way to think about the relation between genome and phenotype, for several reasons: (a) it relates morphogenetic problems, and more general problems of adaptation in biological systems, with the theories of measurement, recognition, classification and discrimination in their broadest sense; (b) it points up vividly that pattern generation and pattern recognition are **coupled processes**; we cannot recognize a pattern in one system without simultaneously generating a pattern in some other system, and conversely. Thus, the kind of genome-directed morphogenetic theory we are envisioning is not a disconnected family of epigenetic models, but rather a theory of genome recognition in the deepest sense.

Bibliographical notes

In this section we shall review some of the historical aspects of the ideas discussed above, and indicate some of the appropriate literature.

1. The analysis of real morphogenetic processes into the "pure" morphogenetic mechanisms of movement, differentiation, and differential birth-and-death, was originally proposed by J. T. Bonner (1952: Morphogenesis. Princeton University Press).
2. An excellent source for the mathematical study of cell sorting models is in a collection of original papers edited by G. D. Mostow (1975: Mathematical Models for Cell Rearrangement. Yale University Press). In this book are to be found the original work of the author and others, the parallel work of Pyatetskii-Schapiro and his school, and related work as well as full background bibliographies. See also the more

recent definitive simulation work of ROGERS & GOEL (1978: "Computer Simulation of Engulfment and Other Movements of Embryonic Tissue". JTB, **71**, 103—140; 1978: Computer Simulation of Cellular Movements: Cell Sorting, Migration through a Tissue and Contact Inhibition. JTB, **71**, 141—166). The application of the sorting models to protein folding was suggested by the problem of "circularization" posed by PYATETSKII-SCHAPIRO cited above; a forthcoming paper by M. YČAW. JACOBSEN, N. GOEL (JTB, **72**, 443—457 (1978)) embodies one line of work in these directions. Viral assembly as a minimal free-energy process is discussed in some classic papers by Caspar & Klug (cf. D. L. D. CASPAR, & A. KLUG, 1962: Physical Principles in the Construction of Regular Viruses. Cold Springs Harbor Symposia # 27).

3. The problem of the establishment of gradients in initially homogeneous tissue, regarded as an instability in a homogeneous steady state of a reaction-diffusion system, was originally treated by RASHEVSKY (N. RASHEVSKY. 1940: An Approach to the Mathematical Biophysics of Biological Self-Regulation and of Cell Polarity. Bull. Math. Biophys. **2**, 15—25). A dozen years later, A. M. TURING (1952: The Chemical Basis of Morphogenesis. Phil. Trans. Roy. Soc. B. **237**, 5—72) developed the same idea in a considerably simpler mathematical format. After another dozen years of neglect, this area has become one of the most active in mathematical biology. On the purely mathematical side, the problem is largely one of bifurcation in coupled reaction-diffusion systems; a typical paper in this area is N. KIPPEL & L. N. HOWARD 1973: Plane Wave Solutions to Reaction-Diffusion Equations. Studies in Applied Math. **52**, 291—328. A good biologically oriented paper is H. G. OTHMER, & L. E. SCRIVEN. 1971: Instability and Dynamic Pattern in Cellular Networks. JTB **32**, 507—537. At the same time. PRIGOGINE and his school were approaching the same problems from the standpoint of irreversible thermodynamics; the reader might consult G. NICOLIS & I. PRIGOGINE 1977: Self-Organization in Nonequilibrium Systems, Wiley, and the extensive bibliography therein. For the ZHABOTINSKII Reaction, and its bearing on such problems, see J. TYSON 1977: The BELOUSOV-ZHABOTINSKII Reaction. Springer Lecture Notes in Biomathematics # 10. The relation of diffusion-reaction systems and network models of excitation is discussed in R. ROSEN 1968: Turing's Morphogens, Two-Factor Systems and Active Transport. Bull. Math. Biophys. **30**, 493—499. The application of diffusion-reaction ideas to the problem of slime-mold aggregation may be found in E. G. KELLER & L. A. SEGAL 1970: Initiation of Slime Aggregation Viewed as an Instability. JTB **26**, 397—415.
4. The definitive discussion of catastrophe theory and its possible biological applications is found in R. THOM 1972: Stabilite Structurelle et Morphogenese, Benjamin. A large number of surveys and potential applications to biology and other sciences has since appeared; cf. the Springer Lecture Notes in Mathematics # 468 and # 525 for these matters and for up-to-date bibliographics. See also the recent collection of papers of C. ZEEMAN: Catastrophe Theory, Benjamin, 1977.
5. The idea of an equation of state, and its detailed application to biological problems, may be found in R. ROSEN 1978: "Dynamical similarity ..." Bull. Math. Biophys. (**40**, 549—579). A detailed development of the formal theory of measurement, which underlies this work, is to appear in a monograph: R. ROSEN 1978: On the Measurement and Representation of Natural Systems. Elsevier, Amsterdam, New York, Oxford. A general survey of morphogenetic problems and their mathematical formulation, in which many of the ideas developed herein were first suggested, may be found in R. ROSEN 1977: The Generation and Recognition of Patterns in Biological Systems. Springer Lecture Notes in Biomathematics # 18 (ed. by D. E. MATTHEWS), 223—341.

2. A discrete stochastic simulation model of the regulation of geneexpression with variable control characteristics

R. Böttner, K. Bellmann, T. Cierzynski

Contents

1. Problems and aims

After enlightening the basic structure of gene expression (DNA → RNA → protein) and exploring molecular mechanisms of gene expression it is necessary and seems possible now to investigate dynamic aspects of genetic information processing, i.e. an important basic process of differentiation, development, and growth.

Within the boundaries of development of the whole organism a cell's DNA releases different parts of stored genetic information in the various phases of cell cycle and in different stages of its differentiation. The mature differentiated cell is able to adapt its gene expression very flexiblely to certain environmental conditions.

For our understanding of differentiation and development and of the mechanisms and principles used by the cell to adapt itself to physiological and pharmocological stimuli it is very important to investigate the regulation of gene expressions:

In which manner does the cell control the temporal course of the numbers of such macromolecules that are most important for the whole process of gene expression? Which importance have synthesis and degradation of the macromolecules in this process? Which functional properties have coupled systems of gene expression and which inter actions between different subprocesses of gene expression are important for the coordination and integration of epigenetic systems?

A further important problem is the investigation of the relationship between the physico-chemical structure of biological macromolecules, the feedback structure of molecular genetic control networks and the function and behaviour of the epigenetic system of the cell. Essential for this problem is the understanding of epigenetic systems dynamic depending on its own internal structure (feed back and feed forward links between nucleinic acids, proteins, metabolits and highly aggregated compounds of the cell). The complexity of the molecular genetic control systems is very high but on the other hand the experimental possibilities are restricted. It is impossible to investigate the system in a functional integrated form. The experiments are reduced to subsystems and only few factors may be controlled. The consequence of the variation of these factors may comprehended only incompletely.

In most cases the behaviour of such strongly interconnected non-linear systems cannot only be recognized by verbal, descriptive analysis.

To bring experimental molecular biology nearer to integrative biology it is needed to develop synthetic methods besides the well established analytical methods.

Mathematical models, e.g. homomorphic maps of the original, are such an integrative tool. By means of mathematical models it becomes possible to get a feeling for dynamic behaviour and investigate basic features of the dynamic behaviour of the non-linear processes of gene expression depending on the structural and functional organization of this information processing system.

Pioneer work in the field of mathematical modelling of epigenetic processes was done by RASHEVSKI (1961), GOODWIN (1963), SUGITA (1961), ROSEN (1972), HEINMETS (1966), and TSANEV, SENDOV (1971). In mathematical modelling of control of gene expression we started from the molecular genetic and biochemical conceptions about differentiation and development (e.g. MOHR and SITTE, 1971) and used non-linear difference equations for the mathematical description (BÖTTNER, 1975; BELLMANN et al., 1974). Non-linearity, and the stochasticity of the equations are the reason why no analytical solutions but only numerical ones by means of computer simulations are possible.

Exceeding earlier models (GOODWIN, 1963; ROSEN, 1972; KNORRE, 1969, 1971; SIMON, 1966; TSANEV, SENDOV, 1971) concerned with transcription, translation, degradation, with simple fixed control characteristics (e.g. MICHAELIS MENTEN shapes) and neglecting internal or external stochasticity we developed a mathematical model additionally representing the following features of real pro- and eucaryotic systems: stochastic behaviour of the subprocesses synthesis and degradation, time depending stabilization and destabilisation of RNA and proteins during differentiation, time consuming transport and processing of informational molecules within and between cells and tissues, optimum and threshold control characteristics on the level of transcription and translation, and multivariable control of transcription, and translation in complex net structures.

The model had to be a multivalent one in studying the dynamic behaviour of epigenetic networks. To enable dynamic analyses of a big class of molecular genetic control systems, the model was developed as a modular program system practicable for the molecular geneticist. By suitable choice of system parameters special pro- and also eucaryotic systems may be mapped without additional programming efforts.

2. Description of the model

2.1. Basic assumptions

1. To reduce the great complexity of the epigenetic system of the control of gene expression we assume the possibility to decompose the whole system in relative autonomous subsystems, interacting together by specific (e.g. regulatory proteins or metabolits) and non specific factors (e.g. nucleotid and aminoacid pools).
   The subdivision of the DNA of the genome in transcription units is a naturally given possibility of decomposition. We comprehend those processes, leading to the synthesis of such a protein coded by a monocistronic RNA (fig. 1, 2) as the most simple unit of gene expression.
   This unit is called an elementary unit of genetic information processing (EUGIP). These EUGIPs serve as elements of more complex gene expression systems, i.e. of epigenetic networks.
2. The whole process of gene expression was subdivided in a sequence of sub-processes, which are linked by exchange of informational molecules (fig. 2).
3. For formal characterization of the single sub-processes we abstract from their special mechanisms and physicochemical details; we only consider their functional features. Each subprocess is described by a transition function, which characterizes the input-output relation of this sub-processes. The inputs of a subsystem are outputs of others.
   In this way one gets a coupled system representing the integrative cooperation of all processes concerned with the gene expression depending on the structural-functional organisation of the whole system.
4. All sub-processes are described by macroscopic parameters and elementary mathematical operations. Experimentally measured figures, characterizing the physicochemical structure of the considered process serve for the computation of the parameters.
5. The regulation·factors that control the different sub-processes are very important for the characterization of the dynamic features of the gene expression systems. In our model the macroscopic parameters depend on these control factors by means of simple algebraic functions representing their effect correctly. A detailed consideration of the stoechiometry of the reactions or of the basic mechanisms of control would lead to very complicated equations, which generate an overall behaviour just as the simple global functions.
6. The activity of a scripton, i.e. the mRNA synthesis rate, may depend on the control factors (regulatory proteins) continuously or in a threshold manner (jump from the active to the inactive state or vice versa).
   The used algebraic control functions are able to represent these two qualitatively different types of control shapes (see 2.2.). It is also possible that the protein synthesis rate depends on the number of mRNA molecules linearly or by means of a saturation process.
7. The protein and mRNA degradation obey a first order kinetics (Greenblatt, 1973; Schimke, Doyle, 1970; Rechcigl, 1971).
8. The total synthesis rate of a certain mRNA is the product of the synthesis rate of this mRNA of a transcription unit and the number of its DNA-copies in the cell. In this way simple processes of phage replication may be desribed approximately.

9. The intracellular conditions (level and composition of the nucleotid and aminoacid pools, concentration and activity of the RNA-polymerase, concentration of binding and releasing factors, ionic milieu, etc.) are assumed not to be limiting for the cellular synthesis processes and to be constant. In the deterministic model variant the macroscopic parameters (i.e. the synthesis and degradations rates) are considered as mean values; in the stochastic one the random fluctuations of both the intra-

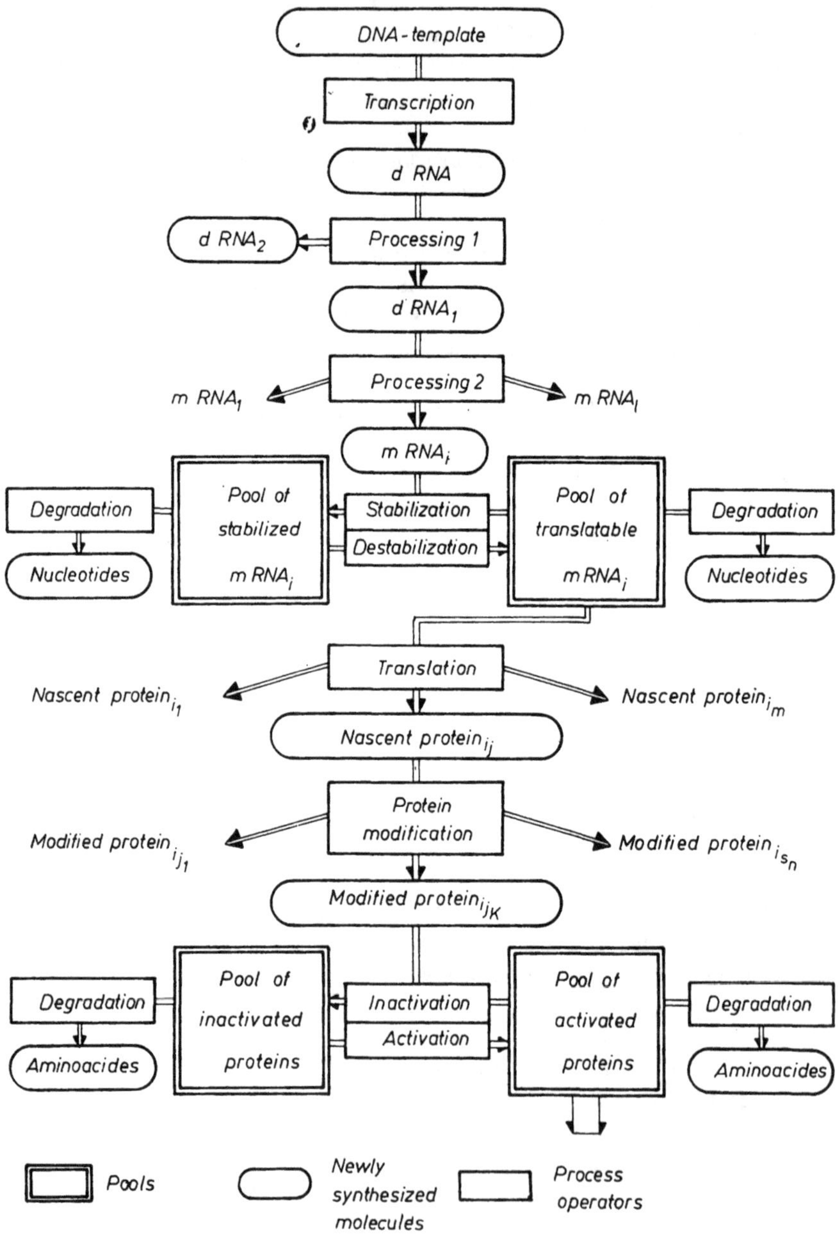

Fig. 1. Scheme of the structure of gene expression

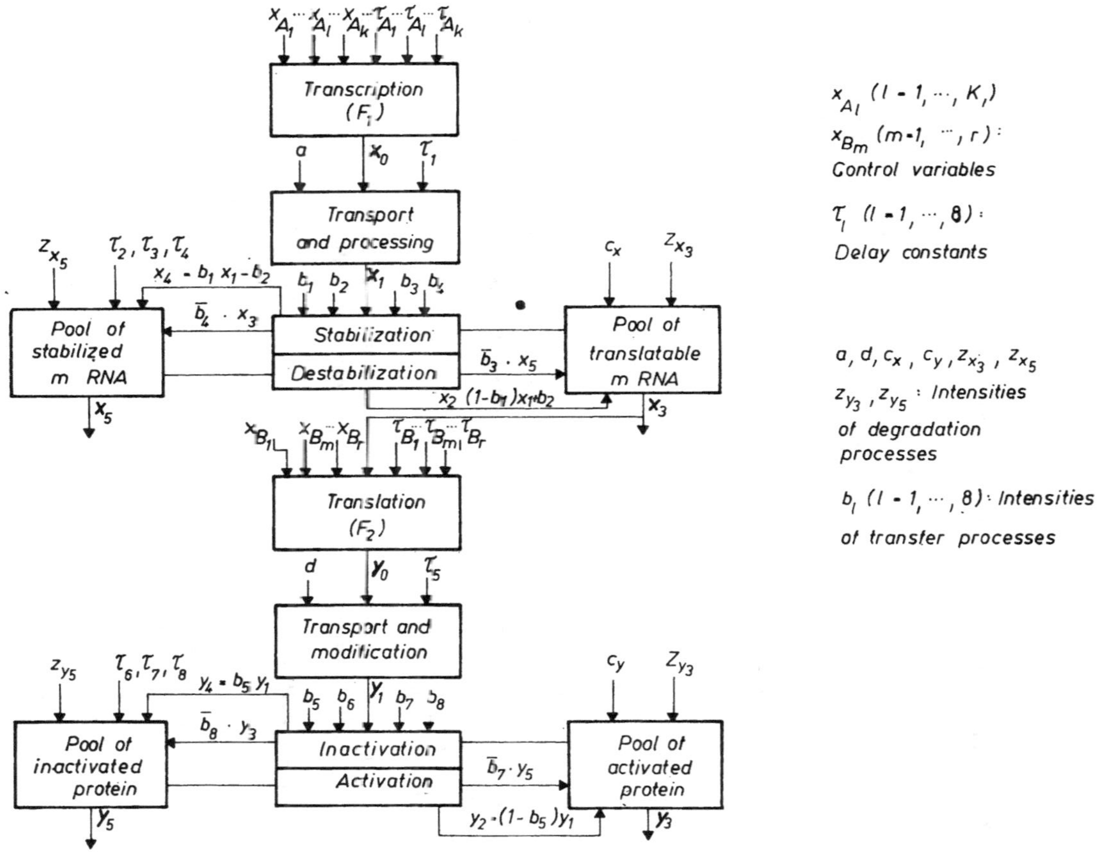

Fig. 2. Block diagram of a single Elementary Unit of Genetic Information Processing (EUGIP) based on the scheme of gene expression.

cellular milieu and the environment as well as the process mechanisms are described by normal distributed (lognormal) random variables.

10. The time is considered as a discrete variable, because the synthesis and degradation processes are discrete events, e.g. the polymerization of the macromolecules occurs indiscrete steps (see contribution NEUMANN, KREISCHER, in this volume). Discretization of the time variable allows a strong reduction of numerical calculation needed.

2.2. Formalization of the main sub-processes of gene expression

The most important species of macromolecules participating in the regulation of gene expression are RNAs and proteins; they are represented by the variables x_i ($i = 0, 1, \ldots, 5$) resp. y_i ($i = 0, 1, \ldots, 5$), where x_i is the concentration of the i-th RNA-pool (dimension: molecules) or the rate (dimension: molecules produced per time step) of the i-th RNA production (in every case to be specified). This holds analoguously for the y_i.

Among the process types the following ones are taken into account

a) Synthesis processes

$$x(t) = \bar{a} \cdot g(t) \left[\frac{\text{nbr. of molec.}}{T}\right] \tag{1}$$

T — time interval in seconds.

$\bar{a}$ is the unmodified, basic mean synthesis rate of molecules per time step. $g(t)$ is the control term modifying $\bar{a}$. $x(t)$ is the real gain of molecules (e.g. RNA molecules in transcription) in the time step t.

b) Transport and delay processes

τ is the characteristic parameter for the delay time and describes the number of time steps needed for transport processes (e.g. from nucleus to cytoplasm).

$$x_\beta(t) = a \cdot x(t - \tau) \tag{2}$$

$x_\alpha(t - \tau)$ designates the number of molecules at the site of their production in the moment $t - \tau$ (e.g. newly synthezised mRNA molecules in the nucleus). $x_\beta(t)$ is the number of molecules arriving at their destination (e.g. cytophasm) after the transport consuming τ time steps. This transport process may be linked with partial degradation (expressed by the constant a, $0 < a \leqq 1$) or with a qualitative modification (that is a question of interpretation of the variable x_β). With this representation in connection with a suitable interpretation of the variables spatial structures of the cell can be considered.

c) Transformation processes

The processes of transformation of a given component in two qualitatively different types (e.g. the change from native protein (x_α) to an active (x_β) and an inactive from (x_γ)) are often observed in gene expression:

$$x_\beta(t) = \left(1 - b_\alpha(t)\right) \cdot x_\alpha(t - \tau) - b_{\alpha'} \tag{3}$$

$$x_\gamma(t) = b_\alpha(t) \cdot x_\alpha(t - \tau) + b_{\alpha'}. \tag{4}$$

This formal representation of the transformation process is interpreted as follows:

$x_\alpha(t - \tau)$ is the number of original molecules in the moment $t - \tau$. The number of the molecule species generated in the transformation process of duration τ is designated by $x_\beta(t)$ and $x_\gamma(t)$. We assume that the sum of the newly synthesized molecules is equal to their original number. The function $b_\alpha(t)$ and the constant $b_{\alpha'}$ define the relation between the numbers of the new molecule species $b_\alpha(t)$ is an impulse function with variable intensity and variable instant of beginning and end of the impulse. With this the control of the beginning, the end and the intensity of the transformation process is possible.

d) Pool formation and degradation processes

The formation of the pool of active protein and the degradation of the protein molecules may be consindered as an example. The integration (storage) of the component y_α synthesized in a single time step into the pool y_β is realized by summing up $y_\alpha(t)$ to $y_\beta(t - 1)$:

$$y_\beta(t) = (1 - z_{y\beta})\, y_\beta(t - 1) + y_\alpha(t) \tag{5}$$

$$z_{y\beta} = 1 - \sqrt[n]{0{,}5} \qquad n = \frac{t_{1/2}\,[\text{seconds}]}{T\,[\text{seconds}]}. \tag{6}$$

A time constant $z_{y\beta}$ represents the degradation rate of the component y_β. $t_{1/2}$ is the half life time of y_β. T is the real time of a time step in sequence.

e) Restoring processes

Restoring is the transition of molecules that belong to a given pool to another pool connected with a change of their state. Such processes can be found in stabilization and destabilization of mRNA or in activation and inactivation of protein molecules. The transition of stabilized (non translatable) mRNA molecules x_γ (informosomes) into the pool of translatable mRNA molecules x_β by destabilization (and possibly transport) may be considered as an example. The transition processes are described by positive and negative terms in the corresponding equations for the acceptor and donator pools resp.

$$x_\beta(t) = (1 - z_{x\beta})\, x_\beta(t - 1) + x_\alpha(t) + b_\gamma(t)\, x_\gamma(t - \tau_R - 1) - b_\beta(t)\, x_\beta(t - 1). \tag{7}$$

The term $b_\gamma(t)\, x_\gamma(t - \tau_R - 1)$ designates the amount of molecules x_γ which come from the donator pool x_γ into the acceptor pool x_γ.

$$b_\gamma(t) = \begin{cases} 0 & \text{for} \quad 0 \leqq t < t_{\gamma a} + \tau_R \\ h_\gamma & \text{for} \quad t_{\gamma a} + \tau_R \leqq t \leqq t_{\gamma e} + \tau_R \\ 0 & \text{for} \quad t_{\gamma e} + \tau_R \leqq t. \end{cases} \tag{8}$$

The impulse function $b_\gamma(t)$ describes the intensity h_γ, the beginning ($t = t_{\gamma a}$), and the end ($t = t_{\gamma e}$) of the transition process. The time constant τ_R designates the duration of the transition process. The negativ term $b_\gamma(t)\, x_\gamma(t - 1)$ describes the release of molecules of type x_γ in another acceptor pool. The impuls function $b_\gamma(t)$ characterizes intensity, beginning, and end of this transition process analoguously to equation (8).

f) Control processes

All processes mentioned above are controllable by extra cellular or intra cellular components, i.e. that the parameters characterizing the single processes are functionally dependent on external factors or on other variables of the systems. This is realized by multiplication of the basic rate of a given process [see equ. (1)] with the control term $g(t)$. The general form of $g(t)$ is:

$$g(t) = \frac{\left[\frac{x_A(t-\tau)}{a_1}\right]^v}{1 + \left[\frac{x_A(t-\tau)}{a_1}\right]^w} \qquad 0 \leqq g(t) \leqq 1 \tag{9}$$

$x_A(t - \tau)$ is the number of molecules of the control factor x_A in the moment $t - \tau$, where τ is a delay constant describing the delay of action of the control factor x_A on the process to be controlled.

The exponents v and w determine the special shape of the control characteristic; the constant a_1 possesses the character of a threshold value of the control variable x_A for high values of v and w ($v, w > 10$).

In figs. 3a, b, c some characteristic control shapes often observed in biochemical systems are shown. They may be generated by the universal control function $g(t)$ [equ. (9)]. In fig. 3a one can obtain different non-linear induction and activation functions resp. $v = w = 1$ is the special case of a hyperbolic induction characteristic of Michaelis Menten type (Michaelis, Menten, 1913). With increasing parameters $v = w$ one gets induction curves with increasingly sigmoidal character.

Analogous control functions can be observed in the case of repression and inactivation control factors ($v = 0$, w increasing). For $w = 1$ (fig. 3b) one can find hyperbolic control functions of Michaelis Menten type and for $w \geqq 2$ sigmoidale repression characteristics and for $w > 10$ threshold behaviour. Fig. 3c shows the control characteristics of the substrat inhibition type, i.e. one can observe an optimal concentration of the control molecules x_A responsible for the maximal intensity of the controlled process (e.g. transcription). Optimum functions are observed if $1 \leqq v \leqq w$. The degree and form of ascent and descent of the curve is determined by v, w, and a_1.

Therefore, by choice of suitable parameters of equ. (9) it is possible to generate non-linear control shapes, important in molecular-genetic control systems. If a sub-process of gene expression is controlled by more than a single control variable more than one

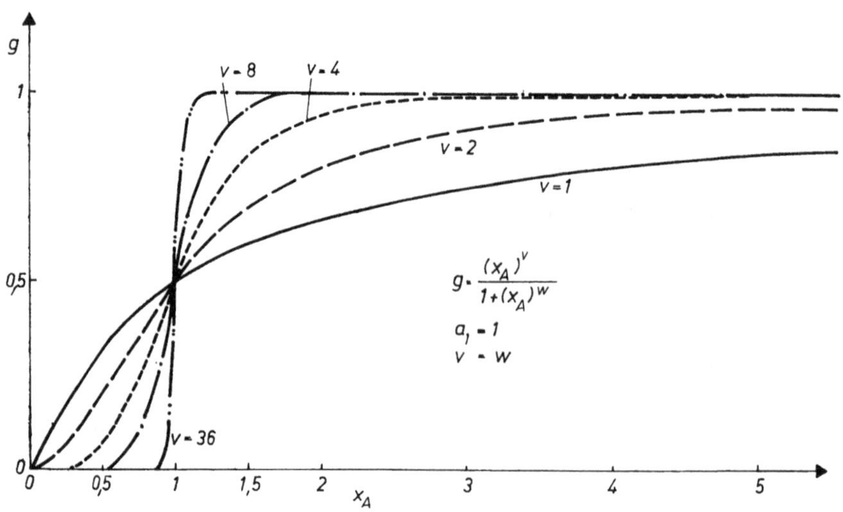

single control function of type (9) is needed. According to the qualitative effects of the different control variables the function $g(t)$ are coupled additively or multiplicatively [see equ. (14/15)].

g) Stochastic fluctuation processes

In the stochastic variant of the model the random effects in transcription, translation, degradation, that may be caused by random fluctuations of the intracellular milieu or stochastic properties of the physico-chemical mechanisms, are mapped by multiplication of the macroscopic parameters (mean synthesis and degradation rates) with random variables (with expected value $E = 1$).

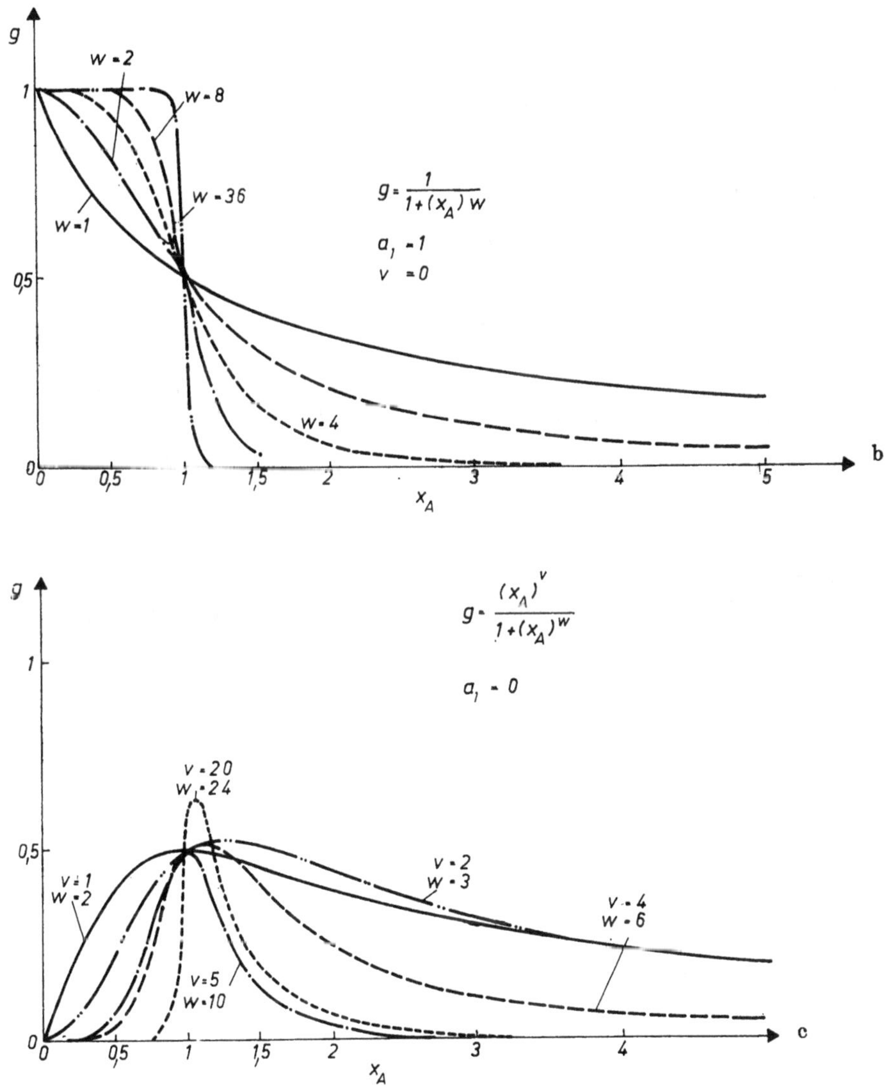

Fig. 3. Used control characteristics
a) induction (activation), b) repression (inactivation), c) optimum control (substrat inhibition).

The random variables $\xi(t)$ are realized step by step by means of a quick random number generator.

E.g. the RNA synthesis with noise is described by

$$x_{as}(t) = \bar{a} \cdot g(t) \cdot \xi(t) \left[\frac{\text{nbr. of molec.}}{T}\right] \tag{10}$$

$x_{as}(t)$ designates the number of molecules of type x_a synthetized in the time interval t. $x_{as}(t)$ is a random variable with $E(x_{a_s}) = \bar{a} \cdot g(t)$. We used normally distributed and log normally distributed random numbers: with the prohability densities

$$p_\xi(x) = \frac{1}{\sqrt{2\pi}\,\sigma} e^{\frac{-(x-\mu)^2}{2\sigma^2}}; \qquad -\infty < x < +\infty \tag{11}$$

with $E_\xi = \mu$, and dispersion $D_\xi^2 = \sigma^2$; and the modus $M_\xi = \mu$
and

$$p_\xi x = \begin{cases} \dfrac{1}{\sqrt{2\pi}\,\sigma \cdot x} e^{\frac{-(\ln x-\mu)^2}{2\sigma^2}} & \text{if} \quad x > 0 \\ 0 & \text{if} \quad x \leqq 0 \end{cases} \tag{12}$$

resp.

In (12) the expected value $E = e^{\mu+\frac{\sigma^2}{2}}$, the dispersion $D_\xi^2 = e^{2\mu+\sigma^2} \cdot (e^{\sigma^2} - 1)$, and the modus $M_\xi = e^{\mu-\sigma^2}$.

Because in the real processes negative rates (of synthesis and degradation) are impossible, we had to secure that always the rates $x_{as}(t) \geqq 0$. This is realized approximately if $\sigma \leqq \frac{1}{3}$ for normally distributed ξ. In this case it is $P(\xi(t) \geqq 0) = 0.99865$; if $\xi(t) < 0$ was realized, we set $x_{as} = 0$. With this probability density of $\xi : N(1, \sigma)$ was changed inconsiderablely. For $\sigma > 1/3$ we used log normally distributed ξ; in this case $x_{as}(t) > 0$ is always realized even in the case of high D_ξ^2.

Unfortunately with increasing D_ξ the difference between $E\xi$ and $M\xi$ increases too, therefore one can observe more frequently realizations $\xi(t) < E\xi$ than realizations $\xi(t) > E\xi$ and rare realizations which are many times higher than $E\xi$. Using the log normally distribution is tantamount to the assumption of irregularly working physico-chemical mechanisms.

2.3. Description of a single EUGIP and of the interactions in a genetic net of EUGIPs

The sub-processes of gene expression mentioned above are series-connected in the following sequence (see fig. 2)

1. Synthesis process.
2. Transport process with time delay and transformation process.
3. Two pool formation processes with degradation, transformation and restoring processes.
4. Synthesis process.

5. Transport process with time delay and transformation process.
6. Two pool formation processes with degradation, transformation and restoring processes.

The processes 1 and 4 are controllable by regulator proteins and external non proteinic signals. The basic unit of gene expression, i.e. an EUGIP, is described by a non-linear difference equation system as follows:

$$
\begin{aligned}
x_0^j(t) &= F_1^j\left(\{x^j_{Al}(t-\tau^j_{Al})\}, H_1^j\right) + a^j_{c1} \qquad (l = 1, 2, \ldots, k^j)\\
x_1^j(t) &= a^j \cdot x_0^j(t-\tau_1^j)\\
x_2^j(t) &= x_1^j(t) - b_1^j(t) \cdot x_1^j(t) - b_2^j\\
x_3^j(t) &= (1 - z^j_{x_3}) \cdot x_3^j(t-1) + x_2^j(t) + \bar{b}_3^j(t) \cdot x_5^j(t-\tau_2^j-1)\\
&\quad - b_4^j(t) \cdot x_3^j(t-1) - c_x^j\\
x_4^j(t) &= b_1^j(t) \cdot x_1^j(t) + b_2^j\\
x_5^j(t) &= (1 - z^j_{x5}) \cdot x_5^j(t-1) + x_4^j(t-\tau_3^j) + \bar{b}_4^j(t) \cdot x_3^j(t-\tau_4^j-1)\\
&\quad - b_3^j(t) \cdot x_5^j(t-1)\\
y_0^j(t) &= F_2^j\left(\{x^j_{Bm}(t-\tau^j_{Bm})\}, H_2^j\right) + a^j_{c2} \qquad (m = 1, 2, \ldots, r^j) \qquad\qquad (13)\\
y_1^j(t) &= d^j \cdot y_0^j(t-\tau_5^j)\\
y_2^j(t) &= y_1^j(t) - b_5^j(t) \cdot y_1^j(t)\\
y_3^j(t) &= (1 - z_3^j) \cdot y_3^j(t-1) + y_2^j(t) + \bar{b}_7^j(t) \cdot y_5^j(t-\tau_6^j-1)\\
&\quad - b_8^j(t) \cdot y_3^j(t-1)\, c_y^j\\
y_4^j(t) &= b_5^j(t) \cdot y_1^j(t)\\
y_5^j(t) &= (1 - z^j_{y5}) \cdot y_5^j(t-1) + y_4^j(t-\tau_7^j) + \bar{b}_8^j(t) \cdot y_3^j(t-\tau_8^j-1)\\
&\quad - b_7^j(t) \cdot y_5^j(t-1)\\
y_6^j(t) &= \frac{y_3^j(t)}{x_3^{p_j}(t)} \qquad (p_j \in \{1, 2, \ldots, n\}).
\end{aligned}
$$

The indices j $(j = 1, 2, \ldots, n)$ of the system variables indicate that n EUGIPs are considered simultaneously. With this all parameters of the model become n-dimensional vectors characterizing the peculiarities of the single EUGIPs. The variables x_i^j $(i = 0, 1, \ldots, 5)$ and y_1^j $(i = 0, 1, \ldots, 5)$ have the following meaning:

$x_0^j(t)$ — number of RNA-molecules synthesized during time step t;
$x_1^j(t)$ — number of RNA molecules arriving at cytoplasm during time step t;
$x_2^j(t)$ — number of translatable mRNA molecules arriving at cytoplasm during time step t;
$x_3^j(t)$ — number of translatable mRNA molecules in the cytoplasm at the end of time step t (concentration of the pool of translatable mRNA molecules);
$x_4^j(t)$ — number of stabilized (nontranslatable) mRNA molecules arriving at cytoplasm during time step t;

$x_5^j(t)$ — number of stabilized mRNA molecules in the cytoplasm at the end of time step t (concentration of the pool of nontranslatable mRNA molecules);

$y_0^j(t)$ — number of protein molecules synthesized during time step t;

$y_1^j(t)$ — number of proteins arriving at their site of employment (e.g. cytoplasm of the own cell and other cells) during time step t;

$y_2^j(t)$ — number of active proteins arriving at their site of employment during time step t;

$y_3^j(t)$ — number of active protein molecules at their site of employment at the end of time step t (concentration of the pool of active protein molecules);

$y_4^j(t)$ — number of inactive proteins arriving at their storage site during time step t;

$y_5^j(t)$ — number of inactive proteins at their storage site at the end of time step t (concentration of the pool of inactive protein molecules);

$y_6^j(t)$ — number of active protein molecules per DNA molecule during time step t (auxiliary variable for the control of transcription in phage replication).

The parameters a^j, d^j, c_x^j, c_y^j, $z_{x_3}^j$, $z_{x_5}^j$, $z_{y_5}^j$ and the functions $b_i^j(t)$ describe the intensities of degradation and restoring processes. The variables $x_{Al}^j(t - \tau_{Al}^j)$ and $x_{Bm}^j(t - \tau_{Bm}^j)$ ($l = 1, 2, \ldots, k^j$; $m = 1, 2, \ldots, r^j$) signify the control factors of transcription and translation, the sets H_1^j and H_2^j contain the parameters to specify the shapes of the control function and threshold values. The functions F_1^j and F_2^j are the main functions of the EUGIP; they describe the control of transcription and translation resp. by means of the control variables x_{Al}^j and x_{Bm}^j.

The algebraic form of these two functions is

$$F_1^j = \sum_{l=1}^{\bar{k}_j} g_{Al}^j + \prod_{l=\bar{k}_j+1}^{k^j} g_{Al}^j \tag{14}$$

or

$$F_1^j = \left(\sum_{l=1}^{\bar{k}_j} g_{Al}^j\right) \cdot \prod_{l=\bar{k}_j+1}^{k^j} g_{Al}^j \tag{15}$$

with the control terms

$$0 \leqq g_{Al}^j\left(x_{Al}^j(t - \tau_{Al}^j)\right) = \bar{a}_{scr_l}^j \cdot \frac{\left[\dfrac{x_{Al}^j(t - \tau_{Al}^j)}{a_{Al}^j}\right]^{v_{Al}^j}}{1 + \left[\dfrac{x_{Al}^j(t - \tau_{Al}^j)}{a_{Al}^j}\right]^{w_{Al}^j}} \leqq \bar{a}_{scr_l}^j \tag{16}$$

$\bar{a}_{scr_l}^j$ — transcription rate (normalized).

The control terms g_{Al}^j are connected additively or multiplicatively (see equ. (14, 15)) according to the control action of the variable x_{Al}^j.

Analogous equations are used for the functions F_2^j, which describe the regulation of translation:

$$F_2^j = \sum_{m=1}^{\bar{r}_j} g_{Bm}^j + \prod_{m=\bar{r}_j+1}^{r^j} g_{Bm}^j \tag{17}$$

or

$$F_2^j = \left(\sum_{m=1}^{\bar{r}_j} g_{Bm}^j\right) \cdot \prod_{m=\bar{r}_j+1}^{r^j} g_{Bm}^j \tag{18}$$

with the control terms:

$$0 \leqq g^j_{Bm}(x^j_{Bm}(t-\tau^j_{Bm})) = \bar{a}^j_{lat_m} \cdot \frac{\left[\dfrac{x^j_{Bm}(t-\tau^j_{Bm})}{a^j_{Bm}}\right]^{v^j_{Bm}}}{1+\left[\dfrac{x^j_{Bm}(t-\tau^j_{Bm})}{a^j_{Bm}}\right]^{w^j_{Bm}}} \leqq \bar{a}^j_{lat_m} \tag{19}$$

$\bar{a}^j_{lat_m}$ — normalized translation rate.

For the dynamic analysis of an EUGIP, e.g. investigation of the transfer behaviour (response to a given time depending input; response means time depending changes of pool concentrations), in the model the following standard test signals may be generated:

(I) $\quad x_{Al}(t) = a_{sl} \cdot \sin(2\pi N_l t) + \bar{a}_{sl}$

a_{sl} and $\bar{a}_{sl}$ determine amplitude and mean value and N_l the frequency of sinusoidal oscillation.

(II) $$x_{Al}(t) = \begin{cases} 0 & \text{if } 0 \leqq t < t_{a_l} \\ c_l & \text{if } t_{a_l} \leqq t < t_{e_l} \\ 0 & \text{if } t_{e_l} \leqq t. \end{cases}$$

In this the control variable $x_{Al}(t)$ is an impulse of intensity c_l and the duration $t_{e_l} - t_{a_l}$.

(III) $$x_{Al}(t) = \begin{cases} a_l t + a_{0_l} & \text{if } 0 \leqq t < t_{e_l} \\ a_l t_{e_l} + a_{0_l} & \text{if } t_{e_l} \leqq t. \end{cases}$$

Here $x_{Al}(t)$ increases linearly with ascent a_l until the moment t_{e_l} and remains constant after this.

The coupling of different units of gene expression is described by coupling matrices $\bar{K}(x^j_{Al})$ and $\bar{K}(x^j_{Bm})$ for the case of transcription variables and translation variables resp. $\bar{K}(x^j_{Al})$ and $\bar{K}(x^j_{Bm})$ are $(13, n)$-matrices at which each row corresponds with a system variable of equ. (13) and each column corresponds with one of the n interacting gene expression units. For example the system variable x_r^s of the s th unit of gene expression is used as control variable x^j_{Al} of the j-th EUGIP, that means the control of transcription of the j-th EUGIP is brought about by the variable x_r^s of the s-th EUGIP if

$$\bar{K}(x^j_{Al})_{o,p} = \begin{cases} 1 & \text{for } o = r, p = s \\ 0 & \text{otherwise} \end{cases} \tag{20}$$

holds. Then it is

$$x^j_{Al}(t) = x_r^s(t).$$

With this it is possible to investigate any net structure of interacting units of gene expression. Each of the 13 variables of an EUGIP from equ. (13) can work as control variable for any other unit.

The model of regulation of gene expression represents a high dimensional system of non-linear stochastic difference equations with time delay. Because of the strong non-linearities an analytical solution of the difference equations is impossible. The only possible way out is computer simulation, e.g. the simultaneous solution of all equations with increasing time parameter. For the stochastic variant of the model random experiments by Monte Carlo simulation (i.e. generation of realizations of the stochastic processes concerned with synthesis and degradation) with subsequent statistic analysis of the results are needed.

The model equations are realized as a program system GENIP (Genetic information processing) implemented for the computer BESM 6 on the basis of ALGOL. With this we have available a universal modular simulation system of regulation of gene expression.

By suitable choice of the system parameters it is possible to investigate the dynamic behaviour of theoretically or practically interesting special cases of molecular genetic networks.

In the next paragraph we present results about some properties of a single EUGIP and about the effects of interaction between two EUGIPs.

By means of GENIP and the Novosibirsk simulation model the complex control-system of early development of the bacteriophage Lambda was investigated. The results are represented in the contribution of Böttner, Bellmann, Tschuraev, Ratner "Modelling of epigenetic networks composed of monogenetic units of gene expression with reference to bacteriophage λ development" in this volume.

3. Application of the simulation system GENIP to some simple systems of regulation of gene expression

3.1. Dynamic analysis of a single EUGIP

3.1.1. Variation of the main parameters

As a simple example for the application of GENIP we depict the investigation of influences of changes of the most important parameters of gene expression to the induction kinetic of the protein pool y_3; these parameters are mRNA synthesis rate a_{scr}, protein synthesis rate a_{lat}, mRNA degradation rate z_{x3}, protein degradation rate z_{y3}.

Inductions of protein synthesis often occur in eucaryotic systems during the terminal phase of differentiation (accumulation of cell specific proteins, Rutter et al., 1968). Induction of protein synthesis may also be obtained for example in mature liver cells. The liver has to adapt to quickly changed conditions (e.g. change of chemical composition of blood after food uptake) to fulfill its task: maintenance of the internal milieu of blood and other tissues. Well investigated are for example the induction of thyrosine amino-transferase (TAT) by glycocorticoids (Lin, Knox, 1956) and by a synthetic corticosteroid (Dexamethason phosphat in tissue cultures, Granner et al., 1970). The kinetic of the protein pools in this experiments show sigmoidal ascent, confirmed by our simulation experiments (see fig. 4). Until to date only the time depending change of concentration of proteins, e.g. of the translation products, are measured in real experiments. Now the question arises, whether it is possible to get information about the sub-processes of gene expression running down before the translation (transcription, degradation of RNA and protein) from the shape of kinetic of the protein pool. Therefore we performed a series of simulation experiments to test the influence of parameter-variations of transcription and degradation on the qualitative shape of induction kinetic of the protein pool. To simplify the analysis we assume that no stabilization of mRNA and inactivation of proteins occur.

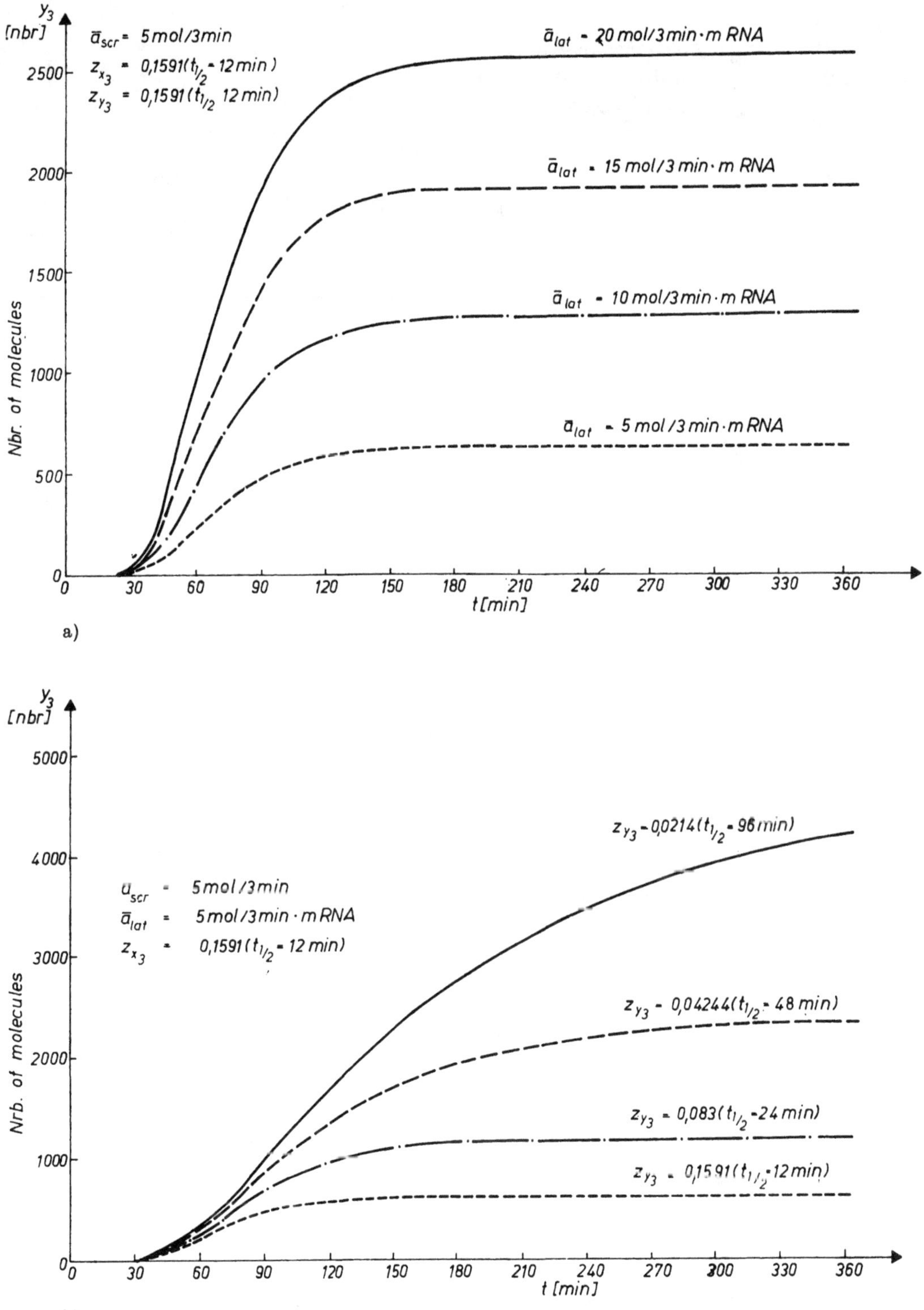

y_3 [nbr]
$\bar{a}_{scr}$ = 5 mol/3 min
z_{x_3} = 0,1591 ($t_{1/2}$ = 12 min)
z_{y_3} = 0,1591 ($t_{1/2}$ 12 min)
$\bar{a}_{lat}$ = 20 mol/3 min · m RNA
$\bar{a}_{lat}$ = 15 mol/3 min · m RNA
$\bar{a}_{lat}$ = 10 mol/3 min · m RNA
$\bar{a}_{lat}$ = 5 mol/3 min · m RNA
Nbr. of molecules
t [min]
a)
y_3 [nbr]
$\bar{a}_{scr}$ = 5 mol/3 min
$\bar{a}_{lat}$ = 5 mol/3 min · m RNA
z_{x_3} = 0,1591 ($t_{1/2}$ = 12 min)
z_{y_3} = 0,0214 ($t_{1/2}$ = 96 min)
z_{y_3} = 0,04244 ($t_{1/2}$ = 48 min)
z_{y_3} = 0,083 ($t_{1/2}$ = 24 min)
z_{y_3} = 0,1591 ($t_{1/2}$ = 12 min)
Nrb. of molecules
t [min]
b)

Fig. 4.

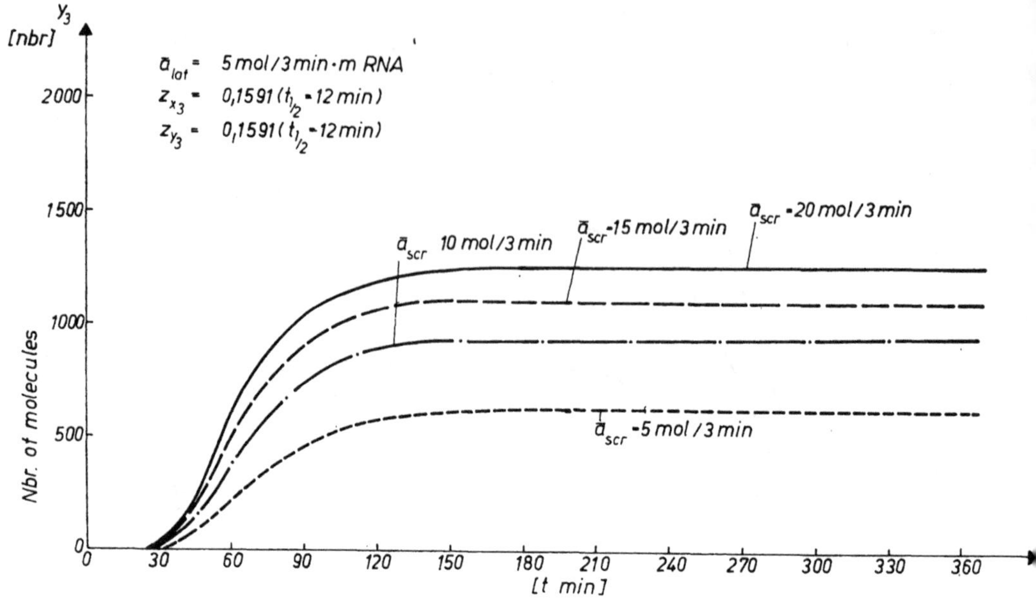

c)

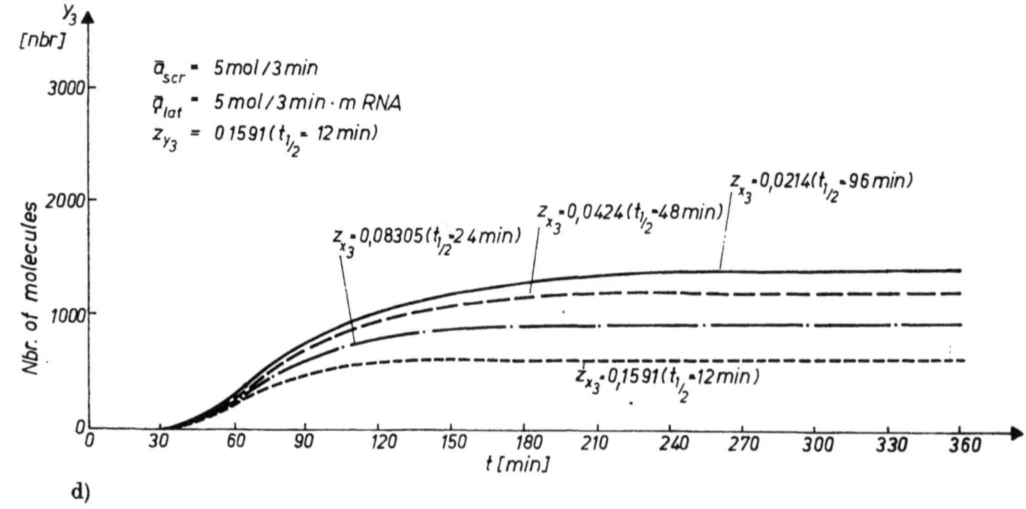

d)

Fig. 4. Induction kinetics of the protein pool for a) different translation rates $\bar{a}_{lat}$ and b) different protein degradation rates z_{y_3}, c) different transcription rates $\bar{a}_{scr}$, d) different mRNA degradation rates Z_{x_3}.

Figs. 4a, b, c, d show different curves of the induction kinetics of the protein pools depending on parameter variations of synthesis and degradation rates, around experimentally observed (BIELKA, 1973; MAHLER, CORDES, 1971; BAKER, YANKOFSKY, 1970; CHAMBERLINE, 1970; RATNER, 1972; ADESNIK, LIVINTHAL, 1970; BERNARD, 1970; WILSON, HOAGLUND, 1967; BOTON, 1971; RECHCIGL, 1971; SCHIMKE, DAYLE, 1970; GREENBLATT, 1973) mean values.

The other unchanged parameters correspond to experimental data too.

Clear qualitative difference between the curve types appear: in doubling of translation rate or of half life time of protein molecules leading to doubling of numbers of protein molecules (figs. 4a, b) in contrast to the effect of doubling of transcription rate or of half life time of the mRNA (figs. 4c, d); increase of the translation and transcription rates lead to nearly equal time instants at which the stationar concentrations of the protein pools are reached (figs. 4a, c) in contrast to the effect of increase of half life time of mRNA and protein (figs. 4b, d).

With this it may be possible to get hints about the target sites of biologically active substances acting on sub-processes of gene expression.

Simulation experiments with stepwise changes of two or few parameters of the model (multiple factor analysis) may help in interpretation of results of experiments conducted to explore the mode of action of pharmaca.

3.1.2. Definition and computation of parameter sensitivities

In order to get a quantitative evaluation of the importance of a given parameter p_k of the model, by means of its effect on the proteinpool concentration y_3 a parameter sensitivity analysis was carried out.

This was done by testing the effect of small variations p_k of the parameter when the system was in its stationary state. The static sensitivity is

$$S_{p_k}^{y_3}(p_1, \ldots, p_k) = \frac{y_3(p_1, \ldots, p_k + \Delta p_k, \ldots, p_n) - y_3(p_1, \ldots, p_k, \ldots, p_n)}{\Delta p_k}. \quad (21)$$

The relative static parameter sensitivity

$$\hat{S}_{p_k}^{y_3}(p_1, \ldots, p_n) = \frac{\dfrac{y_3(p_1, \ldots, p_k + \Delta p_k, \ldots, p_n) - y_3(p_1, \ldots, p_n)}{y_3(p_1, \ldots, p_n)}}{\dfrac{\Delta p_k}{p_k}}$$

$$= S_{p_k}^{y_3}(p_1, \ldots, p_n) \cdot \frac{p_k}{y_3(p_1, \ldots, p_n)} \quad (22)$$

is a measure for the relation between relative change of the number of active protein molecules and relative variation of the parameter considered. Other measures of parameter sensitivities are possible for instance the effect of Δp_k on other macromolecular pools. The sensitivities of equ. (21) and (22) are functions of the vector $\bar{p} = (p_1, \ldots, p_n)$ in n-dimensional parameter space. Because of the high demension of this space (a EUGIP contains 10 main parameters) it is impossible to calculate $S_{p_k}^{y_3}(\bar{p})$ or $\hat{S}_{p_k}^{y_3}(\bar{p})$ for each parameter vector. That is the reason why we have only determined the sensitivity for such vectors $\bar{p}$ that are contained in subspaces corresponding experimental results found in some pro- and eucaryotic systems; we have tried to get a survey on the range and functional dependencies of the sensitivities in experimentally determined subspace. Consequently we have chosen some points $\overline{p}_1, \ldots, \overline{p}_7$ in the experimental parameter subspace.

In these points of the numerical values of the sensitivities $\hat{S}^{y_3}_{p_k}(\overline{p})$ for 6 different parameters p_k ($k = 1, \ldots, 6$) were calculated using the model.

As the "normal" working point $\overline{p}_1$ of an EUGIP the following parameter values were chosen

$$p_1: a_{scr} = 5 \text{ mol/3 min}, \quad p_2: a_{lat} = 5 \text{ mol/3 min} \cdot \text{mRNA}$$

$$p_3, p_4: z_{y_3} = z_{x_3} = 0{,}1591 \text{ (corresponding to half life time } t_{1/2} = 12 \text{ min using time step of } T = 3 \text{ min.)}$$

$$p_5: a_B = 60 \text{ mol}, \quad p_6: w_B = 1.$$

a_B and w_B are parameters for the determination of the control of translation by the number of mRNA molecules [see equ. (19)].

The relative, static parameter sensitivities $\hat{S}^{y_3}_{p_k}(\overline{p})$ were also calculated for the points $\overline{p}_2, \overline{p}_3, \ldots, \overline{p}_7$.

$\overline{p}_2$: parameters equal to $\overline{p}_1$ but $a_{scr} = 15$ mol/3 min

$\overline{p}_3$: parameters equal to $\overline{p}_1$ but $a_{lat} = 15$ mol/3 min · mRNA

$\overline{p}_4$: parameters equal to $\overline{p}_1$ but $z_{x_3} = 0.08299$ ($t_{1/2} = 24$ min)

$\overline{p}_5$: parameters equal to $\overline{p}_1$ but $a_{scr} = 15$ mol/3 min and $w_B = 2$

$\overline{p}_6$: parameters equal to $\overline{p}_1$ but $a_{scr} = 15$ mol/3 min and $a_B = 40$ mol

$\overline{p}_7$: parameters equal to $\overline{p}_1$ but $a_{scr} = 15$ mol/3 min, $a_B = 40$ mol and $w_B = 2$.

The relative parameter sensitivities for the different points $\overline{p}_1, \ldots, \overline{p}_7$ are shown in table 1. At the normal working point $\overline{p}_1$ of the system the relative static sensitivities of the translation rate and the protein degradation rates are the greatest ones ($|\hat{S}^{y_3}_{a_{lat}}| = |\hat{S}^{y_3}_{z_{y3}}| = 1$). The relative static sensitivities of the transcription rate and of the mRNA degradation rate are ± 0.65. The relative static sensitivities of the control function parameters (a_B, w_B) are low.

$|\hat{S}^{y_3}_{a_{lat}}| = |\hat{S}^{y_3}_{z_{y3}}| = 1$ means in the stationary state the number of protein molecules depends linearly on the parameters a_{lat} and z_{y3} resp.

By increasing the transcription rate a_{scr} up to the amount of $a_{scr} = 15$ mol/3 min (point $\overline{p}_2$) the relative static sensitivities of the translation rate and protein degradation rate remain constant, the sensitivities $\hat{S}^{y_3}_{a_{scr}}$ and $\hat{S}^{y_3}_{z_{x3}}$ decrease down to the level $\pm 0{,}39$, while $\hat{S}^{y_3}_{a_B}$ increases (0.60) and $\hat{S}^{y_3}_{w_B}$ even changes its sign (-0.27).

A similar case may be observed in point $\overline{p}_5$ relative to the parameters a_{scr} and z_{y_3}. Summing up one may recognize that the sensitivity of a given parameter depends on the working point in the parameter space. Nevertheless, in general we observe (i) that at the parameter considered the sensitivities of the translation rate and of the protein degradation rate are greater than the sensitivities of the transcription rate and the mRNA degradation rate and (ii) that $\hat{S}^{y_3}_{a_{lat}}$ and $\hat{S}^{y_3}_{z_{y3}}$ remain constant independent on the space point and (iii) that changing a_{lat} (point $\overline{p}_3$) does not lead to a change of sensitivities corresponding to point $\overline{p}_1$.

Table 1. Relative static parameter sensitivities in different points $\bar{p}_1, \ldots, \bar{p}_7$ of the experimentally determined parameter space of an EUGIP.

point of parameter spaces	$\hat{S}^{y_3}_{a_{scr}}$	$\hat{S}^{y_3}_{a_{lat}}$	$\hat{S}^{y_3}_{z_{x3}}$	$\hat{S}^{y_3}_{z_{y3}}$	$\hat{S}^{y_3}_{a_B}$	$\hat{S}^{y_3}_{w_B}$
$\bar{p}_1$	0,65	1,00	−0,66	−1,01	0,34	0,22
$\bar{p}_2$	0,39	1,00	−0,39	−1,01	0,60	−0,27
$\bar{p}_3$	0,66	1,00	−0,66	−1,00	0,34	0,22
$\bar{p}_4$	0,49	1,00	−0,49	−1,01	0,49	0,00
$\bar{p}_5$	−0,42	1,00	0,42	−0,99	0,14	−0,64
$\bar{p}_6$	0,30	1,00	−0,30	−0,99	0,70	−0,60
$\bar{p}_7$	−0,83	1,00	0,83	−0,99	1,68	−1,45

One may observe a strong dependency of sensitivities of the control function parameters w_b and a_B upon given working points. These few examples enlight the complicated parameter-dependensies even in the case of an simple EUGIP.

We find a much more complicated situation in the case of estimation of sensitivities of EUGIPs parameters, when the EUGIPs are elements of a genetic network with non-linear control functions. Here it is only possible to calculate the sensitivities for a given working point. After our experiences it seems nearly hopeless to find a general sensitivity level for a given parameter in a more or less extended subspace. The statement about the sensitivities, and about the system's stability too, are only correct in the boundaries of the model and its assumptions.

3.1.3. Dynamics of gene expression depending on the shape of the control characteristics

As an example we consider the non-linear signal transformation in the control of the transcription using optimum control characteristic (see fig. 3c and equ. (9), $v = 2$, $w = 4$, $a_1 = 10$). We assume that the protein signal (control variable x_A) controlling the transcription is a sinusoidal one with 0 to 40 molecules and a period of oscillation of 96 min. For a time step we have choosen 3 min, furthermore $a_{scr} = 10$ mol/3 min, $a_{lat} = 8$ mol/3 min · mRNA, $z_{x_3} = z_{y_3} = 0.12944$ (corresponding to a half life time of 15 min). As shown in fig. 5 the frequency of the oscillation of the protein pool concentrations is the twofold of the frequency of control variable x_{Aj}. The sinusoidal oscilation of the control variable x_A is strongly distorted by the optimum control characteristic, i.e. if there are control characteristic of the type of substrate inhibition a frequency multiplication in this epigenetic system may be observed. We see this behaviour in connection with the well known physiological phenomenon generally called physiological clock. It may be possible that the cell uses this mechanism in order to generate characteristic time scales, which are the basis for temporal coordination of processes occurring simultaneously during the differentiation and the cell cycle.

With this we realize that signal transformation processes in a single EUGIP or a net of EUGIPs are very complex and that it is impossible to explain the function of molecular genetic control systems only by intuitive considerations about the dynamic events of the expression.

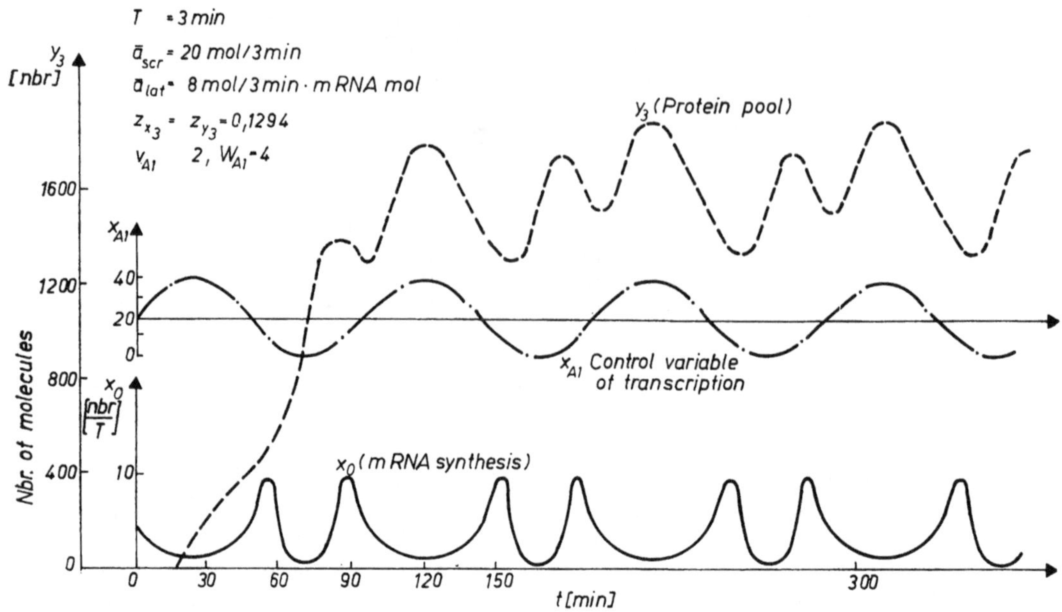

Fig. 5. Dynamics of mRNA synthesis and protein pool concentration depending on sinusoidal control of transcription using a optimum control characteristic of type c) in fig. 3.

3.2. Dynamic analysis of two interacting EUGIPs

3.2.1. Structure of the interaction

We consider two EUGIPs interacting by mutual control of their transcription process: each EUGIP produces a protein acting as a repressor of the transcription of the other EUGIP (see fig. 6). The transcription of EUGIP 1 is controlled by x^1_{A1} (protein of EUGIP 2), the transcription of EUGIP 2 by x^2_{A1} (protein of EUGIP 1). The control function $F_1{}^j$ ($j = 1, 2$) consists only of a single term [see equ. (16)],

$$F_1{}^j = g^j_{A1}\big(x^j_{A1}(t - \tau^j_{A1})\big), \quad j = 1, 2$$

describing the dependence of the transcription on the regulatory protein of the other EUGIP in each case. Because the interaction is repressive, $v^j_{A1} = 0$ [see equ. (16) and fig. 3b] holds for both EUGIP's. The shape of the repression function (hyperbolic, sigmoidal or threshold) is determined by the parameter w^j_{A1}.

The repressive feedback coupling of the two EUGIPs is described by the coupling matrices [see equ. (20)]:

$$\bar{K}(x^1_{A1})_{0,p} = \begin{cases} 1 & \text{if} \quad 0 = 12, p = 2 \\ 0 & \text{otherwise} \end{cases}$$

and

$$\bar{K}(x^2_{A1})_{0,p} = \begin{cases} 1 & \text{if} \quad 0 = 12, p = 1 \\ 0 & \text{otherwise} \end{cases}$$

$$0 = 1, 2, \ldots, 13, p = 1, 2.$$

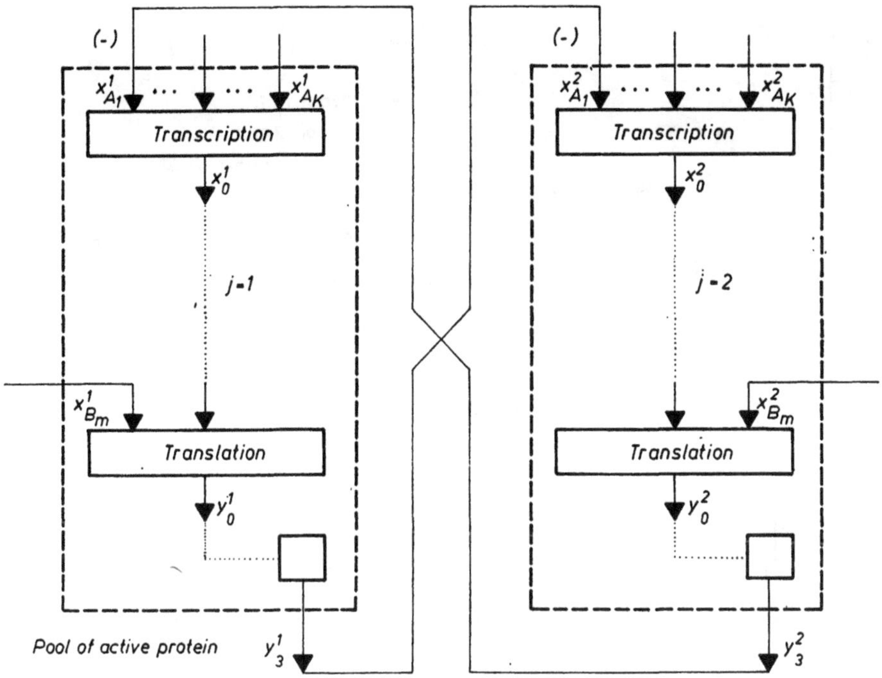

Fig. 6. Feedback structure of a 2-EUGIP-system with transcription mutually controlled by their proteins.

3.2.2. Some characteristic properties of a deterministic two-EUGIP system

With regard to the function of molecular control systems in the cell especially the number and stability of stationary states, their dependency on the parameters and initial values of the variables of the two EUGIPs, the transitions between stationary states and the possible cyclic process behaviour are of main interest. First of all we consider the case that both EUGIPs are parameterized identically. Then the coupled system has three stationary states:

1. EUGIP 1 synthesizes more protein than EUGIP 2 (most extreme case: EUGIP 1 maximal production, EUGIP 2 no production), (stable state),
2. reversed situation (stable state),
3. The two EUGIPs produce an equal amount of protein molecules (unstable state).

Which of the three states is actually reached depends on the initial values of the system variables.

A state 1 is oberserved if, for instance, the initial concentration of protein pool of EUGIP 1 is higher than that of EUGIP 2. State 2 is observed in the reversed case. State 3 is reached if the initial concentrations are equal in both pools. The unstable equilibrium case has some peculiarities: (i) the equal intensity of protein synthesis of both EUGIPs may be constant (fig. 7) after the transition period or (ii) periodic oscillations may be observed (fig. 8). Which of the two possibilities appear depends on the shape of the control characteristic of transcription.

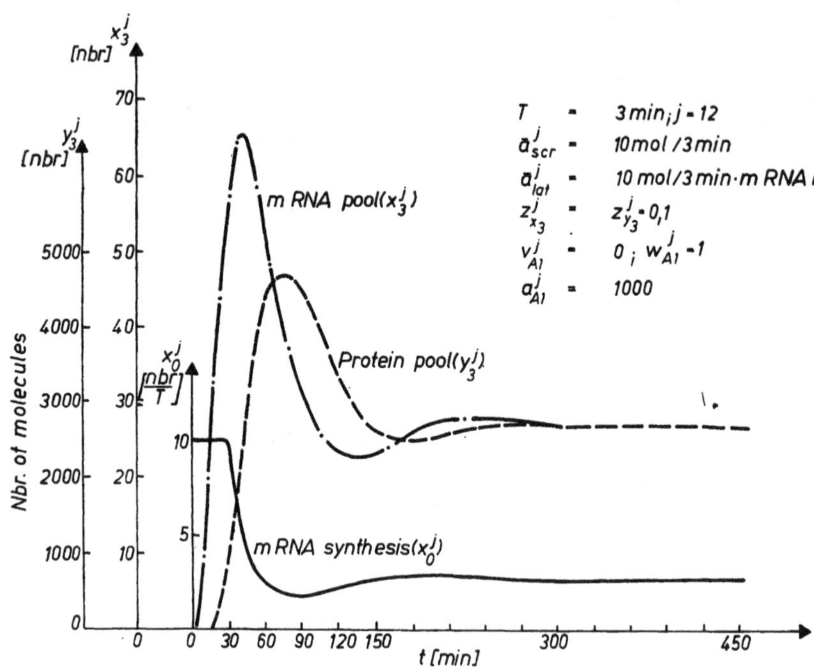

Fig. 7. Dynamics of mRNA synthesis, mRNA pool, and protein pool concentrations for both of the two interacting EUGIPs of fig. 6, with mutually protein controlled transcription by means of a hyperbolic repression control characteristic (see fig. 3b: $v = 0$, $w = 1$).

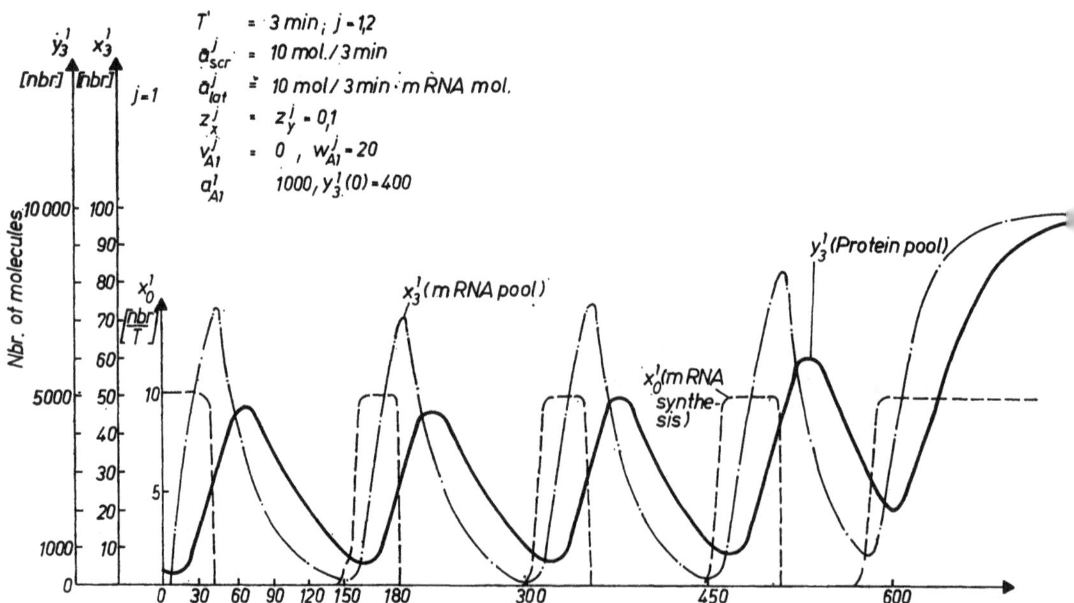

Fig. 8. Dynamics of the 2-EUGIP-system represented in fig. 6; transcription control by means of a sigmoidal characteristic (see fig. 3b: $v = 0$, $w = 4$).

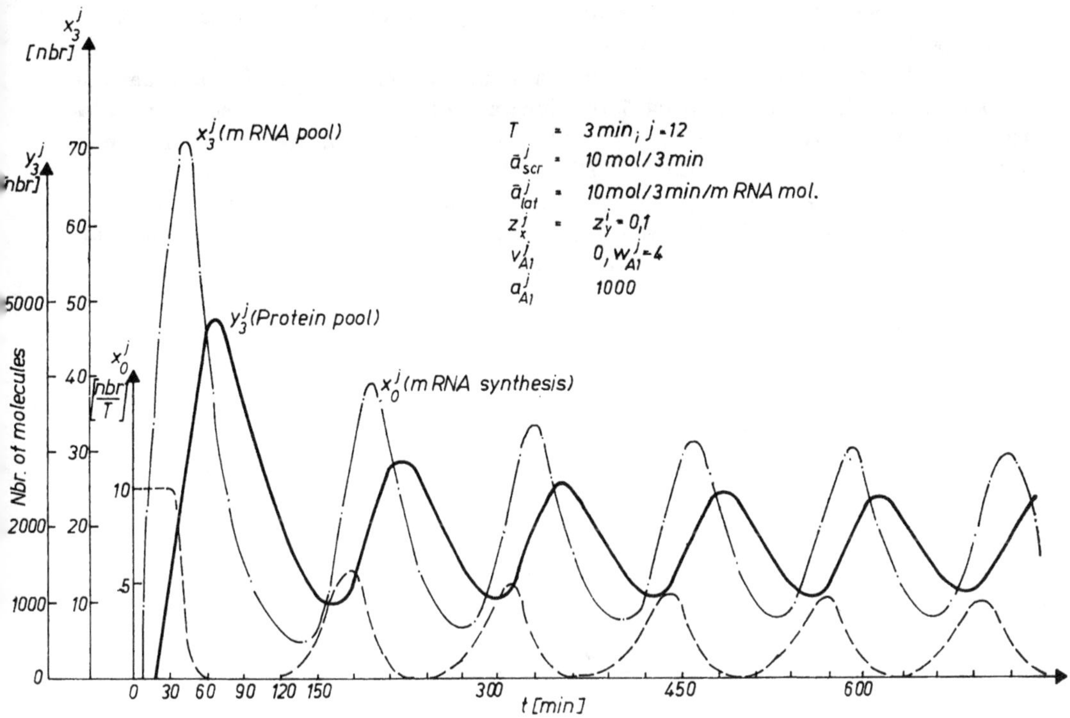

a)

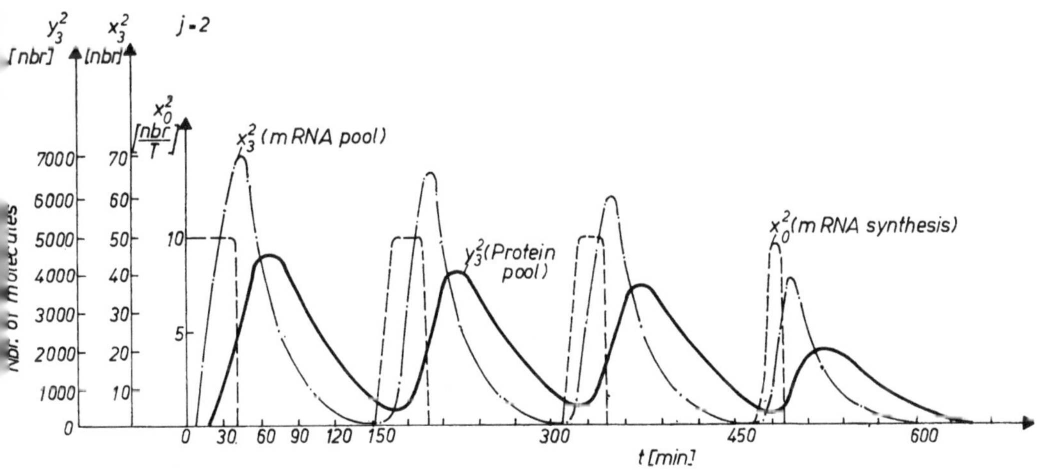

b)

Fig. 9. Dynamics of the 2-EUGIP-system represented in fig. 6; initial protein pool concentration of EUGIP 1 and 2 are 400 and 0 molecules resp.; transcription control by means of a threshold characteristic (see fig. 3b: $v = 0$, $w = 20$). a) and b): dynamics of EUGIP 1 and EUGIP 2 resp.

The condition for (i) is $w^j_{A1} = 1, 2, 3$; $w^j_{A1} = 4$ ($j = 1, 2$) leads to syncronic oscilation of both EUGIP's.

The qualitative shape of the repression characteristic controls strongly the behaviour of both interacting systems. In fig. 7 the dynamics of the important system variables (pool concentrations for $w^j_{A1} = 1$; $j = 1, 2$) is demonstrated. This representation holds for both EUGIPs because both EUGIPs work synchronously. After approximately 300 minutes the system converges against the stationary state, i.e. each EUGIP produces, 2.8 molecules of mRNA in the elementary time interval of 3 min. This fig. holds if the initial concentration of the mRNA pool in both EUGIPs was zero.

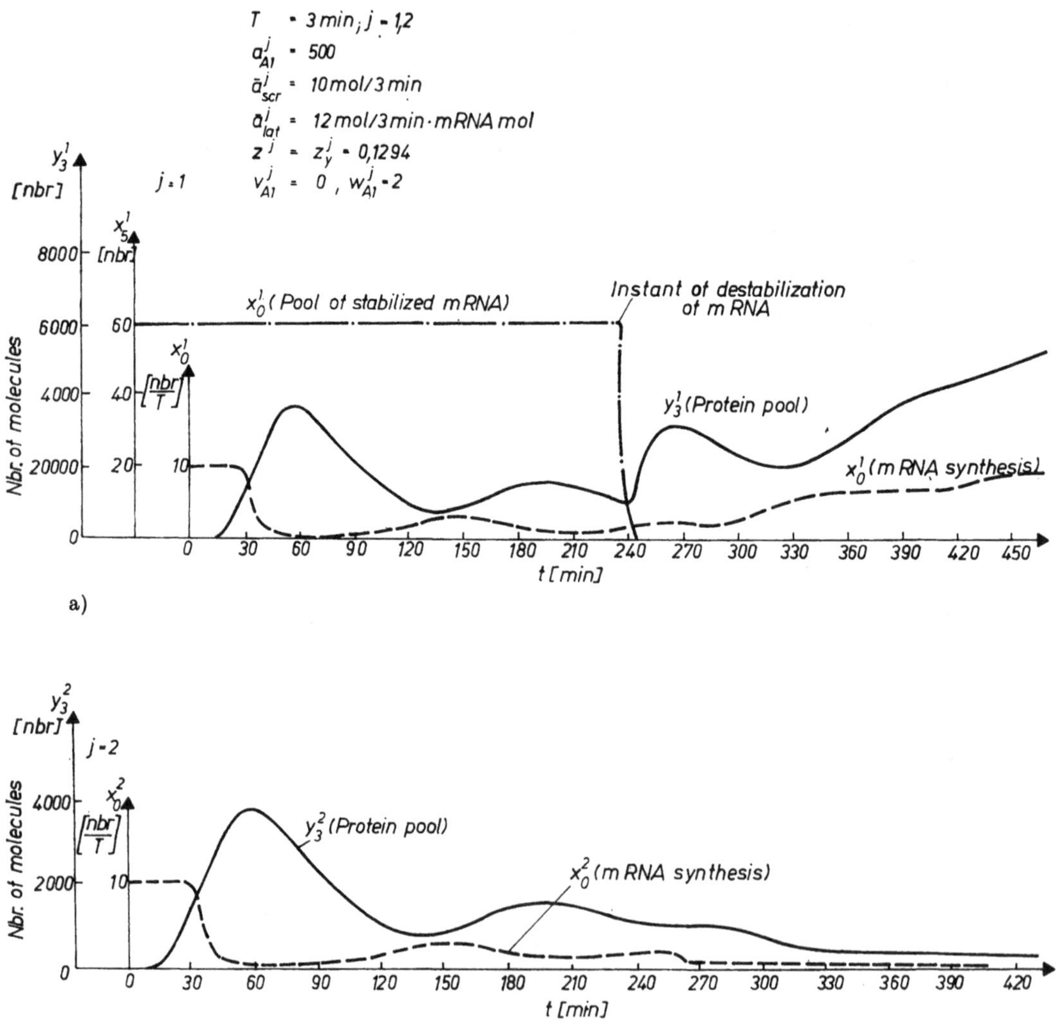

Fig. 10. Effect of destabilization of stabilized mRNA molecules (informosomes) on the dynamics of the 2-EUGIP-system represented in fig. 6; transcription control by means of a sigmoidal repression control characteristic (see fig. 3b: $v = 0$, $w = 2$). a) and b): dynamics of EUGIP 1 and EUGIP 2 resp.

If $w^j_{A1} = 4$ ($j = 1, 2$) undamped oscillations of system variables are observed (fig. 8). Increasing w^j_{A1} ($j = 1, 2$) up to 20, i.e. up to a threshold characteristic, results in an impulse like control of gene expression (see figs. 9a, b). Furthermore the behaviour of the two coupled gene expression units is determined by the initial concentrations of the mRNA and protein pools.

If corresponding initial concentrations of both EUGIP's (e.g. protein pools) are different, in the early phase of the process both systems oscillate and converge to two different levels of mRNA synthesis stationary state 1 resp. 2 mentioned above). see figs. 9a, b with $w^1_{A1} = 20$, $w^2_{A1} = 20$ and $y_{31}(0) = 400$, $y_{32}(0) = 0$. After four oscillations the system turns to a state characterized by full synthesis of EUGIP 1 and no synthesis of EUGIP 2. We emphasize that this convergence behaviour is only the result of a small initial advantage of EUGIP 1: $y_{31}(0) = 400$ but $y_{32}(0) = 0$, e.g. EUGIP 1 possesses already 400 protein molecules repressing the transcription of EUGIP 2. This advantage leads to the successive repression process of the other sub-system. This successive repression process proceeds in a characteristic time interval (depending on the chosen parameters). With this the time structure of a process is determined. One may speculate that such an internal programming of genome depending processes is an important element of regulation and synchronization of processes concerned with control of cell cycle.

The generation of periodic signals by interacting parts of the genome and the phenomenon of frequency multiplication (see fig. 5) could be significant for the coordination of processes concerned with cell differentiation and morphogenesis.

In fig. 10 an example for transition of the two-EUGIP system (with EUGIP 1 and 2) from one state to another is demonstrated. This example represents the situation that during cell differentiation in a given instant the informosomes of EUGIP 1 are activated by an external signal. As a consequence the two operon system changes from the equilibrium state 3 to state 1, e.g. EUGIP 1 increases its synthesis rate drastically with the result that after a successive repression process only EUGIP 1 synthesizes RNA, EUGIP 2 is permanently completely repressed.

Until now we have considered EUGIPs identically parametrisized, e.g. equal parameters of the control functions, of the synthesis and degradation rates. If these system parameters are different in both units no equilibrium state (no undamped oscilation) exists even in the case of equal initial pool concentrations. In every case, the system developes to one of the stable states basically characterized by different RNA synthesis-rates. Changing these parameters for both units leads to qualitatively the same behaviour patterns observable after changing initial pool concentration as mentioned above. For example it is possible to observe the same behaviour as shown in figs. 9a, b if e.g. transcription rates, mRNA degradation, protein synthesis rates, protein degradation rates and/or control parameters are different in both EUGIPs.

Experimental results about periodicities of protein synthesis supporting the theoretically derived possibilities of cyclic behaviour in epigenetic networks are given e.g. by Knorre [30] and Fuller [31].

3.2.3. Dynamic behaviour of a stochastically disturbed two-EUGIP system

It is a well known fact, that biological systems are stable within certain boundaries of environmental conditions, even when their sub-systems are working more or less unreliably. Such variable functioning of sub-systems may be based either on intrinsic stochasticity of the mechanisms or on stochastic fluctuations of environmental conditions or on both. One may ask now, which degree of stochasticity of single processes of the whole system can be still tolerated, i.e. how strongly sub-processes may be stochastically disturbed without loss of stability of the whole system; stability means: no qualitative differences in the time behaviour up to a certain maximal level of stochasticity of the sub-processes. We will demonstrate that this maximal stochastic level depends strongly on the number of system elements, i.e. sub-systems, and the type of their interactions: topology and the shape of the control characteristics. If we consider a genetic net work consisting only of two EUGIPs with a special feedback structure and with special further parametrizations (rates etc.) the maximal stochastic level seems to be much lower than that which is valid for a genetic network composed on more than two operons, as e.g. the control system of the early development of the λ-phage consisting of six operons with a naturally evolved feedback structure and control characteristics, explored in experiments by molecular geneticist.

A broad outline about computer experiments about stability of the λ-system is given in the contribution of Böttner, Bellmann, Tschuraev, and Ratner in this volume.

In the following the results about the effect of stochasticity in a two operon system is demonstrated.

In the system, which behaviour is represented in figs. 9a, b, the sub-processes transcription, translation, degradation of mRNA and protein are considered as to be disturbed stochastically. The real rates which are valid in a given time interval were calculated by equ. (10), paragraph 2.2. g. The used system is equally parameterized to that of figs. 9a, b.

The degree of stochasticity is characterized by the parameter σ, related to the expected rate $E(x_{a_s}) = \bar{a} \cdot g(t)$; $0 \leqq \sigma \leqq 1$.

The computer experiments to test the effect of stochasticity on the system stability were performed with $\sigma = 0.03$ and $\sigma = 0.3$ in comparison with $\sigma = 0$.

In the deterministic case ($\sigma = 0$) one may observe (see figs. 9 and 11) that the system oscillates in a characteristic time interval of 665 min. It is the interval of the successive repression process. After this time only EUGIP 1 is working. It is typical for this case that the transcription works in both systems four times and then only in EUGIP 1.

This characteristic behaviour is qualitatively changed (see table 2) by the weak stochasticity of $\sigma = 0{,}03\%$: In 54.9% of 51 replicated stochastic computer experiments the system behaves as the deterministic one, 31.4% of the cases show only three oscillations, i.e. more than four oscillations are observed in a frequency of about 14%. In fig. 12, a stochastic run with $\sigma = 0.03$ is demonstrated. Examples of stochastic runs in the case of $\sigma = 0.3$ are shown in figs. 13a, b, c. The drastically disturbed rates lead to very abnormal figures (see table 2).

It is typical that in 50% of the runs only two transcription impulses were observed, after that only EUGIP 1 was working. Only 11.9% of the runs show the typical behaviour of the deterministic variant.

The characteristic time interval of the successive repression process of EUGIP 2 by EUGIP 1, (in the determinstic case 665 min) is generally reduced by the tested stochasticity of $\sigma = 0.3$, in about 17% of the runs reduced to 345 min.

Assuming that the protein of EUGIP 1 controls another important process within the cell (e.g. cell division) by a threshold characteristic with a threshold of 5000 EUGIP 1 protein molecules one see that the characteristic time (in the normal case 665 min) needed to reach the threshold is strongly changed in different directions, that may lead to a desorganization of subsequent cell divisions.

This may be a model of thought in the discussion of cancer problems.

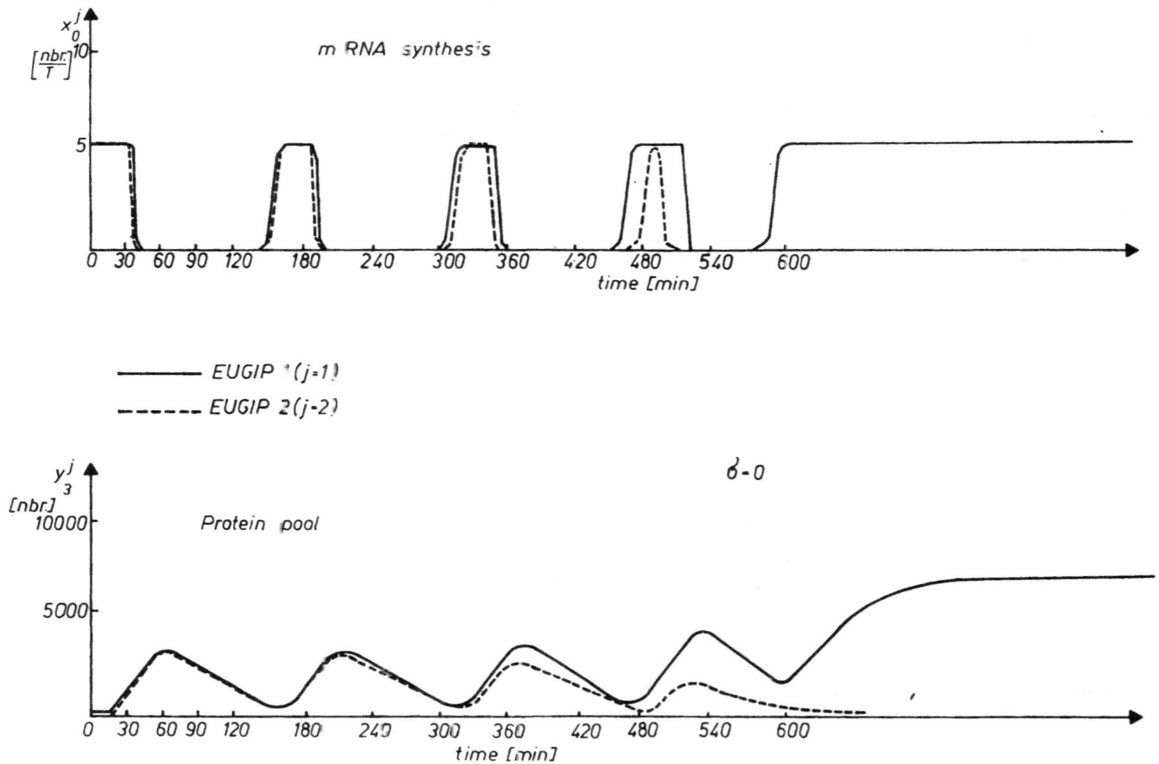

Fig. 11. Dynamics (mRNA synthesis, protein pool concentration) of the 2-EUGIP-system represented in fig. 6; deterministic case: $\sigma = 0$.

57

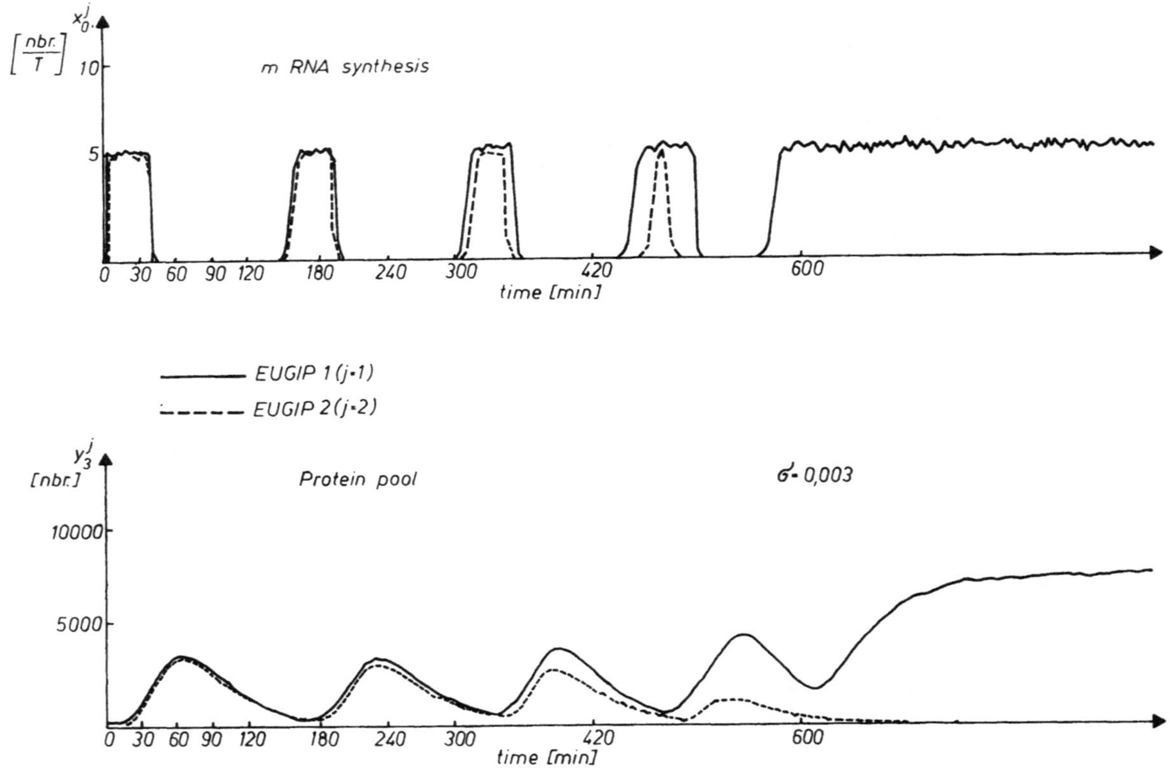

Fig. 12. Effect of weak stochasticity of the sub-processes transcription, translation, mRNA and protein degradation on the dynamics of the 2-EUGIP-system of fig. 11; stochastic case: normally distributed random variables with $\sigma = 0{,}03$. As an example only the variation of the transcription rate (mRNA synthesis) is shown.

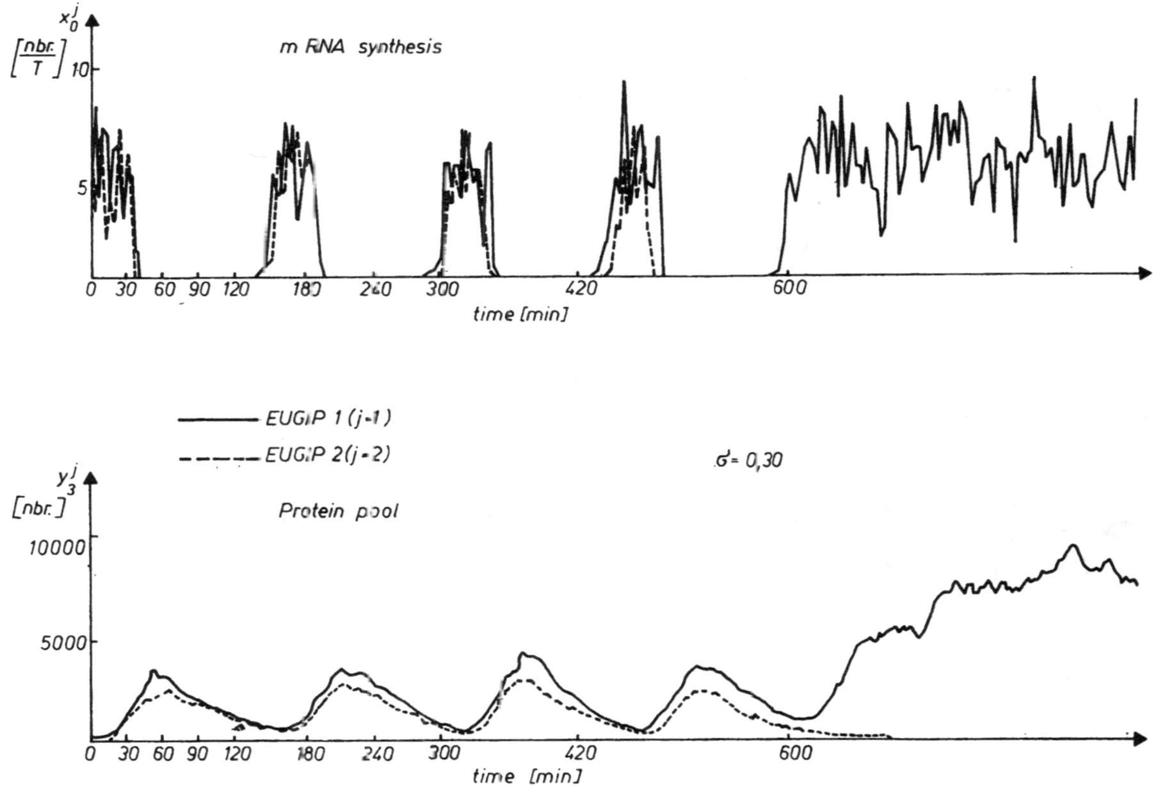
x_0^j
[nbr./T]
10
5
m RNA synthesis
0 30 60 90 120 180 240 300 420 600
time [min]
EUGIP 1 (j=1)
EUGIP 2 (j=2)
σ = 0,30
y_3^j
[nbr.]
Protein pool
10000
5000
0 30 60 90 120 180 240 300 420 600
time [min]

Fig. 13 a)

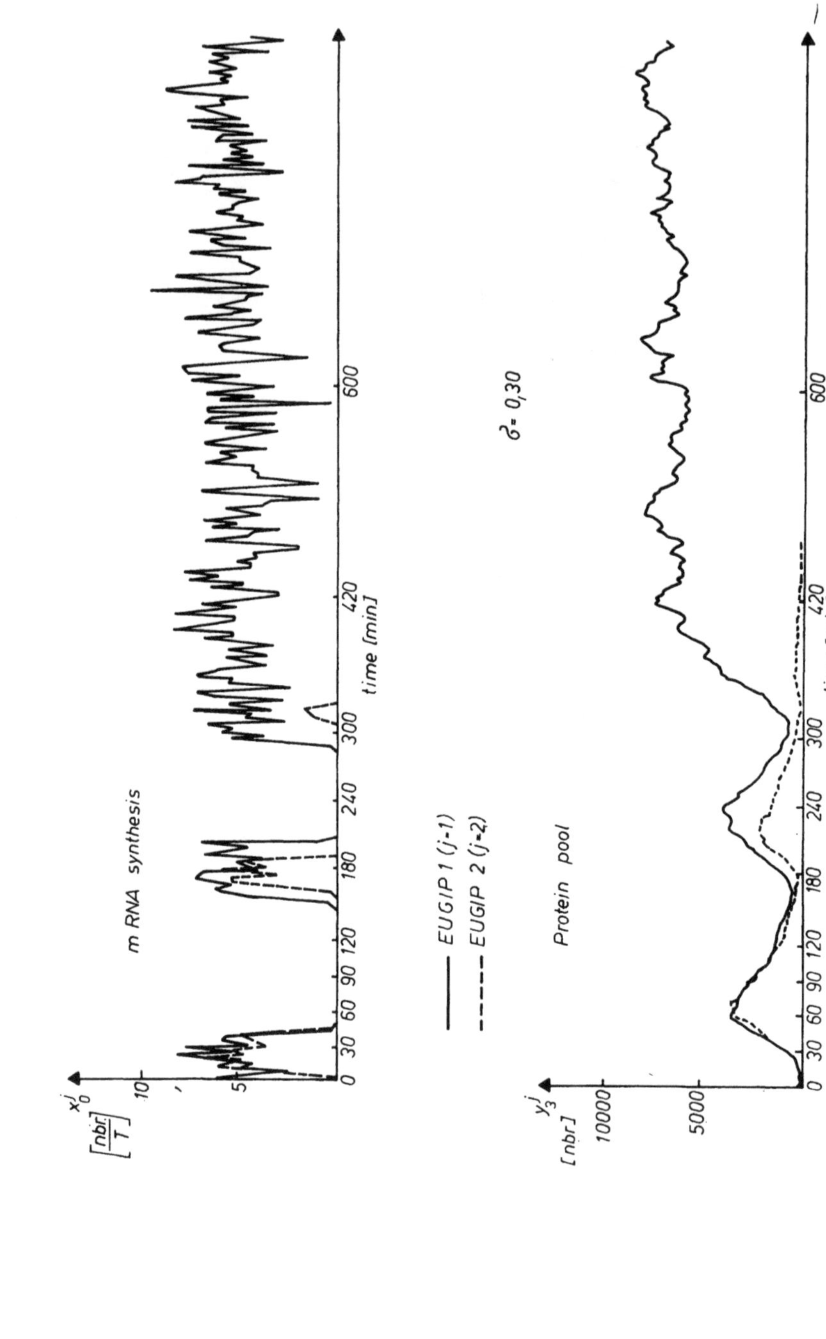

x_0^j
[nbr/T]
10
5
0 30 60 90 120 180 240 300 420 600
time [min]
m RNA synthesis
EUGIP 1 (j-1)
EUGIP 2 (j-2)
Protein pool
σ²= 0,30
y_3^j
[nbr]
10000
5000
0 30 60 90 120 180 240 300 420 600
time [min]

Fig. 13b)

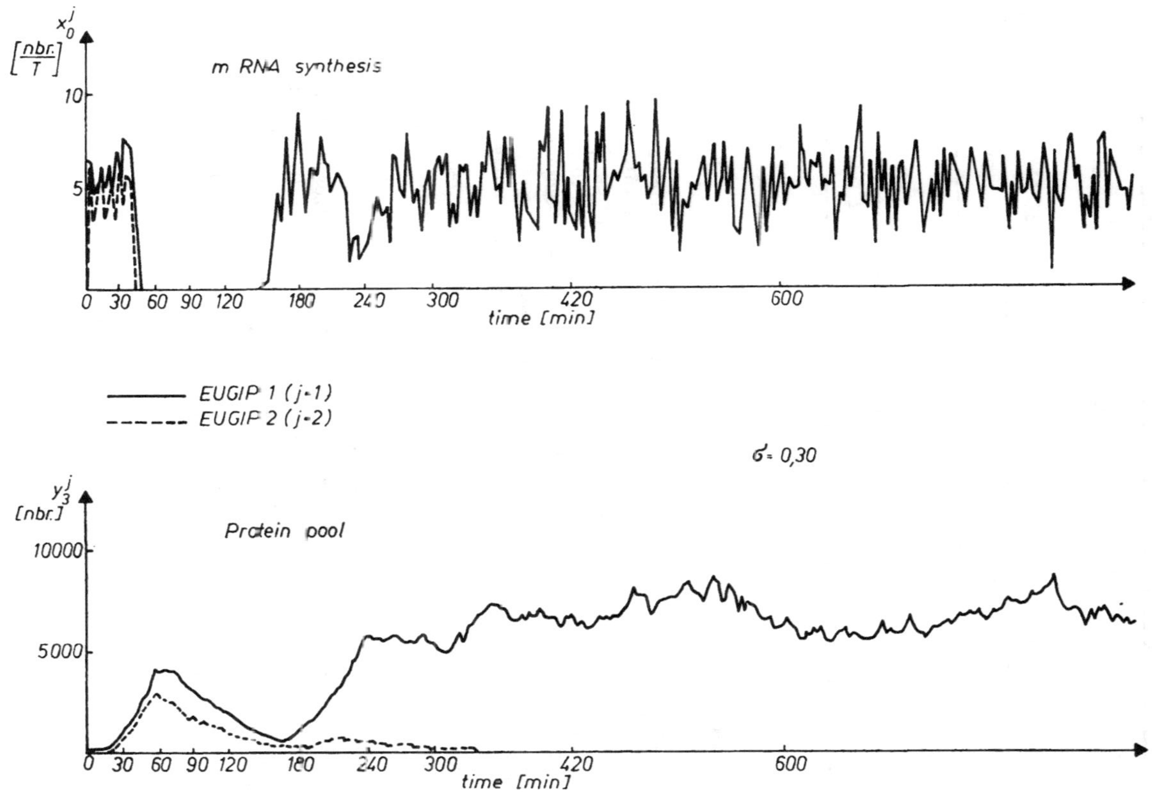

c)

Fig. 13. Effect of strong stochasticity in the 2-EUGIP-system of fig. 11; $\sigma = 0{,}30$.
a) qualitatively normal response: 4 oscillations, b) 2 oscillations, c) 1 oscillation.

Table 2. Effect of stochastic disturbances on the behaviour (number of oscillation periods) of the two-EUGIP-system.

		nbr. of oscillation periods							sum of computer experiments
σ		1	2	3	4	5	6	> 6	
0.03	nbr. of computer experiments	0	0	16	28	5	0	2	51
	%	0	0	31.4	54.9	9.8	0	3.9	100%
0.3	nbr. of computer experiments	7	21	8	5	0	0	1	42
	%	16.7	50.0	19.0	11.9	0	0	2.4	100%

References

ADESNIK, M., and C. LEVINTHAL (1970): The synthesis and degradation of lactose operon messenger RNA in *E. Coli.* Cold Spring Harb. Symp. on Quantit. Biol. **35**, S. 451.

BAKER, R., and C. YANKOFSKY (1970): Transcription initiation frequency for the tryptophan operon of *Escherichia Coli.* Cold Spring Harb. Symp. on Quant. Biol. **35**, S. 467.

BELLMANN, K., R. BÖTTNER, A. KNIJNENBURG and H. NEUMANN (1974): Computer simulation models of gene expression. Symp. Biol. Hung. **18**, pp. 227–256.

BERNARD, W. (1970): Struktur und Funktion in der Proteinsynthese. Nova Acta Leopoldina, Nr. 194, Bd. 35, S. 185.

BIELKA, H. (1973): Molekulare Biologie der Zelle, VEB Gustav Fischer Verlag, Jena.

BOTON LE, A. V. (1971): Protein synthesis in the rat liver acinus after injektion of antinomysin D. Currents in Modern Biology **3**, 353.

BÖTTNER, R. (1976): Mathematische Modellierung der Regulation der Genexpression, Tag.-Ber. Akad. Landwirtsch.-Wiss. DDR, Berlin (1976), **138**, S. 153–174.

CHAMBERLINE, M. (1970): Transcription: a summary, Cold Spr. Harb. Symp. Quant. Biol. **35**, 851.

FULLER, R. W. (1971): Rhytmic changes in enzyme activity and their control. In: RECHCIGL, M. (Ed.), Enzyme synthesis and degradation in mammalian systems. Verlag Karger, Basel.

GOODWIN, B. (1963): Temporal organization in cells, Academic Press, New York.

GRANNER, D. K., E. B. TOMPSON and G. B. TOMKINS (1970): Dexametasone phosphate-induced synthesis of tyrosine aminotransferase in Hepatoma tissue culture cells. J. Biol. Chem. **245**, 1472–1478.

GREENBLATT, J. (1973): Regulation of the expression of the N gene of bacteriophage Lambda. Proc. Nat. Acad. Sci. USA, **70**, 421.

HEINMETS, F. (1966): Analysis of normal and abnormal cell growth. Plenum Press, New York.

KNORRE, W. A. (1968): Oscillations of the rate of synthesis of β-Galactosidase in *E. Coli* ML 30 and ML 308. Biochem. Biophys. Res. Comm. **31**, 812.

KNORRE, W. A. (1969): Mathematische Modelle der Enzymsynthese in Bakterien. studia biophysica **14**, 29–42.

KNORRE, W. A., and R. BÖTTNER (1971): Some Considerations on dynamics of biological systems. Vortrag auf dem 1. Europäischen Biophysik-Kongreß 1971, Baden bei Wien.

LIN, E., and W. E. KNOX (1956): Biochem. Biophys. Acta **26**, 85.

MAHLER, H. R., and E. H. CORDES (1971): Biological Chemistry, Harper u. Row, New York.

MICHALIS, L., and M. L. MENTEN (1913): Biochem. Z. **49**, 1375.

MOHR, H., and P. SITTE (1971): Molekulare Grundlagen der Entwicklung, Akademie-Verlag Berlin.

RASHEVSKY, N. (1961): Mathematical biophysics, Dover, New York.

RATNER, W. A. (1972): Prinzipien der Organisation und Mechanismen molekular-genetischer Prozesse, Verlag Nauka, Novosibirsk (in russ.).

Rechcigl, M. (Ed.) (1971): Enzym Synthesis and Degradation in Mammalian Systems, Verlag Karger, Basel.

Rechcigl, M. (1971): Intracellular protein turnover and the roles of synthesis and degradation, in Rechcigl, M.

Rosen, R. (1972): Foundations of Mathematical Biology, Academic Press, New York.

Rutter, W. J. (et. al.) (1968): Regulation of specific protein synthesis in cytodifferentiation, J. Cell. Physiol. **72** Supplement 1, 1–18.

Schimke, R. T., and D. Doyle (1970): Control of enzyme levels in animal tissues, Ann. Rev. Biochem. **39**, 929.

Simon, Z., and E. Ruckenstein (1966): Regulation and synthesis in the living cell, J. theoret. Biol. **11**, 282–333.

Sugita, M. (1961): Functional analysis of chemical systems in vivo using a logical circuit equivalent, J. theoret. Biol. **1**, 415–430.

Tsanev, R., and B. Sendov (1971): Possible molecular mechanism for cell differentiation in multicellular organisms, J. theor. Biol. **30**, 337.

Wilson, S. H., and M. B. Hoaglund (1967): The stability of messenger ribonucleic acid and ribosomes, Biochem. J. **103**, 556.

3. A continuous approach with threshold characteristics for simulation of gene expression

R. N. Tchuraev and V. A. Ratner

Contents

1. Introduction

This chapter presents an approximate method to describe genetical control systems composed of DNA, mRNA and proteins. The parameters to be analyzed are the concentrations of the components of molecular-genetical systems. In theoretical analysis of such systems, the problem is to describe the function on the basis of a given structure by constructing mathematical models.

The problem of model building can be approached in several ways. One approach is to handle data by the methods of chemical kinetics, founded on the theory of differential equations. Some simple control systems at the molecular level have been simulated by sets of differential equations (Simon, 1965; Simon and Ruckenstein, 1966; Simon, 1970; Romanovsky et al., 1975). In studies of complex molecular systems, this approach is difficult because little is known, so far, about the kinetical details of molecular systems and no efficient methods have been developed for analyzing nonlinear differential equations. Another approach to modelling of molecular control systems, though simple and "robust", rests on methods involved with calculus of predicates and automata theory (Tchuraev, Ratner, 1972; Kauffman, 1974; Tchuraev, 1975). The drawback of this approach is its limited scope of application. Hence, it appeared worthwhile to search for methods of approximate descriptions of the dynamics of molecular-genetical control systems that would combine the essential features of kinetics with a relatively unsophisticated mathematical apparatus.

The method of threshold models seems to meet these requirements. The idea of the method is as follows. The complex and inadequately studied processes of interaction between regulatory proteins (repressors, inducers and others) and sites of specific recognition (operators, terminators, and others) are approximately described by introducing threshold concentrations and logical functions. Changes in the concentrations of each component, with sets logical of variables remaining constant, are described by linear sets of easily solved differential equations. The solutions are "sown" together at the boundaries of regions differing in values of the logical variables. The solution obtained in this way provides an overall description of the dynamics of the complex system.

2. Basic premises

In building threshold models, we proceed from the following basic premises.

1. A sricpton (or its segment) may be either in an active or inactive state. When in an active state, a scripton synthesizes mRNA, when inactive, it does not synthesize RNA.
2. Scripton activity (or the activity of its segments) is unambiguously determined by regulatory substances (repressors or activators).
3. Regulatory substances affect the scriptons state only in the case, when their concentrations exceed certain threshold values.
4. The rate of mRNA synthesis, when the scripton is in an active state, is proportional to the number of scripton copies in the cell.
5. A given protein is synthesized at a rate proportional to the number of mRNA molecules containing the given cistron.
6. The rate of mRNA degradation is proportional to its concentration.
7. Protein disappears as a result of either dilution during sequential cell division or enzymic degradation. In the latter case, the rate of protein degradation is proportional to its concentration.
8. The concentrations of substrates and enzymes needed for the synthesis of DNA, RNA and proteins do not set any limits to transcription nor translation.
9. The translation of mRNA occurs already during its synthesis.

Premises 1 and 2 are natural generalizations of facts established for the regulation of scripton transcription. Premise 3 is a simplification that in some cases agrees with real data (Chadwick et al., 1970; Wu et al., 1972). Premises 4 and 5 follow from the widespread current concepts of transcription and translation processes; the only stimulation is that the concentration of a given mRNA fraction is measured by the number of molecules containing the given cistron. Premise 6 is an inference drawn from our knowledge of the mechanisms of mRNA degradation in microorganisms, according to which this process is effected by nucleases randomly binding to the 5′-end of mRNA molecules (Mosteller et al., 1970). Premise 7 summarizes data concerning protein disappearance, which imply the existence of stable proteins (for example, protein C_I, lac-repressor, a number of enzymes) and unstable proteins (for example, protein N), in the latter case, enzymic degration of proteins conforms to first order kinetics (Greenblatt, 1973; Shimke, Doyle, 1970). Premise 8 rests on the assumption that the pool of the corresponding substrates and enzymes is common and adequate (NORRIS, KOCH, 1972). Premise 9 issues from indirect data, as well as from direct electron microphotographic evidence (MILLER et al., 1970a, b); with respect to phages, however, it is only a semblance of actuality.

3. General description of the threshold model

3.1. Elements of the model

Let us start with the main features of the threshold model. The extent to which molecular systems are described in detail may vary and this reflects upon the choice of the elements of the system. As element of the molecular system, we accept a protein synthesizing block comprising a part of a scripton and mechanisms synthesizing mRNA and the given protein.

Each synthesizing block corresponds to its model functioning in continuous time given by:

binary input variables $e_{ij}(t)$, where j is the index of the block; i is the order number of the variable ($i \in \{1, 2, \ldots, n_1\}$, $j \in \{1, 2, \ldots, n_2\}$);
logical function:

$$S_j(t) = S_j\big(e_{1j}(t), e_{2j}(t), \ldots, e_{n_1-1j}(t)\big);$$

binary variable

$$u_j(t) = \begin{cases} \sigma & \text{if} \quad t < \tau_j, \\ s_j(t - \tau_j) & \text{if} \quad t \geqq \tau_j, \end{cases}$$

where σ is the binary number and τ_j denotes delay; functions $m_j = m_j(t)$, $r_j = r_j(t)$ and $\tilde{r}_j = \tilde{r}_j(t)$; constants P_{lj} ($l \in \{1, 2, \ldots, n_3\}$);
binary output variables

$$v_{lj}(t) = \begin{cases} 1 & \text{if} \quad \tilde{r}_j(t) \geqq P_{lj}, \\ 0 & \text{if} \quad \tilde{r}_j(t) < P_{lj}. \end{cases}$$

The input variables enter through n_1 input channels and the output variables leave through n_3 output channels. Each channel has the same index as the variable assigned to the channel. This model will be referred to as the formal synthesizing block or simply synthesizing block.

The values of the input variable $e_{ij}(t)$ denotes the presence (1) or absence (0) at a given point of time, of a threshold dose of the regulatory substance. The input variable $e_{n_{1j}}(t)$ indicates the presence (1) or absence (0) of a threshold dose of replication initiators, i.e., indicates whether or not replication is occurring. The BOOLEAN function $S_j(t)$ describes the "total" effect of the interaction between regulatory substances and scriptons. The variable $u_j(t)$ (activity of the block) indicates the presence (1) or absence (0) of mRNA synthesis. Functions $m_j(t)$ and $r_j(t)$ describe how the absolute concentrations of mRNA and protein, respectively, change with time depending on the input variables and initial values. Function $\tilde{r}_j(t)$ denotes protein concentration per genome. Constants P_{lj} denote the values of the threshold doses of the protein product of the given block.

3.2. Systems of the formal blocks of proteins synthesis

Once the inputs and outputs of the blocks are connected, we have a system of blocks modelling their respective molecular systems. The connections between the elements of a system are given by a finite oriented graph, in the apices of which are the blocks. The edges of these graphs represent channels along which the symbols are transmitted (see, for example, RATNER, 1977; fig. 42, Chapter 10). The input and output channels, which are not connected with other channels are taken is inputs and outputs of the system.

3.3. Determination of the explicit form of the functions $m_j(t)$ and $r_j(t)$

General for the various blocks is the form of the functions $m_j(t)$ and $r_j(t)$. In fact, to determine these functions, one has to appeal to the basic premises 4—8, which are sufficiently general. The following set of equations is derived from these premises:

$$\frac{dg}{dt} = KJ$$

$$\frac{dm_j}{dt} = u_j a_{1j} g - a_{2j} m_j \qquad (1)$$

$$\frac{dr_j}{dt} = b_{1j} m_j - b_{2j} r_j$$

where g is the cellular concentration of the loci of the j-th protein (it is taken for granted that this value is the same for all the loci); m_j is the concentration of mRNA molecules containing the cistrons of the given protein, r_j is the absolute concentration of the protein.

$$J = \begin{cases} 1, & \text{if the loci replicate,} \\ 0, & \text{if the loci do not replicate.} \end{cases}$$

The binary variable u_j was specified above as indicating the activity of the given locus. K, a_{1j}, and b_{1j} are the constants of the respective replication, synthesis of mRNA (transcription) and protein (translation) of the j-th fractions; a_{2j} and b_{2j} are the degradation constants of the respective j-th fractions of mRNA and protein. The system implies that replication follows linear kinetics and that protein is unstable. It is pertinent to recall that λ-genome replication, e.g. is described by a linear function (STEVENS et al., 1971). At each constant value of the variables J and u_j, the system is linear and consequently, the equations are solved without any difficulty, provided that the binary variables are constant. When at least one of binary variables changes the form of the linear system and its analytical solution are modified, and the initial values of the variables in the new system are equal to the last values before the deviation in binary variables occurred. The solutions of the linear systems are "sown up" to give a general solution of the system.

At various values of the binary variables, the systems are solved as follows.

1) Let $J = 0$ and $u = 0$. This corresponds to the absence of replication and to the inactive state of the locus.

Then

$$g = g_0, \quad m = m_0 e^{-a_2(t-t_0)} \tag{2}$$

$$r = \frac{b_1 m_0}{b_2 - a_2} [e^{-a_2(t-t_0)} - e^{-b_2(t-t_0)}] + r_0 e^{-b_2(t-t_0)}. \tag{3}$$

2) If $J = 0$, $u = 1$ (no replication, the locus is active), then $g = g_0$

$$m = a_1 g_0 - \left(\frac{a_1 g_0}{a_2} - m_0\right) e^{-a_2(t-t_0)} \tag{4}$$

$$r = \frac{a_1 b_1 g_0}{a_2 b_2} [1 - e^{-b_2(t-t_0)}] - \frac{b_1(a_1 g_0 - a_2 m_0)}{a_2(b_2 - a_2)} [e^{-a_2(t-t_0)} - e^{-b_2(t-t_0)}]$$
$$+ r_0 e^{-b_2(t-t_0)}. \tag{5}$$

3) If $J = 1$, $u = 0$ (replication is proceeding, the locus is inactive), then $g = K(t - t_0) + g_0$ and the functions $m(t)$ and $r(t)$ assume the same form as when $J = 0$, $u = 0$.

4) If $J = 1$, $u = 1$ (replication is proceeding, the locus is active), then $g = K(t - t_0) + g_0$

$$m = \frac{a_1 K}{a_2^2} [a_2(t - t_0) - 1] + \frac{a_1 K}{a_2^2} e^{-a_2(t-t_0)} + \frac{a_1 g_0}{a_2} \cdot$$
$$\cdot [1 - e^{-a_2(t-t_0)}] + m_0 e^{-a_2(t-t_0)}, \tag{6}$$

$$r = \frac{[e^{-a_2(t-t_0)} - e^{-b_2(t-t_0)}]}{b_2 - a_2} \left(b_1 m_0 + \frac{a_1 b_1 K}{a_2^2} - \frac{a_1 b_1 g_0}{a_2}\right) +$$
$$+ \frac{[1 - e^{-b_2(t-t_0)}]}{a_2 b_2} \left(g_0 - \frac{K}{a_2}\right) a_1 b_1 + \frac{a_1 b_1 K}{a_2 b_2^2} \cdot$$
$$\cdot [b_2(t - t_0) - e^{-b_2(t-t_0)} - 1] + r_0 e^{-b_2(t-t_0)}. \tag{7}$$

The expressions hold true only when $b_2 \neq 0$. If $b_2 = 0$, which corresponds to the case, when the protein in question is stable and the shape of the function $r(t)$ is changed

5) When $J = 0$, $u = 0$ or $J = 1$, $u = 0$

$$r = \frac{b_1 m_0}{a_2} [1 - e^{-a_2(t-t_0)}] + r_0. \tag{8}$$

6) When $J = 0$, $u = 1$

$$r = \frac{a_1 b_1 g_0}{a_2} (t - t_0) + \frac{b_1}{a_2^2} (a_1 g_0 - a_2 m_0) [e^{-a_2(t-t_0)} - 1] + r_0. \tag{9}$$

7) When $J = 1$, $u = 1$

$$r = \frac{b_1}{a_2} [1 - e^{-a_2(t-t_0)}] \left(m_0 + \frac{a_1 K}{a_2^2} - \frac{a_1 g_0}{a_2}\right) +$$
$$+ \frac{a_1 b_1}{a_2} (t - t_0) \left[\frac{K}{2} (t - t_0) - \frac{K}{a_2} + g_0\right] + r_0. \tag{10}$$

It is mathematically convenient to introduce expressions into these formulae, namely

Formula (2): $m(t) = \varphi_0{}^0(a_1, a_2, g_0, m_0, t)$

Formula (4): $m(t) = \varphi_0{}^1(a_1, a_2, g_0, m_0, t)$

Formula (6): $m(t) = \varphi_1{}^1(a_1, a_2, g_0, m_0, t)$

Formula (3): $r(t) = \psi_0{}^0(a_1, a_2, b_1, b_2, g_0, m_0, t) + r_0 e^{-b_2(t-t_0)}$

Formula (5): $r(t) = \psi_0{}^1(a_1, a_2, b_1, b_2, g_0, m_0, t) + r_0 e^{-b_2(t-t_0)}$

Formula (7): $r(t) = \psi_1{}^1(a_1, a_2, b_1, b_2, g_0, m_0, t) + r_0 e^{-b_2(t-t_0)}$

Formula (8): $r(t) = \hat{\psi}_0{}^0(a_1, a_2, b_1, g_0, m_0, t) + r_0$

Formula (9): $r(t) = \hat{\psi}_0{}^1(a_1, a_2, b_1, g_0, m_0, t) + r_0$

Formula (10): $r(t) = \hat{\psi}_1{}^1(a_1, a_2, b_1, g_0, m_0, t) + r_0$

Here, the top index of symbols φ and ψ is equal to the value of the variable u, and the bottom index to that of the variable J.

After these transformations, the expressions for $m(t)$ and $r(t)$ may be written as

$$m(t) = \begin{cases} \varphi_0{}^0(a_1, a_2, g_0, m_0, t) & \text{if} \quad u = 0; \\ \varphi_0{}^1(a_1, a_2, g_0, m_0, t) & \text{if} \quad J = 0, u = 1; \\ \varphi_1{}^1(a_1, a_2, g_0, m_0, t) & \text{if} \quad J = 1, u = 1. \end{cases} \tag{11}$$

When $b_2 \neq 0$

$$r(t) = \begin{cases} \psi_0{}^0(a_1, a_2, b_2, b_2, g_0, m_0, t) + r_0 e^{-b_2(t-t_0)} & \text{if} \quad u = 0; \\ \psi_0{}^1(a_1, a_2, b_1, b_2, g_0, m_0, t) + r_0 e^{-b_2(t-t_0)} & \text{if} \quad J = 0, u = 1; \\ \psi_1{}^1(a_1, a_2, b_1, b_2, g_0, m_0, t) + r_0 e^{-b_2(t-t_0)} & \text{if} \quad J = 1, u = 1. \end{cases} \tag{12}$$

When $b_2 = 0$

$$r(t) = \begin{cases} \hat{\psi}_0{}^0(a_1, a_2, b_1, g_0, m_0, t + r_0 & \text{if} \quad u = 0; \\ \hat{\psi}_0{}^1(a_1, a_2, b_1, g_0, m_0, t) + r_0 & \text{if} \quad J = 0, u = 1; \\ \hat{\psi}_1{}^1(a_1, a_2, b_1, g_0, m_0, t) + r_0 & \text{if} \quad J = 1, u = 1. \end{cases} \tag{13}$$

4. Some properties of the elements of threshold models

4.1. Behaviour of function $m(t)$

We shall first examine the behaviour of functions $m(t)$ and $r(t)$ depending on the values of the binary variables $u(t)$, $J(t)$ and the initial values. Such properties of these functions as monotony, increase or decrease are of interest. Let us inspect formula (11). If in some interval of time $[t_0, t_1]$ $u(t) = 0$, then either $\frac{dm}{dt} = 0$ (when $m_0 = 0$) or $\frac{dm}{dt} < 0$ (when $m_0 > 0$). This means that, when $u(t) = 0$ in the time interval $[t_0, t_1]$, $m(t)$ remains constantly equal to 0 (when $m_0 = 0$), or monotonously decreases (when $m_0 > 0$). With increasing t in the unlimited semi-interval $[t_0, \infty)$, the value $m(t) \to 0$.

If in the interval of time $[t_0, t_1]$ $u(t) = 0$ and $I(t) = 1$, then, in the case when $m_0 < \frac{a_1 g_0}{a_2}$ $\frac{dm}{dt} > 0$, $m(t)$ will monotonously increase within this interval; when $m_0 = \frac{a_1 g_0}{a_2}$ $\frac{dm}{dt} = 0$, $m(t)$ will remain unchanged; when $m_0 > \frac{a_1 g_0}{a_2}$ $\frac{dm}{dt} < 0$, $m(t)$ will monotonously decrease. When $m_0 \neq \frac{a_1 g_0}{a_2}$, with increasing t in the unlimited semiinterval $[t_0, \infty)$, the value of $m(t)$ will asymtotically tend towards the value $\frac{a_1 g_0}{a_2}$.

Let $u(t) = 1$ and $I(t) = 1$ in the interval $[t_0, t_1]$, then

$$\frac{dm}{dt} = \frac{a_1 K}{a_2} [1 - e^{-a_2(t-t_0)}] + e^{-a_2(t-t_0)}(a_1 g_0 - a_2 m_0). \tag{14}$$

The first term in the right-hand side of the expression is not negative, because $e^{-a_2(t-t_0)} \leqq 1$. Let us restrict ourselves to the case, when $m_0 < \frac{a_1 g_0}{a_2}$. Then the second term is positive. Thus, if interval $[t_0, t_1]$ $u(t) = 1$, $I(t) = 1$ and $m_0 < \frac{a_1 g_0}{a_2}$. Hence, $m(t)$ will grow monotonously within this interval. In this case, with increasing t in the semiinterval $[t_0, \infty)$, $m(t)$ will increase infinitely.

4.2. Behaviour of function $r(t)$

The behaviour of function $r(t)$ will be first analyzed at $b_2 = 0$. Suppose that in interval $[t_0, t_1]$ $u(t) = 0$. Then either $\frac{dr}{dt} = 0$ (when $m_0 = 0$) or $\frac{dr}{dt} > 0$ (when $m_0 > 0$). In the former case, the value of $r(t)$ is constant in the interval $[t_0, t_1]$, in the latter case, increases monotonously. Thus, when $u(t) = 0$, $r(t)$ is nondecreasing in the interval $[t_0, t_1]$. In the semiinterval $[t_0, \infty)$, with increasing t, $r(t)$ tends to the value $r_k = r_0 + \frac{b_1 m_0}{a_2}$.

If $I(t) = 0$, $u(t) = 1$ in the interval $[t_0, t_1]$, then

$$\frac{dr}{dt} = \frac{a_1 b_1 g_0}{a_2} - \frac{b_1}{a_2} (a_1 g_0 - a_2 m_0)\, e^{-a_2(t-t_0)}. \tag{15}$$

Suppose $m_0 < \frac{a_1 g_0}{a_2}$, then $\frac{dr}{dt}$ will increase with t; $\inf \frac{dr}{dt} = b_1 m_0$. Because $m_0 \geqq 0$, in the interval (t_0, t_1) $\frac{dr}{dt} > 0$, i.e., $r(t)$ increases monotonously. If $m_0 \geqq \frac{a_1 g_0}{a_2}$, $r(t)$ will increase monotonously in the interval $[t_0, t_1]$, because $\frac{dr}{dt} > 0$ in this interval. As t increases, so does $r(t)$ infinitely.

Suppose $I(t) = 1$; $u(t) = 1$ in the interval $[t_0, t_1]$, then the first derivative of $r(t)$ will have the form:

$$\frac{dr}{dt} = \left(m_0 + \frac{a_1 K}{a_2{}^2} - \frac{a_1 g_0}{a_2}\right) b_1 e^{-a_2(t-t_0)} + \frac{a_1 b_1 K}{a_2} (t - t_0) - \frac{a_1 b_1 K}{a_2} - \frac{a_1 b_1 K}{a_2{}^2}, \tag{16}$$

or the equivalent form

$$\frac{dr}{dt} = m_0 b_1 e^{-a_2(t-t_0)} + \frac{a_1 b_1 K}{a_2^2} e^{-a_2(t-t_0)} + \frac{a_1 b_1 g_0}{a_2} [1 - e^{-a_2(t-t_0)}] +$$

$$+ \frac{a_1 b_1 K}{a_2} (t - t_0) - \frac{a_1 b_1 K}{a_2^2}. \tag{17}$$

Let us denote

$$A = m_0 b_1 e^{-a_2(t-t_0)} + \frac{a_1 b_1 K}{a_2^2} e^{-a_2(t-t_0)} + \frac{a_1 b_1 g_0}{a_2} \cdot$$

$$\cdot [1 - e^{-a_2(t-t_0)}] + \frac{a_1 b_1 K}{a_2} (t - t_0).$$

Let us consider only the case when $m_0 < \frac{a_1 g_0}{a_2}$.

It is easy to see that, if $\frac{dr}{dt} \geqq 0$ at moment $t = t_0$, then $\frac{dr}{dt} > 0$ when $t > t_0$. In fact, the first derivative of the positive component of the right-hand side of (17) is

$$\frac{dA}{dt} = \frac{a_1 b_1 K}{a_2} [1 - e^{-a_2(t-t_0)}] + b_1 e^{-a_2(t-t_0)}(a_1 g_0 - a_2 m_0). \tag{18}$$

If $m_0 < \frac{a_1 g_0}{a_2}$, $\frac{dA}{dt} > 0$. Since the negative component of the expression (17) is fixed, $\frac{dr}{dt}$ will only increase.

Because $\frac{dr}{dt} = m_0 b_1 \geqq 0$, at moment t_0, $\frac{dr}{dt} > 0$ in the semiinterval $(t_0, t_1]$, i.e., the function $r(t)$ will increase monotonously. Consequently, at any values of the binary values $u(t)$ and $I(t)$, the function $r(t)$ will be nondecreasing when $b_2 = 0$. As to function $\tilde{r}(t)$, when $J(t) = 1$ and $u(t) = 1$, it will increase monotonously starting from some moment.

Let us examine the behaviour of function $r(t)$, when $b_2 \neq 0$. Let $J(t) = 0$, $u(t) = 0$ in interval $[t_0, t_1]$. Then

$$\frac{dr}{dt} = e^{-b_2(t-t_0)} \left\{\frac{b_1 m_0}{b_2 - a_2} [b_2 - a_2 e^{(b_2-a_2)(t-t_0)}] - b_2 r_0\right\}. \tag{19}$$

What sign will the derivative have? The first factor in the right-hand side of the expression is always positive; for this reason, the sign of the derivative only on the second factor:

$$B = \frac{b_1 m_0}{b_2 - a_2} [b_2 - a_2 e^{(b_2-a_2)(t-t_0)}] - b_2 r_0. \tag{20}$$

Quite apparently, in the case when $B \leqq 0$ at moment $t = t_0$, $B < 0$ when $t > t_0$. When $B > 0$ at moment $t = t_0$, then starting from some moment $t > t_0$, the sign of B will be reverse. Acc. to formula (20), if $b_1 m_0 \leqq b_2 r_0$, then $B \leqq 0$ is at moment t_0. Thus, if $b_1 m_0 < b_2 r_0$, the function $r(t)$ will either decrease monotonously, or remain unchanged, when $r_0 = 0$. If $b_1 m_0 > b_2 r_0$, $r(t)$ will start decreasing monotonously only after some moment $t_2 > t_0$.

Suppose $b_2 \neq 0$, $J(t) = 0$, $u(t) = 1$ at time interval $[t_0, t_1]$. Then

$$\frac{\mathrm{d}r}{\mathrm{d}t} = \frac{a_1 b_1 g_0}{a_2} e^{-b_2(t-t_0)} - \frac{b_1(a_1 g_0 - a_2 m_0)}{a_2(b_2 - a_2)} [b_2 e^{-b_2(t-t_0)} - a_2 e^{-a_2(t-t_0)}] - \\ - b_2 r_0 e^{-b_2(t-t_0)}. \tag{21}$$

Let $m_0 < \dfrac{a_1 g_0}{a_2}$; as in the preceding case, it is easy to see that if $b_1 m_0 \geqq b_2 r_0$ at moment $t = t_0$, then $r(t)$ will increase monotonously. If $b_1 m_0 < b_2 r_0$, $r(t)$ will start to grow monotonously only after some moment $t_2' > t_0$. With increasing r in the unlimited semi-interval $[t_0, \infty)$ $r(t) \to \dfrac{a_1 b_1 g_0}{a_2 b_2}$.

Of importance is that, if $r(t)$ increases (or decreases) monotonously in the interval $[t_0, t_1]$, within this interval $\dfrac{m}{r} \geqq \dfrac{b_2}{b_1}$ (when $r(t)$ increases) and $\dfrac{m}{r} < \dfrac{b_2}{b_1}$ (when $r(t)$ decreases).

This follows from the statements given above and from the fact that, at fixed values of variables $u(t)$ and $J(t)$, the system has a single solution. As in the preceding case, it is easy to see that, when $J(t) = 1$, $u(t) = 1$, functions $r(t)$ and $\bar{r}(t)$ monotonously increase in the interval $[t_0, t_1]$, starting from some moment $t \geqq t_0$.

5. Simplest applications of the model

5.1. Cyclic one-operon system

Let us build a threshold model of an one-operon system, in which an operon contains a regulatory gene coding for the unstable repressor of this operon. The block synthesizing the repressor comprises an operon, mechanisms synthesizing mRNA and protein. The formal synthesizing block may be briefly outlined as follows. The input variable $e(t)$ indicates the presence or absence of the threshold dose of the repressor, the effect of which on the operon is desribed by function

$$s(t) = \overline{e(t)}. \tag{22}$$

Then the activity of the operon is given by formula

$$u(t) = \begin{cases} \sigma & \text{if} \quad t < \tau; \\ s(t - \tau) & \text{if} \quad t \geqq \tau. \end{cases} \tag{23}$$

The concentration of mRNA, which contains the cistron of the repressor is described by function

$$m(t) = \begin{cases} \varphi_0{}^0(a_1, a_2, g_0, m_0, t) & \text{if} \quad u(t) = 0, \\ \varphi_0{}^1(a_1, a_2, g_0, m_0, t) & \text{if} \quad u(t) = 1. \end{cases} \tag{24}$$

The concentration of the repressor is described by the following function

$$r(t) = \begin{cases} \psi_1{}^0(a_1, a_2, b_1, b_2, g_0, m_0, t) + r_0 e^{-b_2(t-t_0)} & \text{if} \quad u(t) = 0, \\ \psi_1{}^1(a_1, a_2, b_1, b_2, g_0, m_0, t) + r_0 e^{-b_2(t-t_0)} & \text{if} \quad u(t) = 1. \end{cases} \tag{25}$$

To illustrate, let us write these functions in expanded form:

$$m(t) = \begin{cases} m_0 e^{-a_2(t-t_0)} & \text{if} \quad u(t) = 0, \\ \dfrac{a_1}{a_2} - \left(\dfrac{a_1}{a_2} - m_0\right) e^{-a_2(t-t_0)} & \text{if} \quad u(t) = 1. \end{cases} \tag{26}$$

$$r(t) = \begin{cases} \dfrac{b_1 m_0}{b_2 - a_2} [e^{-a_2(t-t_0)} - e^{-b_2(t-t_0)}] + r_0 e^{-b_2(t-t_0)} & \text{if} \quad u(t) = 0, \\ \dfrac{a_1 b_1 g_0}{a_2 b_2} [1 - e^{-a_2(t-t_0)}] - \dfrac{b_1(a_1 g_0 - a_2 m_0)}{a_2(b_2 - a_2)} \cdot & \text{if} \quad u(t) = 1. \\ \cdot [e^{-a_2(t-t_0)} - e^{-b_2(t-t_0)}] + r_0 e^{-b_2(t-t_0)} & \end{cases} \tag{27}$$

The values of the threshold dose, when the repressor acts on the operon's operators, is denoted by the constant P. It is assumed that $0 < P < \dfrac{a_1 b_1 g_0}{a_2 b_2}$, which is the condition for threshold attainment. The block has a single output variable.

$$v = \begin{cases} 0 & \text{if} \quad r < P, \\ 1 & \text{if} \quad r \geqq P. \end{cases} \tag{28}$$

Let us connect the output channel of the block with the input channel of the same block, i.e., $v = e$. The resulting model for a cyclic operon system is given in fig. 1.

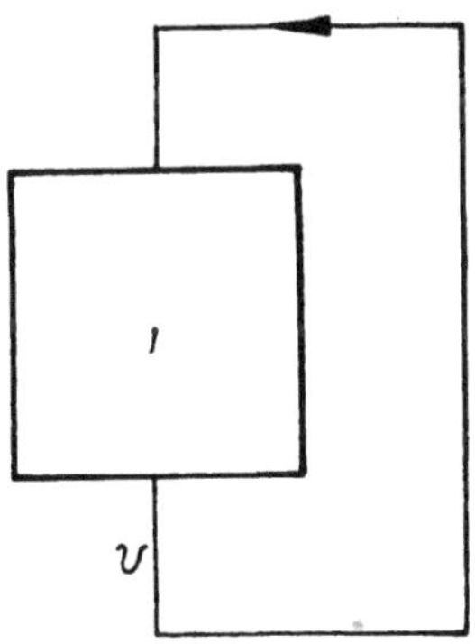

Fig. 1. Scheme of a cyclic operon.

Let us trace the dynamics of this system during an interval between sequential cell divisions. Let $m_0 = 0$ and $r_0 = 0$ at a starting point of time t_0, i.e., $b_1 m_0 \geqq b_2 r_0$. If so, the block works $(u(t) = 1)$, while $m(t)$ and $r(t)$ increase monotonously (see section 3) until $r(t)$ reaches threshold P. After attaining this threshold, the block ceases to work $(u(t) = 0)$ and $m(t)$ starts to decrease. After some time, $r(t)$ will fall below threshold values and the block is active again. Hence, the values of $m(t)$ and $r(t)$ oscillate which suggests that a single-system has a periodic regime. To support this assumption, the dynamics of the system was calculated with a computer at assigned initial values $m_0 = 0$, $r_0 = 0$. It proved that **the cyclic single-operon system with negative feedback has a periodic regime.**

5.2. Cyclic two-operon system

By way of illustration, let us consider a threshold model for a cyclic two-operon system (see fig. 2). This model consists of two repressor synthesizing blocks, differing from the preceding blocks only in the presence of input channels e_{2j} ($j = 1, 2$), corresponding to the channels for effector metabolites and the function $s(t)$ which has the form

$$s_j(t) = \overline{e_{1j}(t)} \vee e_{2j}(t), \tag{29}$$

where $e_{1j}(t)$ and $e_{2j}(t)$ indicate the presence or absence of the threshold doses of the repressor and effector metabolite correspondently, j is the index of the synthesizing block.

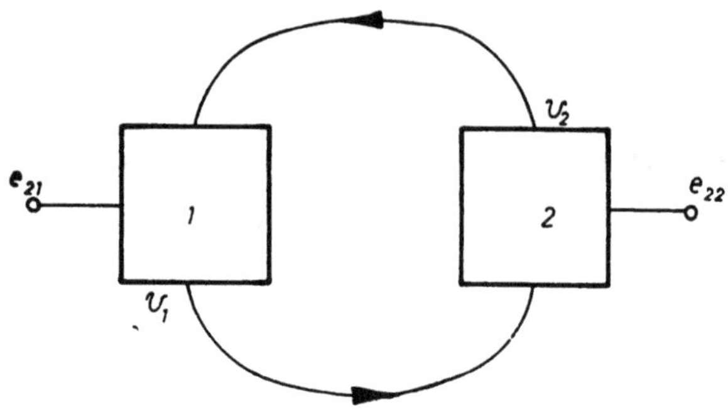

Fig. 2. Scheme of a cyclic two-operon system.

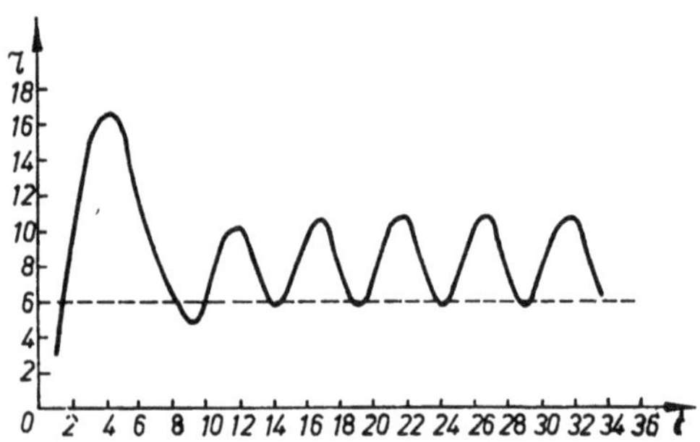

Fig. 3. Periodic regime of a cyclic two-operon system.

The constants P_1 and P_2 denote the values of the threshold doses, when the repressor acts on the operators of the first and second operon. The output channel of each block is connected with the input channel e_1 of the other block: $v_2(t) = e_{12}(t)$ and $v_2(t) = e_{11}(t)$.

The threshold model of a cyclic two-operon system has a trigger regime, so that, while one synthesizing block is active, the other is not, under the condition that $e_{2j}(t) = 0$ for each j.

Indeed, let us analyze the following initial values

$$m_{1(0)} = r_{1(0)} = 0; \qquad P_2 < r_{2(0)} < \frac{a_{12}b_{12}g_0}{a_{22}b_{22}};$$

$$b_{12}m_{2(0)} \geqq b_{22}r_{2(0)}; \qquad m_{2(0)} < \frac{a_{12}g_0}{a_{22}}.$$

Therefore, $u_1(t) = 0$, $u_2(t) = 1$ and the values $r_1(t) = r_{1(0)}(t) = 0$ does not change, whereas function $r_2(t)$ increases monotonously, tending to $\frac{a_{12}b_{12}g_0}{a_{22}b_{22}}$. Owing to the system's symmetry, there exists an alternative state, when $u_2(t) = 1$, $u_2(t) = 0$.

As computer calculations show, **the threshold model for the cyclic two-operon system has a periodic regime** (see fig. 3). This regime may be designated as "synchronous" because both blocks of the system are periodically and synchronously turned on and off, and, at any point of time, the values of the functions $r(t)$ and $m(t)$ of one block are identical to those of the other block. The necessary condition for the existence of the regime is identical characteristics of the blocks.

6. A comparative estimate of the method of threshold models

6.1. General remarks

The very approximate method of threshold models will be further used (see Chapter 10) to describe operon systems of the genetical control of λ-phage development. A distinctive feature of this system (and, perhaps, of a large class of other ontogenetic systems) is that their direct connections and feedbacks are effected through regulatory proteins, not through regulatory metabolites (RATNER, 1975). For this reason, as major dynamic variables are chosen the absolute cellular concentrations of polymers (DNA, mRNA and proteins) and regulatory metabolities are involved only as deterministic extraneous signals.

The kinetics of the template synthesis and degradation, in spite of its complexity (see, for example, RUCKENSTEIN, SIMON, 1966; SIMON, RUCKENSTEIN, 1966), is well described by linear approximation, this raises no difficulties. The problem is to describe the interaction of the regulatory proteins with the specific sites of scriptons and replicons, about which information is fragmentary. What has been established is that regulatory proteins are homomultimers (lac repressor, the protein C_I of the λ-phage and others) possessing allosteric properties. In these cases, the kinetics of the interaction between regulatory proteins and recognition sites are often described by a sigmoid-like curve. Analysis of the kinetical models of such interaction leads to sigmoidal curves representing complex combinations of kinetical parameters.

Thus, the aim of an approximate description of such systems is to make a reasonable choice of such an approximation of the sigmoidal surves that would contain a small number of easily estimated parameters. By introducing a threshold, a sigmoid curve is replaced by a rectangular step with a single parameter, which is equivalent to the introduction of logical variables into the model and keeping linear the system of kinetical equations for the concentrations of macromolecules. Analysis of simple examples demonstrated that this approximation "spares" the main rough features of the system's functioning such as oscillation, trigger regime and other features.

Let us compare the method of threshold models with other approaches used to describe operon systems interacting through regulatory proteins. These include:

a) Mathematical-logical models (SUGITA, 1961, 1963; KAUFFMAN, 1974; TCHURAEV, 1975);
b) Automata models (TCHURAEV, RATNER, 1972);

c) Kinetical models with limiting factors (Koulitchkov, Schamin, 1977 personal communication);
d) Kinetical models (Varshavsky, 1968; Simon, 1971; Romanovsky et al., 1975).

To be more concrete, we shall concentrate our attention on definite links of the system — on the interaction of the repressor with the operator and on the two-operon trigger system. Let us utilize equations of chemical kinetics as giving the fullest descriptions of real processes.

6.2. Interaction between a repressor dimer and an operator

Let us now describe the kinetical model. It is assumed that the regulator cistron produces protomers and that the repressor is a dimer. The processes examined can now be expressed as

$$\begin{aligned} R + R &\underset{K-2}{\overset{K+1}{\rightleftarrows}} RR \\ RR + O &\underset{K-2}{\overset{K+2}{\longleftrightarrow}} ORR \end{aligned} \tag{29}$$

where R are the protomers, RR are dimers, O are operators, and ORR are complexes of operators with dimers. The concentrations are denoted by respective indices in square brackets. We further assume that complexes RR and ORR are in equilibrium. Then, at fixed initial concentrations of the protomers $[R]_0$ and free operators $[O]_0$, the relations between $Z = [ORR]$, $x = [R]_0$ and $y = [O]$ are given by

$$x = \sqrt{\frac{Z}{K_1 K_2 (y - Z)}} + \frac{2Z}{K_2 (y - Z)} + 2Z. \tag{30}$$

Graphically, this expression is described by a sigmoidal curve (Fig. 4).
Although cumbersome, a precise analysis of the curve is feasible provided that two relative constants are estimated $K_1 = \frac{K_{-1}}{K_{+2}}$. and $K_2 = \frac{K_{-2}}{K_{+2}}$. However, the curve may be approximated with more convenient functions. Böttner and Cierzynski (see Chapter 4) have chosen curves conforming to

$$Z = \frac{x^n}{1 + x^n}. \tag{31}$$

Here, too, computer calculations are liable to be lengthy, but the results are good.
Koulichkov and Schamin (1977) have proposed linear approximation

$$Z^* = \begin{cases} 0 & x < x_1 \\ c(x - x_1) & x_1 \leqq x \leqq x_2 \\ Z_0 & x > x_2. \end{cases} \tag{32}$$

The parameters x_1 and x_2, as well as the angle coefficient $c = \frac{Z_0}{(x_2 - x_1)}$, are estimated from the minimization requirements of a reasonable functional characterizing the distance between the kinetical and approximating curves under the additional condition that $c(x_2 - x_1) = Z_0$. It should be noted that the choice of linear approximation is in its essence, similar to POLETAEV's method of limiting factors (see GILDERMAN et al., 1970). Indeed, on the zone $x_1 \leqq x \leqq x_2$, the number of complexes linearly depends on the initial concentrations of repressor protomers $x = [R]_0$. The limiting factor here is the concentration of repressor protomers. x_1 and x_2 are the border points of this limiting zone. When $x < x_1$ the process is turned off, when $x > x_2$ limiting is the number of free operators $Z_0 = y = [O]_0$. Computer calculations involved with linear approximation are also tedious.

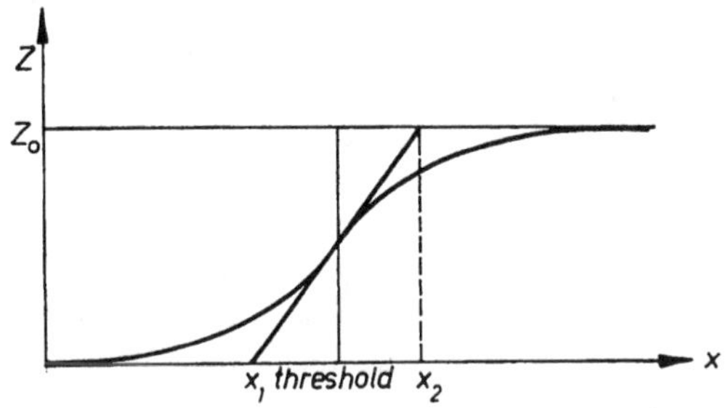

Fig. 4. Sigmoidal connection between concentration of the operator-repressor-dimer complex z concentration of the protomers x.

Application of the threshold method results in rough approximation, with the sigmoid curves being approximated by a rectangular step line with a single parameter x^*

$$Z^* = \begin{cases} 0 & \text{when} \quad x < x^*, \\ Z_0 & \text{when} \quad x \geqq x^*. \end{cases} \tag{33}$$

The threshold value of x^* is definable under the condition of the minimization of a reasonable functional characterizing the distance between kinetical and threshold values. For example, if the functional chosen has the form

$$L = \int_0^\infty [f(x) - Z^*]^2 \, dx = \int_0^{x^*} f^2(x) \, dx + \int_{x^*}^\infty [f(x) - Z_0]^2 \, dx. \tag{34}$$

The condition for functional minimum

$$\frac{dL}{dx^*} = f^2(x^*) - [f(x^*) - Z_0]^2 = f^2(x^*) - f^2(x^*) + 2Z_0 f(x^*) - Z_0^2 = 0$$
$$f(x^*) = \frac{Z_0}{2} \tag{35}$$

i.e., the threshold x^* corresponds to the value $f(x)$ equal to half of the value $\sup f(x) = Z_0$.

Here, the application of the methods of **automata theory** entails rejection of a continuous quantitative description of the concentrations of repressors, operators and complexes. Two possible states of the repressor-operator system are then examined — open and close. The repressor appears or disappears in definite doses, while the operators may fall into one of the two states, open (1) or close (0).

Finally, methods of **mathematical logics** exclude all quantitative descriptions of concentrations. Only threshold concentrations of repressor doses are taken into account; 1 — denotes threshold dose, 0 — denotes no threshold dose. These correspond to the state of the operator: 0 — turned off, 1 — turned on.

6.3. Two-operon trigger system

We shall discuss what can be achieved by each method in reverse order.

In terms of **mathematical logics**, a trigger is described by a set of BOOLEAN functions

$$\begin{aligned} u_1 &= \bar{u}_2 \vee e_1, \\ u_2 &= \bar{u}_2 \vee e_2, \end{aligned} \tag{37}$$

where u_1, u_2 are the binary output variables of the operons, e_1 and e_2 are the binary input variables of the operons corresponding to the metabolite effectors of the first and second operons (TCHURAEV, 1975). In the case, when $e_1 = e_2 = 0$ which corresponds to the absence of external signals, this system has two solutions: $u_1 = 0$, $u_2 = 1$ and $u_2 = 1$, $u_2 = 0$. Each solution corresponds to a stationary state of the trigger. Therefore, a mathematical logical model of a trigger merely indicates that it may fall into one of the alternative stationary states. This approach is the roughest.

Methods of **automata theory** have been first extended to a two operon trigger by TCHURAEV, RATNER (1972) and also by ŠOSKA (personal communication). The procedure consists in matching each operon-repressor pair of a trigger system with a discrete automaton V, which functions in discrete time intervals. Each automaton V has there input channels e, r, w and one output channel v. The binary signals e and r enter into the input channels e and r. The signal $e = 1$ denotes the presence of a portion of the specific metabolite, the signal $r = 1$ denotes the presence of a portion of repressor activity acting on the operon; 0 denotes the absence of a corresponding portion. Binary symbols are also transmitted along channel w. If symbol 1 enters the channel at some moment, then the loci of the regulatory gene and operon will double at the next moment.

Let us describe the automaton V. It has a set of states $Q = \{q_0, q_1, \ldots, q_m, \ldots, q_{2m}, \ldots q_M\}$. Each state corresponds to the number of repressor molecules per operon locus. q_0 denotes repressor absence, q_M denotes maximum portion of the repressor per locus.

$$\begin{aligned} &\text{Input alphabet } A = \{0, 1\}. \\ &\text{Output alphabet } B = \{0, 1\}. \\ &\text{Funktion of transition } q(t+1) = \varphi\big(q(t), r(t+1), w(t)\big). \\ &\text{Funktion of output } v(t+1) = \psi\big(q(t), e(t)\big). \end{aligned} \tag{38}$$

The functioning of the automaton is described by

$$\varphi(q_i, 0, 0) = q_i, \qquad \varphi(q_i, 0, 1) = \begin{cases} \dfrac{i}{2} & \text{if } i \text{ is an even number;} \\[2ex] \dfrac{i-1}{2} & \text{if is an uneven number;} \end{cases}$$

$$\varphi(q_i, 1, 0) = q_{i+1},$$

$$\varphi(q_i, 1, 1) = \begin{cases} \dfrac{i}{2} & \text{if } i \text{ is an even number;} \\[2ex] \dfrac{i+1}{2} & \text{if } i \text{ is an uneven number.} \end{cases}$$

$$\psi(q_i, 0) = \begin{cases} 1 & \text{if} \quad i = 0, \\ 0 & \text{if} \quad i > 0, \end{cases} \qquad \psi(q_i, 1) = 1.$$

The automaton introduced is called "formal operon". In the model of the trigger, the input r of each operon is connected with the output v of the other operon. The results obtained are as follows.

Besides alternative stable states, the model simulates the discrete dynamics of the establishement of these states and the transition from one state to another. Furthermore, another dynamic trigger system regime was found possible with the same parameters: the so called "**synchronous regime**" with periodic turning on and off of both operons. When applied to more complex systems, analysis of automaton models proves to be rather complex and amenable, mainly, to computer treatment.

The threshold model of a two-operon trigger was described above. This model enables to define one of the alternative trigger states and also reproduces the approximate dynamics of the factions of the major macromolecular components of the system, that of mRNA and proteins. This approach has many advantages: the linear sets of equations are solved analytically, the number of parameters is small and computer calculations are not very time consuming. Moreover, thanks to this approach, one can define all the regimes of the functioning and transmission between the states established earlier by methods of automata theory.

The dynamic model incorporating **limiting factors** (KULITCHKOV, SCHAMIN, 1977) provides a more precise description of the dynamics of macromolecular fractions in the middle portion of the sigmoidal curves of interaction. However, models with limiting factors are associated with lengthy computer calculations.

Finally, **kinetic models** of a trigger analyzed by ROMANOVSKY et al. (1975), SIMON et al. (1975) make possible the calculation of phase portraits of a system, i.e., to examine its dynamics qualitatively. With these models, the existence of two alternative stable states is always established, although the characteristics of the kinetic phase curves depend on the choice of the kinetic parameter processes. In addition, many kinetic constants are to be known.

It may be concluded that the approximate method of threshold models, when applied to operon systems with sigmoidal kinetics of interaction between regulatory proteins and control sites, ensures relative mathematical simplicity and the presentation of the major features of interacting mechanisms, i.e., this method may be efficiently applied to the analysis of the dynamics of linked operon systems.

References

CHADWICK, P., V. PIRROTTA and R. STEINBERG (1970): The λ and 434 phage repressors. Cold Spring Harbor Symp. Quant. Biol., **35**, 283.

GILDERMAN, YA. I., K. N. KUDRINA and I. A. POLETAEV (1970): Models of L-systems (systems with limiting factors). In: "Studies on cybernetics" "Sov. Radio", Moskow, 165—210.

GREENBLATT, J. (1973): Regulation of expression of the *N* gene of bacteriophage lambda. Proc. Nat. Acad. Sci. USA, **70**, 2, 421.

KAUFFMAN, S. (1974): The large scale structure and dynamics of gene control circuite: an ensemble approach. J. Theor. Biol., **44**, I, 167.

MILLER, O. L., B. R. BEATTY, B. A. HAMCALO and C. A. THOMAS (1970): Electron microscopic visualization of transcription. Cold Spring. Harb. Symp. Quant. Biol.. **35**, 505.

MOSTELLER, R. D., J. K. ROSE and C. YANOFSKY (1970): Transcription initiation and degradation of trp-mRNA. Cold Spring Harb. Symp. Quant. Biol., **35**, 461.

NORRIS, T. E., and A. L. KOCH (1972): Effect of growth rates on the relative rates of synthesis of messenger, ribosomal and transfer RNA in *Escherihia coli*. J. Mol. Biol., **64**, 3, 633.

RATNER, V. A. (1977): Molekulargenetische Steuerungssysteme (ed. K. Bellmann)., Akademie-Verlag, Berlin and Gustav-Fischer-Verlag, Stuttgart, New York.

ROMANOFSKI, YU. M., N. V. STEPANOVA and D. S. CHERNAVSKI (1975): Mathematical modelling in biophysics. Publ. House "Fizmatgis", Moskow.

RUCKENSTEIN, E., and Z. SIMON (1966): Regulation and synthesis in the living cell. I. Kinetics of ribonucleic asid synthesis. J. Theor. Biol., **11**, 2, 282.

SHIMKE, R. T., and D. DOYLE (1970): Control of enzyme levels in animal tissues. Ann. Rev. Biochem., **39**, 929.

SIMON, Z. (1965): Multi-study — stage for cell differentiation. J. Theor. Biol., **8**, 2, 258.

SIMON, Z. (1970): Triggers of interrelated operons and gene activation degrees in animal cells. Stud. Biophys., **20**, 3, 139.

SIMON, Z., and E. RUCKENSTEIN (1966): Regulation and synthesis processes in living cell. II. Kinetics of protein synthesis. J. Theor. Biol., **11**, 2, 299.

SIMON, Z., and E. RUCKENSTEIN (1966): Regulation and synthesis processes in living cell. III. Interrelated operon triggers as elements of the cell automaton. J. Theor. Biol., **11**, 2, 314.

STEVENS, W., S. ADHYA and W. SZYBALSKI (1971): Origin and biderectional orientation of DNA replication in coliphage lambda. In: "Bacteriophage Lambda" (A. D. HERSHEY, ed.), Cold Spring Harbor Laboratories, Cold Spring Harbor, 515—533.

SUGITA, M. (1961): Functional analysis of chemical systems in vivo using a logical circuit equivalent. J. Theor. Biol. **1**, 4, 415.

SUGITA, M. (1963): Functional analysis of chemical systems in vivo using a logical circuit equivalent. II. The idea of molecular automaton. J. Theor. Biol., **4**, 2, 179.

TCHURAEV, R. N. (1975): Mathematic-logical models for molecular regulatory systems. In: "Stu-, dies on mathematical genetics" (V. A. RATNER, ed.), Inst. Cyt. Genet. Press, Novosibirsk 67—77 (Russ.).

TCHURAEV, R. N., and V. A. RATNER (1972): Modelling of the molecular genetic systems of control by means of automata theory language. I. Operons and operon systems. In: "Studies on theoretical genetics" (V. A. RATNER, ed.), Inst. Cyt. Genet. Press, Novosibirsk, 210—228 (Russ.).

VARSHAVSKI, A. YA. (1968): The regulation of synthesis of genetic repressors in bacteria. J. Mol. Biol., **2**, I (Russ.).

WU, A. M., S. GHOSH, H. ECHOLS and W. G. SPIEGELMAN (1972): Repression by the C_I protein of phage: in vitro inhibition of RNA synthesis. J. Mol. Biol., **67**, 3, 407.

4. Modelling of epigenetic networks composed of monogenic units of gene expression, with reference to bacteriophage lambda development

R. Böttner, K. Bellmann, R. N. Tchuraev, V. A. Ratner

Contents

1. Problems and aims

The preceding chapters deal with the major current approaches to the simulation of operon systems controlling gene expression. In this chapter, we consider a concrete multioperon system, which controls developing bacteriophage λ, as well as portrait models simulating the function of this phage. The building of portrait models is a method making possible experimental designing and predictions aimed at solving fundamental biological problems. On the other hand, by comparing models with experimental data one can achieve close agreement between the similarities portrayed.

There is hardly any sense in building a portrait model of a system about which everything is known. What a formidable and futile task it would be modelling a system about which nothing is known. Between these two extremes there are real situations where there is some point in portrait modelling. Thus, basing on the information already available about a system's general behaviour, one can estimate or limit some of its unknown parameters and then deduce how the system would behave as a whole if our guesses were true. Thus, we have a method to explain the dynamic behaviour of the system on the basis of the structure of its molecular genetic control system. By means of this model we can probe hypotheses on the λ-control system and give ideas about new hypotheses and their experimental check. With respect to the λ-phage, we attempted to answer the following general questions.

1. Is the control system of the λ-phage adequately described to account for the main stages of its development?
2. How does the dynamics of the synthesis of the regulating proteins, genome DNA and mRNA fractions change with time?
3. How do some gene mutations or other influence from the system environment affect the dynamics of the system?
4. To what extent are the regimes of the λ-phage sensitive to chance variations of the parameters of the molecular-genetic-control system?

A special purpose of the paper is to compare the two methods for construction of models of molecular genetic control systems which are described in the contributions 2 and 3 of this volume, i.e. continuous and discrete approach.

By comparing models of the same system obtained by various methods, one can see how efficient the methods applied are. Furthermore, these comparisons help to define those properties of the given system that are insensitive to the modelling method.

2. A qualitative description of the structure and function of the lambda-genome

2.1. The scriptons

The phage λ-genome has five major scriptons (fig. 1). The scripton has been defined as a unit of transcriptions having one, at least, promoter and a definite orientation (Szybalski et al., 1970). The scripton may contain several cistrons coding for various proteins and cis-dominant sites for negative and positive controls. The various segments of scriptons are not necessarily read-off in a coordinate manner, though the transcription of the segments distal to the promoter always depends on the initiation in the autonomous promoter. The various scriptons may overlap. As fig. 1 shows, the phage λ has three leftward oriented scriptons (L_1, L_2, L_3) and two rightward oriented ones (R_1, R_2).

The scripton L_1 comprises the promoter P_c, the operator O_c, which is specific to the product of the cistron tof, the cistrons C_I and rex, and the hypothetical terminator t_c (Szybalsky, 1971; Spiegelman et al., 1970; Kumar et al., 1970). The cistron C_I codes for the structure of the repressor C_I, which acts on the operators O_L and O_R (Ptashne, Hopkins, 1968). The protein C_I seems to be also the activator of the transcription of the scripton L_1 (Szybalski, 1971; Kourilsky et al., 1970). The cistron rex produces a product repressing the development of the phage *T4rII* (Mark, Szybalski, 1973).

The scripton L_2 may be subdivided into there segments:

1) $P_L - O_L - O_L' - S_L - N - t_L$;
2) $t_L - C_{III} - \gamma - B -$ exo $-$ xis $-$ int $- aa'$;
3) $aa' - b_2 - \cdots - 5$.

The transcription of the first segment is effected by the host RNA-polymerase immediately after the operator element O_L is free from the repressor C_I (SZYBALSKI et al., 1970). At the point t_L, in the absence of the protein N, polymerase is stopped by the host ϱ-factor (ROBERTS, 1970).

The operator O_L is specific to the product of the cistron *tof* which represses the transcription of the scripton L_2. The promoter P_L and the operators O_L and O_L' (i.e., a segment about 200 nucleotide pair long) is not transcripted and polymerization commences at the starting point S_L (BLATTNER, DAHLBERG, 1972; BLATTNER et al., 1972).

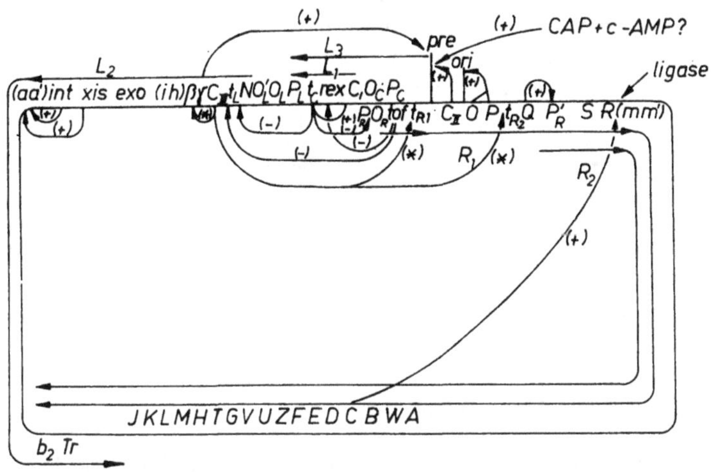

Fig. 1. Genetic map of λ-phage (SZYBALSKI, 1971, 1974) with indication of scriptons, cistrons, punctuation marks and interactions between them: (—) — negative repression, (+) — positive activation, (*) — positive antitermination.

The second segment of the scripton L_2 is transcribed provided that termination at the site t_L is prevented by the product of the cistron N (SZYBALSKI, 1971; HUMAR et al., 1970; HEINEMANN, SPIEGELMAN, 1970). This zone contains the cistron C_{III}, needed for the establishment of the lysogenic state, the cistrons *xis* — *int*, the products of which are necessary for the integration of the phage λ genome into the bacterial genome or for the excision of the latter from the former (DOVE, 1968; SZYBALSKI, 1970; FOLKMANIS, FREIFELDER, 1972). The transcription of the third segment of the scripton L_2 is possible only when it is physically attached to the second segment, as in the case of the circular phage λ genome or tandem prophage order. The third segment seems to be devoid of influential genes, because the deletion of this segment does not interfere with the development of the λ phage (SZYBALSKI et al., 1970).

The scripton L_3 contains genes $pre - \overline{tof} - C_I - rex - t_c$; it is activated by the products of the cistrons C_{II} and C_{III} (ECHOLS, GREEN, 1971). This cistron provides a pulse-like increase of the molecules of the protein C_I, needed for the establishment of the lysogenic state of the phage (the locus $\overline{tof}$ is complementary to the cistron *tof*).

The scripton R_1 consists of the following segments: 1) $P_R - O_R - tof - t_{R_1}$; 2) $t_{R_1} - C_{II} - O - P - t_{R_2}$; 3) $t_{R_2} - Q - S - R - mm'$; 4) $mm' - A - \cdots - I - b_2$. The first segment contains an operator O_R specific to the protein C_I (PTASHNE, HOPKINS, 1968), the cistron *tof*, coding for the antirepressor *tof* (NIJKAMP et al., 1971; OPPENHEIM et al., 1970) and the ϱ-dependent terminator t_{R_1} (HEINEMANN, SPIEGELMAN, 1970). The second segment of the scripton R_1 contains the cistron C_{II}, which codes for the protein necessary for lysogeny establishment (ECHOLS, GREEN, 1971); the products of which (replication initiators) are necessary for the replication of λ DNA synthesis starting at the replicator ori (STEVENS et al., 1972). The third segment contains the cistron Q, coding for the product activating the scripton R_2 (HERSKOWITZ, SIGNER. 1970; NIJKAMP et al., 1970; COUNTURIER et al., 1973). The transcription of the second and third segments is possible only when the protein N is present in amounts sufficient for preventing termination at t_{R_1} and t_{R_2} (HERSKOWITZ, SIGNER, 1970). The fourth segment contains the structural cistrons, essential for the formation of the tail and the head of the phage, and the cistrons S and R, which are necessary for bacterial lysis (DOVE, 1968; SZYBALSKI, 1970). The transcription of this segment depends on the fusion of the m and m' ends of the λ genome; in the prophage and in cyclic λDNA, these cohesive ends are closed, while in the mature λ particle they are open (SZYBALSKI et al., 1970).

The scripton R_1 may not provide the phage particles with the products of the cistrons S, R and of a group of cistrons from A to J. R_2 is the only scripton that provides products in sufficient amounts. This scripton starts from the promoter P_R', which is active only in the presence of a product of the cistron Q. The promoter P_R' is 10 times more efficient than the promoter P_R of the scripton R_1 (SZYBALSKI, 1971). The scripton R_2 overlaps with the scripton R_1. Its end presumably lies on the left part of the segment b_2.

2.2. Description of the control region

Let us examine the control zone of the phage λ genome. This zone is represented by the scriptons L_1, L_3, the first two segments of the scripton L_2, which are subdivided by the promoter P_R, the terminator t_L, the integration site segments of the scripton R_1, which are flanked by the promoter P_R, the terminators t_{R_1} and t_{R_2} the site of phage circularization (mm').

The functional relations between the scriptons of the λ phage are depicted in fig. 1. The designations are as follows: (+) — positive activation, (—) — negative repression, (*) — antitermination. The scriptons function only in the presence of certain regulatory substances and in the absence of others. Thus, the scripton L_1 is transcribed solely in the presence of the protein C_I and in the absence of the protein *tof* (KUMAR, CALEF, SZYBALSKI, 1970; KOURILSKY et al., 1970; NIJKAMP et al., 1970). The promoter P_c is weak: when activated, it produces approximately 3.7 protomers of protein C_I per min (REICHARDT, KAISER, 1971); the active form of the protein C_I is dimeric (CHADWICK, et al., 1970). This protein is a repressor, specific of the operators O_L and O_R, and efficiently represses transcription starting at the promoters P_L and P_R (PTASHNE, HOPKINS, 1968; CHADWICK et al., 1970). The interaction of the repressor C_I with the operator is described by an S-shaped curve (CHADWICK et al., 1970). The repressor protein is stable, and it disappears after synthesis termination as a result of sequential cell divisions (OGAWA, TOMIZAWA, 1967).

The transcription of the scripton L_2 is repressed by the repressors C_I and *tof*, the former being more efficient (HAMPACHEROVA et al., 1973). The promoter P_L is strong (SZYBALSKI et al., 1970). No special activator is needed for the transcription of the first segment of the scripton L_2, whereas the transcription of the scripton's second segment occurs only in the presence of the antiterminator-protein, which seems to hinder the recognition of the terminator by the ϱ-factor, thereby permitting RNA-polymerase to proceed through the terminator (SZYBALSKI, 1971; KUMAR et al., 1970). The protein N is unstable and degrades with a half-life value of 2 min after its synthesis is over (SCHWARTZ, 1970; GREENBLATT, 1973).

For the transcription of the scripton L_3, the products of the cistrons C_{II} and C_{III} (ECHOLS, GREEN, 1970; ECHOLS, 1972) and, possibly, the catabolic acceptor protein (CAP) and cAMP complex are requisite (GRODZICKER et al., 1972). The promoter *pre* of the scripton L_3 is 7 times more efficient than the promoter P_c. When L_3 is active, 28 protomers per min of the protein C_I are produced, on the average (REICHARDT, KAISER, 1971).

The protein C_I and, probably, partly the product of the cistron *tof* repress the transcription of the scripton R_1 (SZYBALSKI, 1972; ECHOLS, 1972). For the transcription of the second and third segments of this scripton, termination at the terminators t_{R_1} and t_{R_2} is to be stopped by the protein N. The second segment contains the cistrons O and P, which code for the proteins necessary for the autonomous replication of the λ phage genome. Replication can proceed only in the absence of the protein C_I (ECHOLS, 1972; GREEN et al., 1967).

The transcription of the scripton R_2 is effected by the product of the cistron Q, which appears to be a σ-like subunit of the host RNA-polymerase (NAONO, TOKUYAMA, 1970). The subunit (σ') replaces the host σ subunit and, thus, ensures the specificity of the scripton R_2 for the promoter P_R'.

a) The lytic regime

As already noted, the λ-phage has two developmental variants, namely lytic growth and lysogenization. Each variant can be effected by prophage interaction or induction (ECHOLS, 1972; RATNER, 1974, 1975). The developmental stages of a phage are:

1. **the immediate early stage,** when only RNA polymerase, not the products of phage cistrons, are involved in the transcription of the phage genome;
2. **the delayed early stage,** when the protein-antiterminator is concerned with transcription;
3. **the late stage,** when transcription is effected with the participation of the product of the cistron Q (SZYBALSKI, 1971).

Based on convincing experimental data, one might conceive the lytic variant proceeding as follows. Upon infection, phage particles inject linear DNA molecules, open at their cohesive termini (*mm'*) into bacteria (DOVE, 1968). Promptly thereafter, there occurs a complementary fusion of the termini, an "annealment" of the bonds by bacterial ligases; as a result, the phage genome forms a circle (WAKE et al., 1972). At **the immediate early stage**, the first segments of the scriptions L_2 and R_1 are transcribed and the synthesis of the proteins N and *tof* is initiated. At the two first stages, the concentration of the protein *tof* is too low to repress the transcription of the scripton L_2, whereas that of the protein N attains in 2—3 min a level sufficiently high to repress termination at the terminators t_L, t_{R_1}, and t_{R_2} (SZYBALSKI, 1971). It is also likely that antitermination consists in the additional initation of transcription after the terminators.

As a consequence, there begins the transcription of the cistrons C_{III} — *int* in the scripton L_2, and of the cistrons C_{II}, O, P in the scripton R_1, then that of the cistrons Q, R, S in the third segment. Thus, the loci of vegetative recombination (*exo*, β, γ), integration (*int* — *xis*), repression of bacterial metabolism (*kil*), replication (O, P) and the activators of the scripton L_3(C_{II} and C_{III}) start to function at the delayed early stage. At about min 4, the phage genome starts to replicate, probably on the cell wall (STEVENS et al.,

1971; KORN, THOMAS, 1971). It is at the delayed early stage that the choice is made between the two alternative variants of development; the proteins C_{II} and C_{III}, the CAP and cAMP complex are involved in this "decision-making". If the conditions in the cell direct phage development along the lytic pathway, replication proceeds, and the σ-like product of the cistron Q initiates the transcription of the cistrons of phage morphogenesis and that of the cistron R and S in the scripton R_2. The product of one of these cistrons (A) participates in the opening of the cohesive termini (mm') of the phage genome during the packing in the phage particle.

Approximately by 10—15 min after infection, the protein *tof* turns on the transcription of the scripton L_2 (ECHOLS et al., 1973; RATNER, 1974, 1975). This brings about a resumption of termination at t_L, t_{R_1}, and t_{R_2}, beause the protein antiterminator N rapidly decays. Lytic growth ends by a rupture of the bacterial membrane under the effect of a lytic enzyme produced by the cistron S.

It may be that the CAP and cAMP complex is in some way implicated in the developmental pathway chosen at the immediate early stags. The level of cAMP reflects the amounts of energy stored in the cell. According to GRODZICKER et al. (1970), if cAMP level is high in a phage-infected bacterial culture, the phage will lysogenize; if high, the phage will give preference to lytic growth. Parenthetically, one should not exclude some other, more indirect mode of action of these factors (ROLFE et al., 1973; BELFORT, WULFF, 1974).

b) The lysogenic regime

The lysogenic variant of development may be envisaged as follows. The products of the cistrons C_{II} and C_{III}, possibly together with the CAP and cAMP complex, activate the promoter *pre* in the scripton L_3. The synthesis of the protein C_I starts 5—6 min after infection (REICHARDT, KAISER, 1971). This protein represses the scriptons R_1 and L_2 and, possibly, activates transcription in the scripton L_1 (SZYBALSKI, 1971; KOURILSKY et al., 1970). This results in the cessation of the synthesis of the protein N and in function resumption at the terminators t_L, t_{R_1}, and t_{R_2}. Replication ceases in the presence of the protein C_I, even though the proteins O and P may be present (GREEN et al., 1967; ECHOLS, 1972). Eventually, the transcription of all the scriptions except that of L_1, is turned off (SZYBALSKI et al., 1970). Intensive transcription of the scripton L_3, which takes about 10 min, is required only for the passage into the lysogenic state; weak transcription ($\sim$ 3.7 protomers per min) maintains this state (ECHOLS, GREEN, 1970; REICHARDT, KAISER, 1971). It should be noted that the onset of the synthesis of the repressor C_I never precedes the appearance of the products of the cistron int, these products are required for the recombinational exchange at the site aa' leading to the integration of the λ phage genome into the bacterial chromosome. L_1 is the sole scripton functioning in the lysogenic state (SZYBALSKI, 1971).

The decision to follow either the lytic or lysogenic regime is not a predetermined event, so that some of the infected bacteria in a culture give a lytic response, the others respond lysogenically. The relative proportion of these responses depends, among other factors, on the numbers of multiinfections (BERTANI, 1960), the state of the arabinose operon system of the host cell. It will be disregarded here that the choice is made stochastically, cAMP concentration in the cell is accepted as the determined signal for the choice of the appropriate development at regime.

It is beyond the scope of this review to do justice to some of the minor ontogenetic events. However, it should be mentioned that the antirepressor *tof* partly represses the cistron *tof*-containing operon R_1 (TAKEDA et al., 1975); there is evidence for the existence of a sixth operon R_2' with a weak promoter between *tof* and t_{R_1} (SZYBALSKI, 1971); data indicate that low concentrations of the C_I induce the transcription of mRNA of this repressor in the operon L_1 and high C_I concentrations inhibit it. Interesting as these details may be, they will be omitted from the model suggested.

3. The models

3.1. The threshold model for the regulatory system of lambda phage development (TH-model)

Let us now consider the system controlling the development of the λ phage (see fig. 1). The method of threshold models has been outlined in contribution 3 in this volume (see also TCHURAEV, RATNER, 1975; RATNER, TCHURAEV, 1977). The basic idea of the method is to describe the threshold models of blocks synthesizing definite proteins with each block including a part of a scripton, the mechanisms synthesizing mRNA and the given protein, then the inputs and the outputs of the formal synthesizing blocks are connected according to the scheme of the functional relations of the system simulated (fig. 2a). The first step is to describe the blocks synthesizing the main regulatory proteins C_I, N, C_{II}, *tof*, C_{III}, O, P, and Q and then to proceed to the building of threshold models. The functions $m(t)$ for each synthesizing block will have the following form (for designations see contribution 3)

$$m_i(t) = \begin{cases} \varphi_0{}^0(a_{1i}, a_{2i}, g_0, m_{i(0)}, t), & \text{if} \quad u_i(t) = 0; \\ \varphi_0{}^1(a_{1i}, a_{2i}, g_0, m_{i(0)}, t), & \text{if} \quad I(t) = 0,\ u_i(t) = 1; \\ \varphi_1{}^1(a_{1i}, a_{2i}, g_0, m_{i(0)}, t), & \text{if} \quad I(t) = 1,\ u_i(t) = 1. \end{cases} \qquad (1)$$

$i \in \{1, 2, \dots 7\}$ the number of mRNA fractions exceeds the number of proteins examined, because the cistron C_I is contained in two fractions of mRNA; one corresponds to the scripton L_1, the other to the L_3). For this reason, these functions will not be used in extended form in the descriptions of the blocks.

a) The block synthesizing to protein C_I

This block comprises parts of the scriptons L_1 and L_3, mechanisms underlying the synthesis of mRNA in both fractions (the scriptons L_1 and L_3) and the protein C_I.

The regulatory substances, which act on this block, are of two kinds: the proteins of positive control (C_I, C_{II}, C_{III} the CAP and cAMP complex) and the protein of negative control — the antirepressor *tof*. The involvement of two mRNA fractions in the synthesis of the protein C_I somewhat modifies the model of this block, as compared with the general scheme given in the preceding paragraph. The relevant parameters of the model developed for this block will be: the input variables $e_{11}(t)$, $e_{21}(t)$, $e_{31}(t)$, $e_{41}(t)$, $e_{51}(t)$, and $I(t)$ denote the respective presence or absence of the threshold doses of the proteins C_I, *tof*, C_{II}, C_{III}, of the CAP and cAMP complex, and replication. Based on data on the interaction of the regulatory substances with the scriptons L_1 and L_3, the two BOOLEAN func-

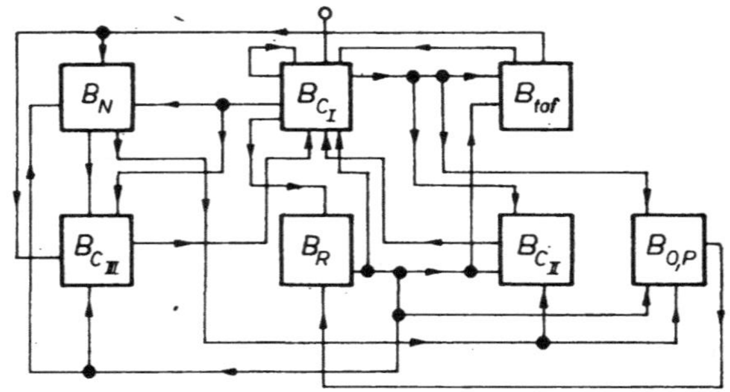

a)

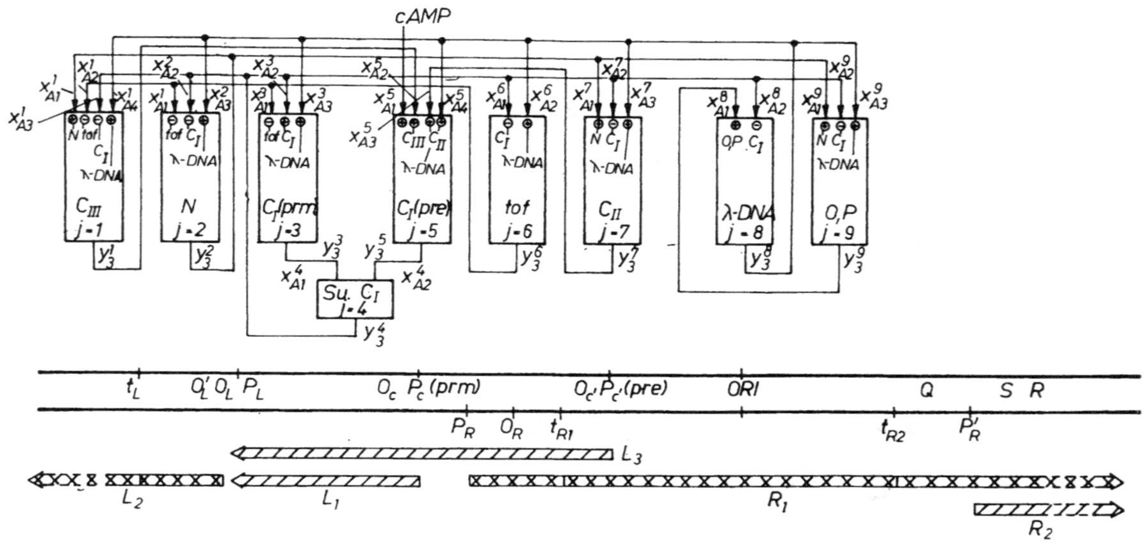

b)

Figs. 2a, 2b. a) Scheme for the threshold model of regulatory system of λ-phage development. Squares: blocks of regulatory protein synthesis. B_R — replication block. b) Scheme for stochastic difference equation model of regulatory system of phage development. λ-DNA-replication unit.

tions may be written as

$$S_1(t) = e_{11}(t) \,\&\, \overline{e_{21}(t)}, \tag{2}$$

$$S_2(t) = e_{31}(t) \,\&\, e_{41}(t) \,\&\, e_{51}(t). \tag{3}$$

Function $S_1(t)$ describes the interaction of the regulatory proteins with the scripton L_1 and the function $S_2(t)$ that with the scripton L_3. Because there is a lag specific of each scripton, in the action of these substances on scripton activity, the variables $u_1(t)$ and $u_2(t)$ describing the activity of the scriptons L_1 and L_3 will have the following form

$$u_1(t) = \begin{cases} \sigma_1 & \text{if} \quad t < \tau_1; \\ S_1(t - \tau_1) & \text{if} \quad t \geqq \tau_1. \end{cases} \tag{4}$$

$$u_2(t) = \begin{cases} \sigma_2 & \text{if} \quad t < \tau_2; \\ S_2(t - \tau_2) & \text{if} \quad t \geqq \tau_2. \end{cases} \tag{5}$$

The concentration of each of the mRNA fractions for the C_I will be described by the functions $m_2(t)$ and $m_2(t)$ (see formula 1).

Since the differential equation for the concentration of the protein C_I is given by

$$\frac{dr_1}{dt} = b_{11}m_1 + b_{12}m_2 \tag{6}$$

($b_2 = 0$ because the protein C_I is stable), the function $r_1(t)$ will be expressed by the formula (See chapter **3** of this volume).

$$r_1(t) = \begin{cases} \hat{\psi}_0{}^0(a_{11}, a_{21}, g_0, b_{11}, m_{1(0)}, t) + \hat{\psi}_0{}^0(a_{12}, a_{22}, g_0, b_{12}, m_{2(0)}, t) + r_{1(0)} \\ \qquad \text{if} \quad u_1 = 0, \quad u_2 = 0; \\ \hat{\psi}_0{}^0(a_{11}, a_{21}, g_0, b_{11}, m_{1(0)}, t) + \hat{\psi}_0{}^1(a_{12}, a_{22}, g_0, b_{12}, m_{2(0)}, t) + r_{1(0)} \\ \qquad \text{if} \quad u_1 = 0, \quad u_2 = 1, \\ \hat{\psi}_1{}^0(a_{11}, a_{21}, g_0, b_{11}, m_{1(0)}, t) + \hat{\psi}_1{}^1(a_{12}, a_{22}, g_0, b_{12}, m_{2(0)}, t) + r_{1(0)} \\ \qquad \text{if} \quad u_1 = 0, \quad u_2 = 1, \quad J = 1; \\ \hat{\psi}_0{}^1(a_{11}, a_{21}, g_0, b_{11}, m_{1(0)}, t) + \hat{\psi}_0{}^1(a_{12}, a_{22}, g_0, b_{12}, m_{2(0)}, t) + r_{1(0)} \\ \qquad \text{if} \quad u_1 = 1, \quad u_2 = 1, \quad J = 0; \\ \hat{\psi}_1{}^1(a_{11}, a_{21}, g_0, b_{11}, m_{1(0)}, t) + \hat{\psi}_1{}^1(a_{12}, a_{22}, g_0, b_{12}, m_{2(0)}, t) + r_{1(0)} \\ \qquad \text{if} \quad u_1 = 1, \quad u_2 = 1, \quad J = 1; \\ \hat{\psi}_0{}^1(a_{11}, a_{21}, g_0, b_{21}, m_{1(0)}, t) + \hat{\psi}_0{}^0(a_{12}, a_{22}, g_0, b_{12}, m_{2(0)}, t) + r_{1(0)} \\ \qquad \text{if} \quad u_1 = 1, \quad u_2 = 0, \quad J = 0; \\ \hat{\psi}_1{}^1(a_{11}, a_{22}, g_0, b_{11}, m_{1(0)}, t) + \hat{\psi}_0{}^0(a_{12}, a_{22}, g_0, b_{12}, m_{2(0)}, t) + r_{1(0)} \\ \qquad \text{if} \quad u_1 = 1, \quad u_2 = 0, \quad J = 0. \end{cases} \tag{7}$$

The constants P_1, P_2, P_3, and P_4 refer to the threshold doses of the protein C_I, when this protein acts on the sites P_c, O_L, O_R, and replication. The variables v_1, v_2, v_3, and v_4 will indicate whether or not the concentration of the protein C_I per genome has reached threshold values

$$v_1 = \begin{cases} 0 & \text{if} \quad \tilde{r}_1(t) < P_1; \\ 1 & \text{if} \quad \tilde{r}_1(t) \geqq P_1. \end{cases}$$

$$v_2 = \begin{cases} 0 & \text{if} \quad \tilde{r}_1(t) < P_2; \\ 1 & \text{if} \quad \tilde{r}_1(t) \geqq P_2. \end{cases}$$

$$v_3 = \begin{cases} 0 & \text{if} \quad \tilde{r}_2(t) < P_3; \\ 1 & \text{if} \quad \tilde{r}_1(t) \geqq P_3. \end{cases}$$

$$v_4 = \begin{cases} 0 & \text{if} \quad \tilde{r}_1(t) < P_4; \\ 1 & \text{if} \quad \tilde{r}_1(t) \geqq P_4; \end{cases} \quad \text{where} \quad \tilde{r}_1(t) = \frac{r_1(t)}{g(t)}.$$

b) The synthesizing block of the protein N

This molecular block consists of the first segment of the scription L_2 and of the mechanisms underlying the synthesis of the protein N and of the respective mRNA. Let us describe the model for this block according to the scheme given in paragraph 2. The input variables $e_{12}(t)$ and $e_{22}(t)$ denots the presence or absence of the threshold doses of the proteins C_I and *tof* at the moment t. The variable $J(t)$ is the same for all the blocks. Since the proteins C_I and *tof* are responsible for the negative control, the function $S_3(t)$, which describes their interaction with the operators O_L und O_L' of the scripton L_2 is written in the form

$$S_3(t) = \overline{e_{12}(t)} \;\&\; \overline{e_{22}(t)}. \tag{12}$$

The variable $u_3(t)$, which characterizes the activity of the first segment of the scripton L_2, is expressed as

$$u_3(t) = \begin{cases} \sigma_3 & \text{if} \quad t < \tau_3; \\ S_3(t - \tau_3) & \text{if} \quad t \geqq \tau_3. \end{cases} \tag{13}$$

The concentration of the cistron N-containing mRNA is described by the function $m_3(t)$ [see formula (1)].

The protein N is unstable and decays according to first order kinetics; consequently, the function $r_2(t)$ obeys the relation (see paragraph 2 for designations):

$$r_2(t) = \begin{cases} \psi_0{}^0(a_{13}, a_{23}, g_0, b_{12}, b_{22}, m_{3(0)}, t) + r_{2(0)}e^{-b_{22}(t-t_0)} & \text{if} \quad u_3 = 0, \quad J = 0; \\ \psi_0{}^1(a_{13}, a_{23}, g_0, b_{12}, b_{22}, m_{3(0)}, t) + r_{2(0)}e^{-b_{22}(t-t_0)} & \text{if} \quad u_3 = 1, \quad J = 0; \\ \psi_1{}^1(a_{13}, a_{23}, g_0, b_{12}, b_{22}, m_{3(0)}, t) + r_{2(0)}e^{-b_{22}(t-t_0)} & \text{if} \quad u_3 = 1, \quad J = 1. \end{cases} \tag{14}$$

The constants P_5, P_6 and P_{12} indicate the value of the threshold doses of the protein N, when the terminators t_L, t_{R_1} and t_{R_2} are under protein action. The respective output variables are given by the equations

$$v_5(t) = \begin{cases} 0 & \text{if} \quad \tilde{r}_2(t) < P_5; \\ 1 & \text{if} \quad \tilde{r}_2(t) \geqq P_5. \end{cases}$$
$$v_6(t) = \begin{cases} 0 & \text{if} \quad \tilde{r}_2(t) < P_6; \\ 1 & \text{if} \quad \tilde{r}_2(t) \geqq P_6. \end{cases} \tag{15}$$

c) The synthesizing block of the protein C_{III}

Because the protein C_{III} synthesizing block comprises the second segment of the scripton L_2, it is under the negative control of the proteins C_I and *tof*, as in the preceding block. The activity of this block is also under the positive control of the protein N. The parameters involved in the respective formal block of synthesis are the input variables $e_{13}(t)$, $e_{23}(t)$, and $e_{33}(t)$ indicating whether or not the doses of the proteins C_I, *tof*, and N are super threshold. The function $S_4(t)$ is defined by

$$S_4(t) = \overline{e_{13}(t)} \,\&\, \overline{e_{23}(t)} \,\&\, e_{33}(t). \tag{16}$$

The variable $u_4(t)$, describing the activity of the cistron C_{III} is expressed by the formula

$$u_4(t) = \begin{cases} \sigma_4 & \text{if} \quad t < \tau_4; \\ S_4(t - \tau_4) & \text{if} \quad t \geqq \tau_4. \end{cases} \tag{17}$$

The concentration of the C_{III}-cistron containing fraction of mRNA is described by the function $m_4(t)$.

Until very recently, it was unknown whether the protein C_{III} is stable or not. For this reason, two variants of the model of the C_{III} synthesizing block will be examined. If one assumes that $b_{23} = 0$, then $r_3(t)$ is described by formula (13), otherwise formula (12) is to be used (see contribution 3). The constant P_7 will denote the value of the threshold, when C_{III} acts on the promoter pre, in which case the output variable becomes

$$v_7(t) = \begin{cases} 0 & \text{if} \quad \tilde{r}_3(t) < P_7; \\ 1 & \text{if} \quad \tilde{r}_3(t) \geqq P_7, \end{cases} \tag{18}$$

which exerts a negative control over this segment.

d) The synthesizing block of the protein *tof*

This block includes the first segment of the scripton R_1. Its activity is determined by the protein C_I, which exerts a negative control over the segment. A description of the formal unit follows. The input variable $e_{14}(t)$ indicates whether or not the protein C_I has attained the threshold value; the effect of the C_I is expressed by the function

$$S_5(t) = \overline{e_{14}(t)}. \tag{19}$$

Then $u_5(t)$ will be expressed by

$$u_5(t) = \begin{cases} \sigma_5 & \text{if} \quad t < \tau_5; \\ S_5(t - \tau_5) & \text{if} \quad t \geqq \tau_5. \end{cases} \tag{20}$$

The concentration of the *tof* cistron containing mRNA is described by the function $m_5(t)$. In view of the presumptive instability of the protein *tof* (Gros et al., 1971), the function $r_4(t)$ is given by the formula of the form: (12) (see contribution 3).

The constants P_8 and P_9 denote the threshold values attained, when the protein *tof* acts in the operators O_c and O_L'. The output variables of this block have the following standard form

$$v_8(t) = \begin{cases} 0 & \text{if} \quad \tilde{r}_4(t) < P_8; \\ 1 & \text{if} \quad \tilde{r}_4(t) \geqq P_8. \end{cases} \tag{21}$$

$$v_9(t) = \begin{cases} 0 & \text{if} \quad \tilde{r}_4(t) < P_9; \\ 1 & \text{if} \quad \tilde{r}_4(t) \geqq P_9. \end{cases} \tag{22}$$

e) The synthesizing block of the protein C_{II}

The block comprises a part of the second segment of the scripton R_1. Besides the protein C_I, this block is controlled by the protein N. In the respective formal synthesizing unit, the presence (or absence) of the threshold doses of these proteins are described by the input variables $e_{15}(t)$ and $e_{25}(t)$.

Their point action of the activity of the block is described by the functions

$$S_6(t) = \overline{e_{15}(t)} \,\&\, e_{25}(t) \tag{23}$$

$$u_6(t) = \begin{cases} \sigma_6 & \text{if} \quad t < \tau_6; \\ S_6(t - \tau_6) & \text{if} \quad t \geqq \tau_6. \end{cases} \tag{24}$$

The function $m_6(t)$ defines the concentration of the C_{II} cistron containing mRNA.

The concentration of the protein C_{II} and that of the C_{III} of the respective block is expressed either by an equation of the form: (12) or (13) (see contribution 3). The value of the threshold dose, when the protein C_{II} acts on the promoter pre is designated by the constant P_{10}. This block has only one output variable

$$v_{10}(t) = \begin{cases} 0 & \text{if} \quad \tilde{r}_5(t) < P_{10}; \\ 1 & \text{if} \quad \tilde{r}_5(t) \geqq P_{10} \end{cases} \tag{25}$$

where

$$\tilde{r}_5(t) = \frac{r_5(t)}{g(t)}.$$

f) The synthesizing block of the proteins O and P

Having suggested that the same parameters characterize the synthesis of the proteins O and P, we shall treat them as one process. Because the cistrons of these proteins are located in the same segment as the cistron C_{II}, the description of the respective protein synthesizing block will differ from the machinery synthesizing the protein C_{II} in the constant P_{11}, the parameter indicating the threshold dose for the proteins O and P which are required for replication and initiation in the output variable

$$v_{11}(t) = \begin{cases} 0 & \text{if} \quad \tilde{r}_6(t) < P_{11}; \\ 1 & \text{if} \quad \tilde{r}_6(t) \geqq P_{11} \end{cases} \tag{26}$$

and also, possibly, in the form of the function $r_6(t)$. The form of this function will be modified only, if the proteins C_{II} and O, P disappear in different manners.

g) The block controlling replication

The next step is to introduce a formal block of replication with input variables $e_{17}(t) = v_4(t)$ and $e_{27}(t) = v_{11}(t)$ and the output variable

$$J(t) = \begin{cases} \overline{e_{17}(t-\tau_7)} \,\&\, e_{27}(t-\tau_7) & \text{if} \quad t \geqq \tau_7; \\ \sigma_7 & \text{if} \quad t < \tau_7. \end{cases} \tag{27}$$

The output variable of this block indicates the presence (or absence) of the replication of the λ phage genome.

h) The synthesizing block of the protein Q

The block synthesizing the protein Q incorporates a part of the third segment of the scripton R_1. The activity of this block is under the control of the proteins C_I and N. The formal synthesizing block may be outlined as follows. The input variables $e_{18}(t)$, $e_{28}(t)$, and $e_{38}(t)$ indicate whether or not the doses of the proteins C_I and N are threshold. Their effect on the activity of the block is described by the functions

$$S_8(t) = \overline{e_{18}(t)} \,\&\, e_{28}(t) \,\&\, e_{38}(t) \tag{28}$$

$$u_8(t) = \begin{cases} \sigma_8 & \text{if} \quad t < \tau_8; \\ S_8(t-\tau_8) & \text{if} \quad t \geqq \tau_8. \end{cases} \tag{29}$$

The concentrations of the cistron Q-containing mRNA and the protein Q are described by the functions $m_9(t)$ and $r_8(t)$. The constant P_{13} denotes the value of that threshold at which the protein Q affects the activity of the fourth segment of the scripton R_1. The output variable is expressed as

$$v_8(t) = \begin{cases} 0 & \text{if} \quad \tilde{r}_8(t) < P_{13}; \\ 1 & \text{if} \quad \tilde{r}_8(t) \geqq P_{13}. \end{cases} \tag{30}$$

i) Linking together the blocks; simulation of the total system of λ phage regulation

In this formal analysis, each synthesizing block is represented by a square with side lines corresponding to the input and output channels. Let us connect the inputs and outputs of the formal blocks complying with the description of the regulatory zone of the λ phage given in section 1 (fig. 1). The threshold model of the system of λ phage regulation is pictured in fig. 2. The set of equations derived from this scheme are:

$$\begin{aligned} e_{11}(t) &= v_1(t), \\ e_{21}(t) &= v_8(t), \\ e_{31}(t) &= v_{10}(t), \\ e_{41}(t) &= v_7(t), \\ e_{12}(t) &= e_{13}(t) = v_2(t), \\ e_{22}(t) &= e_{23}(t) = v_9(t), \end{aligned} \tag{31}$$

$$e_{33}(t) = v_5(t),$$

$$e_{14}(t) = e_{15}(t) = e_{16}(t) = v_3(t), \tag{31}$$

$$e_{25}(t) = e_{26}(t) = v_6(t).$$

Clearly, in the BOOLEAN function $S_i(t)$ of each block the arguments are the output variables of the other block connected with the given block. The threshold model of the phase development regulatory system (PDRS) resides in the specification of the scheme of relations between the blocks and the descriptions of the composing blocks.

3.2. The discrete, stochastic model (DS model) with variable control characteristics for the regulatory system of lambda-phage development

Now we model the system controlling the development of λ-phage (see fig. 1) by means of the program system GENIP outlined in contribution 2 (see also BÖTTNER, 1974, BELLMANN, BÖTTNER et al., 1976, BÖTTNER, 1976, 1977). The basic idea is, as with the threshold model, to describe the parts of the λ scriptions by elementary units of gene expression, which are described by the euqations (13) in contribution 2). Then we can use the possibilities of this modular program system to build a genetic net of the molecular genetic control system of λ-phage and investigate its dynamic and stochastic properties. The inputs and outputs of the basic units are connected according to the functional interactions of the parts of the molecular control system of λ-development. This is accomplished in the model system GENIP by the coupling-matrices each related to a single control variable of a unit of gene expression.

Since the interaction between the regulator molecules and the control sequences of the DNA lead to threshold-like control characteristics, we choose the parameters v^j_{Al} and w^j_{Al} in the control terms g^j_{Al} [see equ. (16) in contribution 2] of the transcription regulating functions $F_1{}^j$ [equ. (14), (15) in contribution 2] in the following way:

$$\left.\begin{aligned} v^j_{Al} &= 25 \\ w^j_{Al} &= 25 \end{aligned}\right\} \quad \text{if the control variable } x^j_{Al} \text{ is an inducing protein (positive controlling) (see fig. 3a in contribution 2)}$$

$$\left.\begin{aligned} v^j_{Al} &= 0 \\ w^j_{Al} &= 25 \end{aligned}\right\} \quad \text{if the control variable } x^j_{Al} \text{ is a repressor protein (negative controlling; see fig. 3b in contribution 2)}$$

The parameters a^j_{Al} in the control terms g^j_{Al} [equ. (16) in contribution 2] are the threshold values, which the control variable x^j_{Al} must reach to unfold its effect.

According to the experimental knowledge on the λ-phage the action of control factors of transcription is conjunctive, that is the control functions $F_1{}^j$ of equ. (14) of contribution 2 are specialized to

$$F_1{}^j = \prod_{l=1}^{k^j} g^j_{Al}, \qquad \bar{k}_j = 0. \tag{32}$$

Since till today no control factor of λ-development at the level of translation has been found the functions $F_2{}^j$ of equs. (17, 18) of contribution 2 are choosen to

$$F_2{}^j \equiv 1 \cdot \bar{a}^j_{lat_l} \tag{33}$$

by an appropriate choice of parameters from the set $H_2{}^j$.

The processes of mRNA- and protein-transport, -processing, -stabilization and -inactivation lack in procaryotic systems. Therefore we turn out these processes by choosing the parameters

$$b_1{}^j = b_2{}^j = b_3{}^j = \bar{b}_3{}^j = b_4{}^j = \bar{b}_4{}^j = b_5{}^j = b_7{}^j = \bar{b}_7{}^j = b_8{}^j = \bar{b}_8{}^j$$
$$= \tau_1{}^j = \tau_5{}^j = 0$$

in the equations (13) of a unit of gene expression in contribution 2 and reflect in this way the tight coupling of the transcription and translation processes in procaryotes.

The other parameters of a unit of gene expression, especially the rate of transcription $\bar{a}^j_{scr}$, the rate of translation $\bar{a}^j_{lat}$, the rate of mRNA degradation $z^j_{x_3}$ and the rate of protein degradation $z^j_{y_3}$ ha to be chosen for every unit separately.

For the construction of a model of the regulation of the development we need 9 units of gene expression (see fig. 2b). Some describe the formation of the regulator proteins:

$j = 1$, c_{III}-protein; $j = 2$, N-protein; $j = 3$, c_I-protein, transcription initiated at the promoter prm; $j = 5$, c_I-protein transcription initiated at the promoter pre; $j = 6$, *tof*-protein; $j = 7$, c_{II}-protein; $j = 9$, *O-P*-proteins. The unit $j = 4$ serves as a formal summation unit of the c_I repressor proteine synthesized from the promoters pre and prm, while the unit $j = 8$ serves as a λ-DNA synthesis unit. In the protein synthesizing units for calculating the number of protein molecular per λ-DNA the parameter p_j of equs. (13) of contribution 2 is set to $p_j = 8$, since the variable $x_3{}^8$ represents the number of λ-DNA molecules (see description of unit 8).

Now we describe the single gene expression units and their coupling to the λ control system.

a) The c_{III}-gene expression unit

This cistron belongs to the second part of the scription L_2. Its transcription is functionally dependent on the activating control factor x^1_{A1}, which represents the number of molecules of antiterminator protein N, per λ-DNA molecule, on the repressing control factors x^1_{A2} and x^1_{A3}, representing the number of antirepressor *tof* per λ-DNA molecule resp. the number of repressor c_I molecules per λ-DNA molecule. A fourth control variable x^1_{A4} represents the number of λ-DNA molecules in the cell and the control term

$$g^1_{A4} = \frac{\left[\dfrac{x^1_{A4}(t - \tau^1_{A4})}{1}\right]^1}{1 + \left[\dfrac{x^1_{A4}(t - \tau^1_{A4})}{1}\right]^0} \cdot 2 \tag{34}$$

describes the linear increases of the total rate of transcription of c_{III}-cistron as a consequence of the increase of the number of λ-DNA molecules caused by replication. This control term is used from all protein generating units and reflects the experimental finding, that the total rate of transcription of a scripton is approximately proportional to the number of λ-DNA copies in the cell.

The parameters a^j_{Al} [see equ. (16) in contribution 2] are in the case of v^j_{Al} and/or w^j_{Al} values greater than 20 the threshold values of the control variable x^j_{Al}. That means, the control variable x^j_{Al} must exceed or fall below the value a^j_{Al} to fulfil an induction or repression function. So the parameter a^1_{A1} corresponds to the threshold value P_5 of the threshold model from section 3.1, the parameter a^1_{A2} corresponds to the threshold dose P_9 and a^1_{A3} corresponds to the threshold value P_2.

The control function of transcription $F_1{}^1$ for the c_{III}-unit of gene expression is thus a product of 4 control terms.
(with $\bar{a}^1_{scr_l} = 1, l \neq 1, 4, \bar{k}^1 = 0, k^1 = 4$) [see equ. (14) contribution 2].

$$F_1{}^1 = \prod_{l=1}^{4} g^1_{Al} \tag{35}$$

The coupling matrices $\bar{K}(x^j_{Al})_{o,p}$ ($o = 1, 2, \ldots, 13; p = 1, 2, \ldots, 9$) of the control variables x^j_{Al} are [see equ. (20)]

$$\bar{K}(x^1_{A1})_{o,p} = \begin{cases} 1 & \text{for} \quad o = 13, \quad p = 2 \\ 0 & \quad \text{otherwise} \end{cases} \tag{36}$$

$$\bar{K}(x^1_{A2})_{o,p} = \begin{cases} 1 & \text{for} \quad o = 13, \quad p = 6 \\ 0 & \quad \text{otherwise} \end{cases} \tag{37}$$

$$\bar{K}(x^1_{A3})_{o,p} = \begin{cases} 1 & \text{for} \quad o = 13, \quad p = 4 \\ 0 & \quad \text{otherwise} \end{cases} \tag{38}$$

$$\bar{K}(x^1_{A4})_{o,p} = \begin{cases} 1 & \text{for} \quad o = 4, \quad p = 8 \\ 0 & \quad \text{otherwise.} \end{cases} \tag{39}$$

b) The N-gene expression unit

The first part of the L_2-scripton is the cistron N. It is controlled by the repressing variables x^2_{A1} (represents the number of antirepressor *tof* molecules per λ-DNA) and x^2_{A2} (represents the number of c_I-protein molecules per λ-DNA). The third control variable x^2_{A3} represente the number of λ-DNA molecules and its control term g^2_{A3} looks analogous as equ. (34). Thus the control function of transcription is

$$F_1{}^2 = \prod_{l=1}^{3} g^2_{Al} \tag{40}$$

with $\bar{a}^2_{scr_l} = 1$, $(l \neq 1{,}3)$, $\bar{k}^2 = 0$, $k^2 = 3$.

The coupling matrices for the control variables x^2_{Al} are

$$\bar{K}(x^2_{A1})_{o,p} = \begin{cases} 1 & \text{for} \quad o = 13, \quad p = 6 \\ 0 & \quad \text{otherwise} \end{cases} \tag{41}$$

$$\bar{K}(x^2_{A2})_{o,p} = \begin{cases} 1 & \text{for} \quad o = 13, \quad p = 4 \\ 0 & \quad \text{otherwise} \end{cases} \tag{42}$$

$$\bar{K}(x^2_{A3})_{o,p} = \begin{cases} 1 & \text{for} \quad o = 4, \quad p = 8 \\ 0 & \quad \text{otherwise.} \end{cases} \tag{43}$$

The parameters a^2_{A1} and a^2_{A2} correspond to the threshold doses P_9 resp. P_2 of the threshold model.

c) The $C_I(prm)$ — gene expression unit

This cistron is the essential part of the scripton L_1. Its transcription is controlled by the antirepressor *tof* (repressing control variable x^3_{A1}) and, its own gene product C_I (acitvating control variable x^3_{A2}) and the number of λ-DNA molecules (control variable x^3_{A3} with a control term g^3_{A3} corresponding to equ. (34)). The control function of transcription of this gene expression unit is ($\bar{a}^3_{scr_l} = 1, l \neq 1, 3; \bar{k}^3 = 0, k^3 = 3$)

$$F_1{}^3 = \prod_{l=1}^{3} g^3_{Al}. \tag{44}$$

The coupling matrices are

$$\bar{K}(x^3_{A1})_{o,p} = \begin{cases} 1 & \text{for} \quad o = 13, \quad p = 6 \\ 0 & \text{otherwise} \end{cases} \tag{45}$$

$$\bar{K}(x^3_{A2})_{o,p} = \begin{cases} 1 & \text{for} \quad o = 13, \quad p = 4 \\ 0 & \end{cases} \tag{46}$$

$$\bar{K}(x^3_{A3})_{o,p} = \begin{cases} 1 & \text{for} \quad o = 4, \quad p = 8 \\ 0 & \text{otherwise.} \end{cases} \tag{47}$$

The parameters a^3_{A2} and a^3_{A1} correspond to the threshold values P_1 and P_8.

d) Summation unit for repressors synthesized from scripton L_1 and L_3

This unit serves as a formal summation unit. Its input variables x^4_{A1} and x^4_{A2} are the number of repressor molecules synthesized from polymerases started at the promoter prm resp. pre. That means the coupling matrices of the variables x^4_{Al} are

$$\bar{K}(x^4_{A1})_{o,p} = \begin{cases} 1 & \text{for} \quad o = 10, \quad p = 3 \\ 0 & \text{otherwise} \end{cases} \tag{48}$$

$$\bar{K}(x^4_{A2})_{o,p} = \begin{cases} 1 & \text{for} \quad o = 10, \quad p = 5 \\ 0 & \text{otherwise.} \end{cases} \tag{49}$$

The function $F_1{}^4$ has the form

$$F_1{}^4 = \sum_{l=1}^{2} g^4_{Al} = \sum_{l=1}^{2} x^4_{Al} \tag{50}$$

[see equs. (14, 15) with $\bar{k}^4 = 2$, $k^4 = 0$, $\bar{a}^4_{scr_l} = 2$, $a^4_{Al} = 1$, $v^4_{Al} = 1$, $w^4_{Al} - 0$, $l = 1, 2$ in contribution 2].

By an appropriate choice of the parameters of the fourth unit of gene expression ($\bar{a}^4_{lat} = 1$, $z^4_{x3} = z^4_{y3} = \tau_1{}^4 = \tau_5{}^4 = 0$, (see equ. (13) and (19) of contribution 2) it is obtained that the output variable $y_3{}^4$ equals the total number of existing C_I — repressor molecules. The variable $y_6{}^4$ then can be used as control variable in the units, which are affected by the C_I repressor.

e) The $C_I(pre)$ — gene expression unit

The synthesis of C_I-repressor, whose mRNA-transcription is initiated at the promotor *pre* (scripton L_3) is described by the unit $j = 5$. It is controlled by four control variables: x^5_{A1} — the activating CAP and cAMP complex (this is an exogenous variable), x^5_{A2} — the activating C_{III} — protein molecules per λ-DNA, x^5_{A3} — the activating C_{II} — protein molecules per λ-DNA and x^5_{A4}, the number of λ-DNA with a control term g^5_{A4} corresponding to equ. (34).

The coupling matrices are:

$$\bar{K}(x^5_{A2})_{o,p} = \begin{cases} 1 & \text{for} \quad o = 13, \quad p = 1 \\ 0 & \text{otherwise} \end{cases} \tag{51}$$

$$\bar{K}(x^5_{A3})_{o,p} = \begin{cases} 1 & \text{for} \quad o = 13, \quad p = 7 \\ 0 & \text{otherwise} \end{cases} \tag{52}$$

$$\bar{K}(x^5_{A4})_{o,p} = \begin{cases} 1 & \text{for} \quad o = 4, \quad p = 8 \\ 0 & \text{otherwise} \end{cases} \tag{53}$$

and the control function of transcription has the form ($\bar{k}^5 = 0$, $k^5 = 4$; $\bar{a}^5_{scr_l} = 1$, $(l \neq 1, 4)$) in equs. (15, 16) of contribution 2.

$$F_1{}^5 = \prod_{l=1}^{4} g^5_{Al}. \tag{54}$$

The parameters a^5_{A2}, a^5_{A3} correspond to the threshold values P_7 and P_{10}.

f) The *tof* — gene expression unit

The cistron is the first part of the scripton R_1. Its control variables are: the number of C_I repressor-molecules per λ-DNA (repressing variable x^6_{A1}) and the number of λ-DNA molecules in the cell (control variable x^6_{A2} with a control term g^6_{A2} analogous to equ. (34). The control function of transcription has the form ($\bar{k}^6 = 0$, $k^6 = 2$) in equ. (14) of contribution 2).

$$F_1{}^6 = \prod_{l=1}^{2} g^6_{Al} \tag{55}$$

and the coupling matrices are

$$\bar{K}(x^6_{A1})_{o,p} = \begin{cases} 1 & \text{for} \quad o = 13, \quad p = 4 \\ 0 & \text{otherwise} \end{cases} \tag{56}$$

$$\bar{K}(x^6_{A2})_{o,p} = \begin{cases} 1 & \text{for} \quad o = 4, \quad p = 8 \\ 0 & \text{otherwise.} \end{cases} \tag{57}$$

Parameter a^6_{A1} corresponds to the parameter P_3 of the threshold model.

g) The C_{II} — gene expression unit

This cistron belongs to the second part of the scripton R_1. Its transcription is possible only in the presence of the antiterminator protein N. The number of antiterminator N-molecules per λ-DNA therefore forme the positive controlling factor x^7_{A1}. The other control variables are the number of C_I-repressor-molecules per λ-DNA (repressing control variable x^7_{A2}) and the number of λ-DNA molecules in the cell (control variable x^7_{A3} with a control term g^7_{A3} analogous to equ. (34). The control function of transcription of the unit seven has the following form: $\bar{k}^7 = 0$, $k^7 = 3$, $\bar{a}^7_{scr_l} = 1$ $(l \neq 1, 3)$ in equs. (14, 16) of contribution 2.

$$F_1^7 = \prod_{l=1}^{3} g^7_{Al} \tag{58}$$

and the coupling matrices are:

$$\bar{K}(x^7_{A1})_{o,p} = \begin{cases} 1 & \text{for} \quad o = 13, \quad p = 2 \\ 0 & \text{otherwise} \end{cases} \tag{59}$$

$$\bar{K}(x^7_{A2})_{o,p} = \begin{cases} 1 & \text{for} \quad o = 13, \quad p = 4 \\ 0 & \text{otherwise} \end{cases} \tag{60}$$

$$\bar{K}(x^7_{A3})_{o,p} = \begin{cases} 1 & \text{for} \quad o = 4, \quad p = 8 \\ 0 & \text{otherwise} \end{cases} \tag{61}$$

The threshold parameters a^7_{A1} and a^7_{A2} correspond to the parameters P_6 and P_3.

h) The λ-DNA — replication unit

This unit serves for the formal description of the replication of λ-DNA. A new interpretation of the variables and functions of the basic unit of gene expression is necessary for this purpose [equ's. (13) and (14—19) of contribution 2]. The variable x_0^8 now means the number of λ-DNA molecules synthesized per time step and the variable x_3^8 means the pool of λ-DNA molecules in the cell. This quantity will be used from the other protein-producing gene expression units for calculating the number of protein molecules per λ-DNA [equ. (13) of contribution 2] and as a control variable, describing approximately the influence of the replication of the λ-DNA on the gene expression. The function F_1^8 describes the control of the λ-DNA replication through the activating control factor x^8_{A1} (number of replication-activating O,P-protein molecules per λ-DNA) and of the repressing control factor x^8_{A2} (number of C_I-molecules per λ-DNA) and has the following form ($\bar{k}^8 = 0$, $k^8 = 2$, $\bar{a}^8_{scr_2} = 1$)

$$F_1^8 = \prod_{l=1}^{2} g^8_{Al}. \tag{62}$$

The coupling matrices are

$$\bar{K}(x^8_{A1})_{o,p} = \begin{cases} 1 & \text{for} \quad o = 13, \quad p = 9 \\ 0 & \text{otherwise} \end{cases} \tag{63}$$

$$\bar{K}(x^8_{A2})_{o,p} = \begin{cases} 1 & \text{for} \quad o = 13, \quad p = 4 \\ 0 & \text{otherwise} \end{cases} \tag{64}$$

and the parameters a^8_{A1} and a^8_{A2} correspond to the threshold parameters P_{11} and P_4.

i) The O, P — gene expression unit

For the same reason as in the threshold model we shall treat the synthesis of the proteins O, P as one process and also notice that the control of the cistron O, P is completely the same as the control of the cistron C_{II}, because both cistrons belong to the second segment of the R_1-scripton. Therefore the coupling matrices and the threshold parameters of the control variables x^9_{A1}, x^9_{A2}, x^9_{A3} and the control function $F_1{}^9$ are chosen completely analogous as in the case of the cistron C_{II} ($j = 7$).

4. Results

4.1. A qualitative analysis of the TH-model

With these preliminaries, we approach directly to the threshold model of the PDRS. Let $m_{i(0)} = 0$, $r_{j(0)} = 0$ ($i \in \{1, 2, \ldots 7\}$, $j \in \{1, 2, \ldots 6\}$), $\sigma_i = 0$ at a starting point t_0, which corresponds to the PDRS state promptly after infection. After delays τ_3 and τ_5, the blocks 2 and 4 (each block index has the same number as each section of its description in the text) are activated ($u_3 = 1$ and $u_5 = 1$), while the remaining blocks are still inactive. It follows that from this point the concentrations of $r_2(t)$ and $r_4(t)$ of the proteins *N* and *tof* rise monotonously. It is taken for granted that **the thresholds P_5 and P_6 are reached earlier than the P_9.** After the attainment of the thresholds P_5 and P_6, blocks 3, 5, and 6 are turned on, and the concentrations $r_3(t)$, $r_5(t)$, $r_6(t)$ of the proteins C_{III}, C_{II}, O and P start to rise. Two functional variants are possible, if we concede to the assumption that the thresholds P_7, P_{10} and P_{11} are attained earlier than the P_9. If $e_{51} = 0$ in the given time interval, block 1 will be inactive during the interval; if $e_{51}(t_y) = 1$, where t_y is the point when both thresholds P_7 and P_{10}, are attained, concentration of the protein C_I, $r_1(t)$ will start to rise with a delay τ_1.

Let us examine the first possibility. According to formula (27) $r(t) = 1$ after the P_{11} is attained, and the functions $r_l(t)$ ($l \in \{2, 3, \ldots, 6\}$) previously expressed as (5) or (9) will assume the form (7) or (10) (See contribution 3). The functions r_l will continue to rise, and so will functions $\tilde{r}_l$, although they may decrease for a time. After some time after this turning on, $r_4(t)$ (the concentration of *tof* protein) will be at the threshold P_9. This occurs with delays τ_3 and τ_4 and inactivates the proteins 2 and 3 (see formulas 12, 16). As a result, the function $r_2(t)$ will be described by formula (3). (See contribution 3), and some time after this turning on, $r_2(t)$ will decrease and tend to 0; concomitantly, there is a point of time t_η, when $\tilde{r}_2(t) < P_5$, $\tilde{r}_2(t) < P_6$. This renders blocks 5 and 6 inactive. Changes in the concentrations of the proteins O and P, expressed by the function $r_6(t)$, deserve attention. The behaviour of this function depends very much on how the proteins O and P disappear. Should these proteins be unstable ($b_2 \neq 0$) $r_6(t)$ decreases and rapidly tends to 0. The decrement of $r_6(t)$ proceeds still faster. As a consequence, $\tilde{r}_6(t)$ falls below the threshold P_{11} and brings replication to a stop. **The stability of the proteins O and P** is a reasonable inference from the fact that a lytically developing phage producing maximum offspring number is conferred populational advantages. If O and P were stable, $r_6(t)$ would be a nondecreasing function (see preceding item), and $\tilde{r}_6(t)$ would decrease much more slowly than in the alternative case, when $b_2 \neq 0$.

Let us now turn to the lysogenic variant of development, when $e_{51} = 1$. After the proteins C_{II} and C_{III} have attained the thresholds P_{20} and P_7, the concentration of the protein C_I starts to grow. After some time interval, the thresholds P_1, P_2, P_3, and P_4 are reached, this causes replication to cease (see formula (27)) and the functions $r_2(t)$, $r_3(t)$, $r_4(t)$, $r_5(t)$, and $r_6(t)$ to assume the form (3) or (8) —, as shown in contribution 3). Let us examine the formulae describing the functions $r_3(t)$ and $r_5(t)$. The threshold P_4 is attained rapidly, because replication seams to be over soon after the onset of the synthesis of the protein C_I (Reichardt, Kaiser, 1971). The concentrations of the proteins C_{II} and C_{III}, when $b_2 = 0$, will not fall to thresholds below the P_{10} and P_7 for the reason that $\tilde{r}_3(t)$ and $\tilde{r}_5(t)$ are nondecreasing functions when $J(t) = 0$. This being so, the scripton L_3 will persist to function, contrary to the observation that it is inactive at min 15—20 (Reichardt, Kaiser, 1971). Obviously, $r_3(t)$ and $r_5(t)$ should decrease when $u_4(t) = 0$ and $u_6(t) = 0$, which is feasible only when $b_2 \neq 0$. The assumption made is that **the proteins C_{II} and C_{III} (or one of them) are unstable as a result of enzymic degradation.** This assumption has now received experimental support (Reichardt, 1975).

It is pertinent to note that blocks 2 and 3 may also be in inactivated by the attainment of the threshold P_9 by the protein *tof*. After the C_I has attained the P_8, the function $r_4(t)$ may be described either by formula (3) or formula (8) given in contribution 3. If formula (3) is applicable, both repressors, C_I and *tof*, will be present on the cell, until either of the two is diluted by sequential divisions of the host cell. Block 1 will remain inactive, until $r_4(t)$ becomes less than the P_8 which takes more than a cell cycle. If formula (3) is applicable block 1 will be reactivated during a period shorter than a cell cycle. The only block remaining active will be block 1.

From this qualitative analysis of the PDRS it follows that the inaccessability of threshold values for a regulatory protein means obstruction of one of the developmental variants. Once a developmental regime has established, development will proceed in an orderly manner. Thus, under the lytic regime, the cistrons N and *tof* start to function first (the immediate early stage), then the cistrons C_{II} and C_{III}, O and P (the delayed early stage), those are followed by the cistron Q and by the morphological cistrons (the late stage). Under the lysogenic regime (the delayed early stage), the protein C_I starts to be synthesized and prevents the onset of the third stage. The control of λ phage development is such that **one cannot alter the sequence of the developmental stages only by modulating the kinetical parameters one has to modify also the scheme of interrelations.** This stams from the preceding analysis. This feature seems to provide **the stability of λ phage ontogenesis, i.e., low sensitivity of developmental regimes to chance fluctuations of parameters.** The plausibility of this suggestion is confirmed by the results of machine experiments based on the threshold model and on the stochastic difference equation model (see section 4.3).

4.2. The dynamics of the lambda-phage regulatory system under the lytic and lysogenic regimes

4.2.1. General considerations

With very weak limitations imposed on its parameters, the models of the λ-system has two regimes, lytic and lysogenic. An appropriate choice of the unknown parameters of in the λ system ensures a reasonably close qualitative and quantitative agreement between the system's behaviour and real-life processes. How will the system function? To answer this question, we set up computer programs corresponding to the description of

Table 1[1]. Designations of dynamic variables and parameter values for blocks of the PDRS model (version 1).

	Blocks synthesizing regulatory proteins, their promoters and order numbers (The order numbers and dynamic variables in parentheses refer to the stochastic difference equation model)								Replication block
	C_I		N	C_{III}	*tof*	C_{II}	O & P	Q^+	
	prm	*pre*	P_L		P_R				*ori*
Block numbers	(3) 1 (4)	(5)	2 (2)	3 (1)	4 (6)	5 (7)	6 (9)	7	8 (8)
Concentration designations: m RNA	$(x_3{}^3)$ $m_1(t)$	$(x_3{}^5)$	$(x_3{}^2)$ $m_2(t)$	$(x_3{}^1)$ $m_3(t)$	$(x_3{}^6)$ $m_4(t)$	$(x_3{}^7)$ $m_5(t)$	$(x_3{}^9)$ $m_6(t)$	$m_7(t)$	
absolute concentrations of proteins	$r_2(t)$ $(y_3{}^4)$		$r_2(t)$ $(y_3{}^2)$	$r_3(t)$ $(y_3{}^1)$	$r_4(t)$ $(y_3{}^6)$	$r_5(t)$ $(y_3{}^7)$	$r_8(t)$ $(y_3{}^9)$	$r_7(t)$	
protein concentrations per genome	$\tilde{r}_I(t)$ $(y_6{}^4)$		$\tilde{r}_2(t)$ $(y_6{}^2)$	$\tilde{r}_3(t)$ $(y_6{}^1)$	$\tilde{r}_4(t)$ $(y_6{}^6)$	$\tilde{r}_5(t)$ $(y_6{}^7)$	$\tilde{r}_8(t)$ $(y_6{}^9)$	$\tilde{r}_7(t)$	
Constants of synthesis and degradation: (numbers of molecules/min.):									
$a_1(a^j_{scr_1})$	0,4	2,8	4	4	2	2	2	2	
$a_2(z^j_{x3})$	0,3	0,3	0,4	0,2	0,2	0,4	0,4	0,4	
$b_1 a^j_{lat})$	*I*	*I*	3	3	4	4	4	4	
$b_2(z^j_{y3})$	0	0	0,3	0,3	0,3	0,3	0	0	
Thresholds (Number of molecules of regulatory protein/genome):									
Protein C_I P_1	8								
Protein C_I P_2			8	8					
Protein C_I P_3					8	8	8	8	
Protein C_I P_4									10
N Protein P_5				10					
N Protein P_6						10	10		
Protein C_{III} P_7	15								
Protein *tof* P_8	4								
Protein *tof* P_9			35	35					
Protein C_{II} P_{10}		9							
Proteins O and P P_{11}									8
Protein N P_{12}								10	
Protein Q P_{13}								50	

Explanation of tables 1 and 2

[1] Protein synthesizing block Q was used only for testing the parametric stability of the model. Threshold P_{12} is connected with the effect of protein N on the terminator t_{R_1}. Threshold P_{13} is connected with the effect of protein Q on the late operon R_2. Because corresponding protein synthesizing blocks were disregarded in the model, the threshold values are given in the same column of block Q. **Note:** The choice of coefficient values was mainly based on indirect considerations. Coefficients a_{1i} were determined by the binding frequencies of RNA — polymerase to the promoters. Here, the coefficients chosen are close to the respective coefficients established for bacterial operons (Baker, Yanofsky, 1972). Reichardt and Kaiser (1971) have suggested that the promoter of scripton L_3 is 7 times more efficient than the promoter of the scripton L_I.

Table 2. Parameter values for the blocks of te PDRS model (version 2).

	Blocks synthesizing regulatory proteins, their promoters and order numbers								Replication block
	C_I		N	C_{III}	*tof*	C_{II}	O & P	Q^+	
	prm	*pre*	p_L			p_R			*ori*
Block numbers	(3)	1 (4) (5)	2 (2)	3 (1)	4 (6)	5 (7)	6 (9)	7	8 (8)
Constants of synthesis and degradation (number of molecules/min.)									
$a_1(a^j_{scr})$	0,4	2,8	6	6	4	4	4	4	
$a_2(z^j_{x3})$	0,3	0,3	0,2	0,4	0,2	0,2	0,2	0,2	
$b_1(a^j_{lat})$	1	1	6	6	6	6	6	6	
$b_2(z^j_{y3})$	0	0	0,3	0,2	0,3	0,3	0	0	
Thresholds (numbers of molecules of regulatory protein/genome):									
Protein C_I P_I	8								
Protein C_I P_2			8	8					
Protein C_I P_3					8	8	8	8	
Protein C_I P_4									10
Protein N P_5				20					
Protein N P_6						20	20	20	
[illegible]tein P_7		30							
Protein *tof* P_8	10								
Protein *tof* P_9			190	190					
Protein C_{II} P_{10}	20								
Proteins O and P P_{11}									3
Protein N P_{12}								20	
Protein Q P_{13}								450	

The coefficients $a_{2i} = \ln 2/t^{1/2}$ (where $t_{1/2}$ is half-life) taken into account that the half-life of bacterial mRNA is several minutes. Choosing the coefficients b_{1i}, it was implied that the binding frequencies of ribosomes to templates is low. The coefficient b_{23} for protein N has been determined experimentally (b_{23} = ($n2/2$ min.) (GREENBLATT, 1973). The other coefficients b_{2i} for unstable proteins were chosen under the assumption that their degradation proceeds similarly. All the constants were expressed as numbers of molecules per min. In case of the stochastic difference equation model the constants were expressed as number of molecules per 0,2 min.

P_1, P_2, P_3 and P_4 — values of the threshold doses of protein C_I when it acts on sites p_c, O_L, O_R and on replication. P_5 and P_6 — values of the threshold doses of protein N, when it acts on terminators t_L and t_{RI}. P_7 — value of the threshold dose of protein C_{III} when it acts on the promoter *pre*. P_8 and P_9 — value of the threshold dose of protein *tof*, when it acts on the operators O_c and O_L'. P_{10} — value of the threshold dose of the protein C_{II} when it acts on promoter *pre*. P_{11} — value of the threshold dose of proteins O and P needed for replication initiation. Thresholds were expressed as numbers of protein molecules per genome. The threshold values were chosen by formulas describing the respective synthesizing blocks, taking into consideration the real time of the turning on and — off of scripton transcription.

the λ system. We then carried out computer studies of λ system behaviour assigning various values to the parameter sets. We introduced several assumptions concerning protein stability: O and P were accepted as stable, C_{II} and C_{III} as unstable; we also assumed that the protein *tof* is unstable (GROS et al., 1971). The cistrons are turnes on and off with delays less than 1 min (with transcription rate amounting to 42 bases per sec, according to ADESNIK and LEVINTHAL (1970)), thanks to the proximity of the cistrons to the promoters (SZYBALSKI, 1971). Since the concentrations of the proteins and mRNA with the threshold model were calculated with an interval of 1 min, it was supposed that

$$\tau_1 = \tau_2 = \cdots = \tau_8 = 1. \tag{65}$$

In the case of the stochastic difference equation model we chose 12 sec as the length of a time step in accordance with the time constants in the system and the nonlinearities in the control of the transcription processes. The corresponding delay constants therefore must be chosen to $\tau^j_{Al} = 3$ and $\tau_5{}^j = 2$, ($j = 1, 2, 3, 5, 6, 7, 9$).

To illustrate this let us see how the model's functions with various values assigned to parameter sets (Tables 1 and 2). The block synthesizing the protein Q is descounted, because the output of this block is not connected with the other ones.

4.2.2. The lytic regime

The analysis of the lytic regime starts with the assumption that $e_{51} = 0$ in the case of the threshold model (TH-model) and $x^5_{A1} = 0$ in the case of the stochastic difference equation model. (SD-model). The function of the λ system under this regime is shown diagramatically in figs. 3a, b, 4a, b, and 5a, b. a) are the results with the TH-model and b) are the results with the SD-model). The zero reference point stands for the moment of cell infection. 1 min later there follows a rise in the concentrations of the mRNA containing the cistrons N, *tof*, and of their proteins N and *tof*. At min 2, the concentrations of the protein N reaches the thresholds P_5 and P_6 and, as a result the concentration of the mRNA, containing the cistrons C_{II} and C_{III}, as well as that of O, P and of the respective proteins start to rise at min 3. The phage genome begins to replicate. At first, this brings about a brief decrease in the relative concentrations (estimated per genome) of the proteins N and *tof*. At min 7, the absolute and relative concentrations of proteins and mRNA increase rapidly. At min 9, the concentration of the protein *tof* attains the threshold P_9 and this causes a decrease in the concentration of the mRNA, containing citrons N and C_{III}. The absolute concentration of the protein N keeps growing to min 10 and that of the protein C_{III} to min 11 (see fig. 5). Then the concentration of the respective mRNA and proteins smoothly decline to zero. At min 16, the relative concentration of the protein N falls below the thresholds P_5, P_6 and entails an exponential decrease to zero of the concentrations of mRNA, containing the cistrons C_{II}, O, and P. At min 24, the concentration of the unstable protein C_{II} rapidly approaches zero and the absolute concentration of the proteins O and P forms a plateau. Owing to the continuing replication, the relative concentration of the proteins O and P starts to diminish at min 19. The concentration of these proteins at min 40 is still about sevenfold higher than the threshold P_{11}. This makes possible the replication of the phage genome during the interval from min 4 to min 40. The acceptance of the stability of the protein Q lad us to the supposition that its concentration changes in the same way as the concentration of O and P, remaining above threshold for a long time and ensuring the activity of the

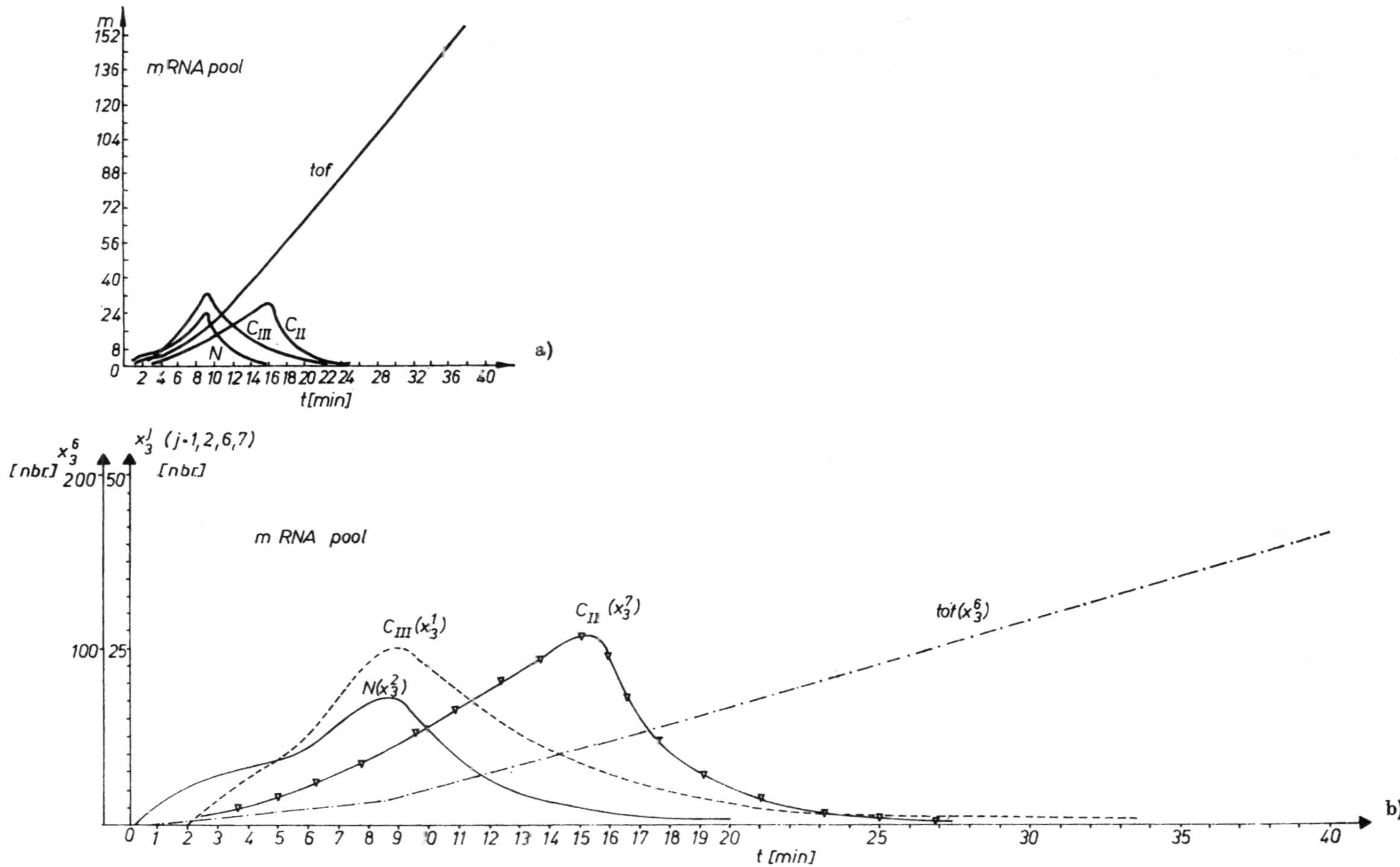

Fig. 3a, b. Lytic regime based on parameter set N I: kinetic curves for mRNA concentrations containing regulatory protein cistrons; m: number of mRNA molecules in the cell, t: time after cell infection (min).
a) results with the threshold model (TH). b) results with the stochastic difference equation model (SD).

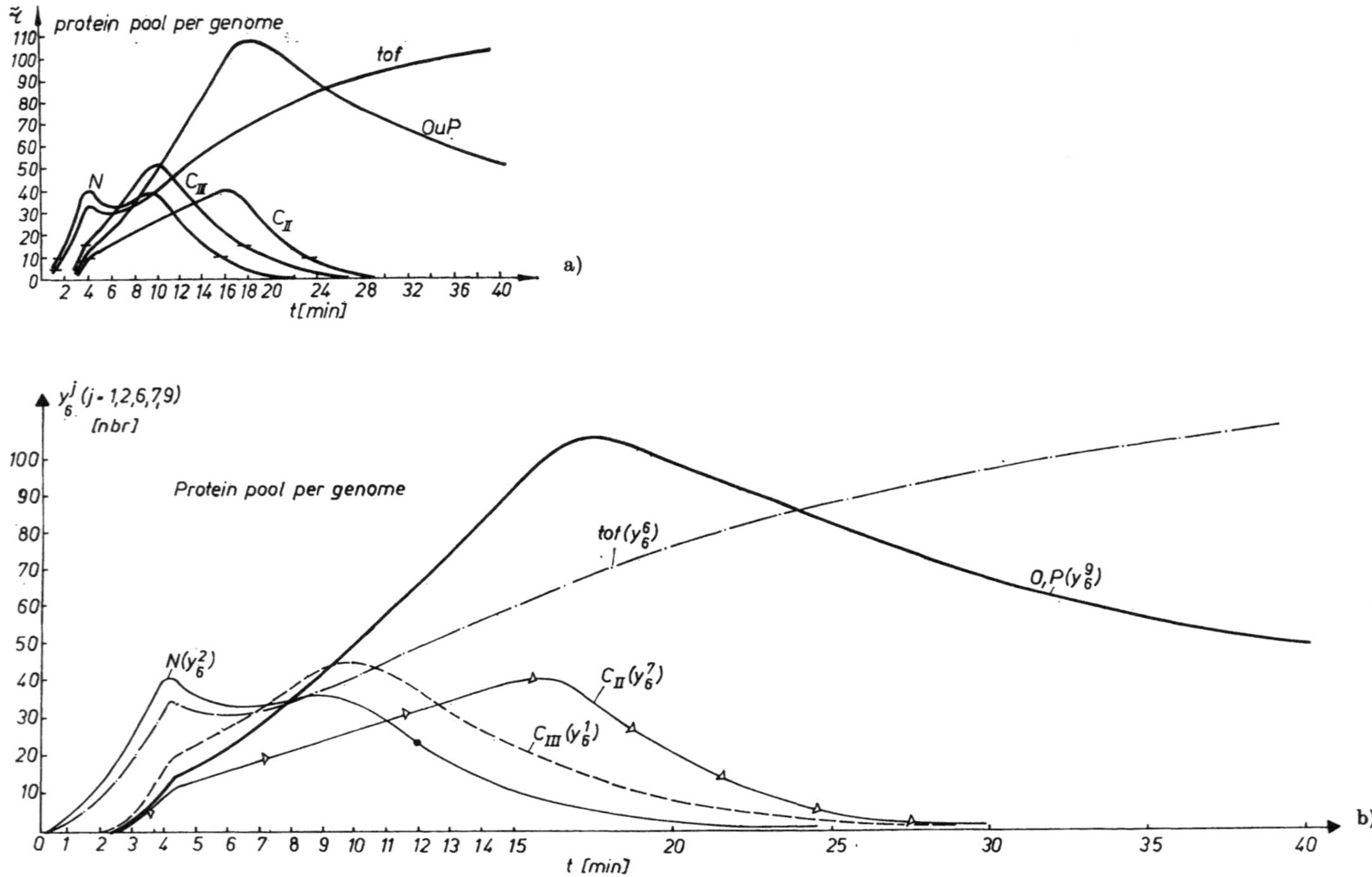

Fig. 4a, b. Lytic regime based on parameter set N I: kinetic curves for *relative* concentrations of regulatory proteins, r: number of protein molecules per genome, t: time elapsed after cell infection (min). dashes on curves: threshold concentrations. a) TH, b) SD

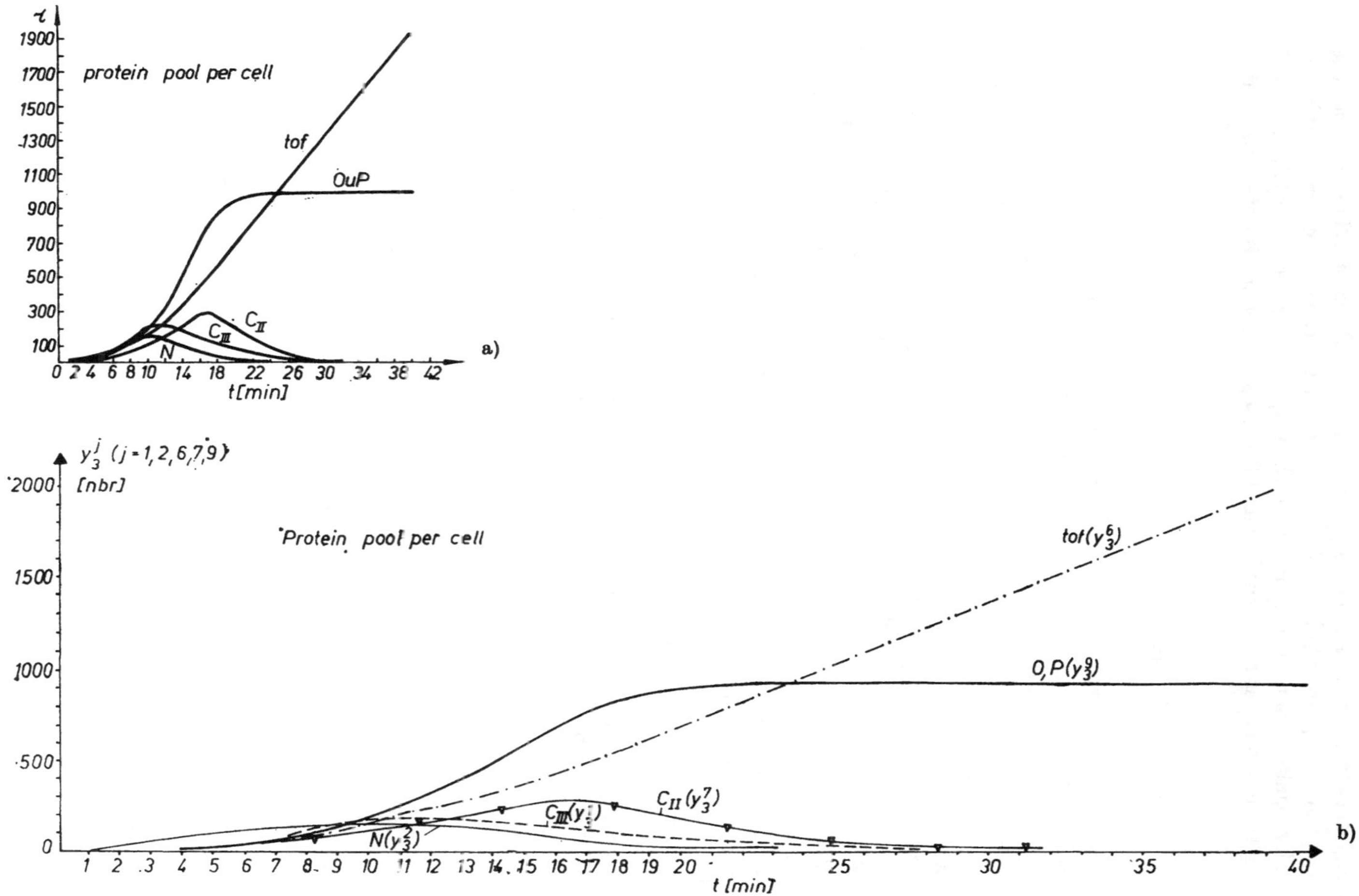

Fig. 5a, b. Lytic regime based on parameter set N I: kinetic curves for ***absolute*** concentrations of regulatory proteins. r: number of protein molecules per cell. t: time elapsed after cell infection (min). a) TH, b) SD.

morphological cistrons of the scripton R_2. It is to be stressed that a fall in the concentration of the proteins C_{III} and C_{II} below P_7 and P_{10} makes the lytic regime irreversible from min 18. This signifies that, even in the case when $e_{51} = 1$ (resp. $x^5_{A1} > a^5_{A1}$), block 1 (5) containing the cistron C_I is not turned on and the lysogenic regime is not established.

4.2.3. The lysogenic regime

Suppose that $e_{51} = 1$ (resp. $x^5_{A1} > a^5_{A1}$) at a starting point from which the system functions to min 5 as earlier (see figs. 6—8a, b). Fig. 7c shows results, computed with the stochastic difference equation model with sigmoid induction and repression control characteristics to evaluate the influence of the shape of the control characteristics on the dynamic properties of the λ-phage regulatory system (see section 5 discussion). The concentration of the proteins C_{II} and C_{III} are at the thresholds P_{10} and P_7 at min 4. Because $e_{51} = 1$ (resp. $x^5_{A1} > a^5_{A1}$), from min 5 onwards, there occurs a rise in the centration of the protein C_I and of the C_I-containing mRNA. This mRNA corresponds to the scripton L_3, the activity of the scripton L_1 is repressed by the protein *tof*. At min 8, the concentration of the protein C_I is at the thresholds P_2 and P_3 causing an exponential decrease in the concentration of the mRNA containing the cistrons N, C_{III}, C_{II}, *tof*, and O, P. The absolute concentrations of the respective proteins rise transiently and, in some cases, later fall to zero. At min 9, the concentration of the protein C_I attains the threshold P_4; this stops replication and makes constant the concentrations of the stable proteins. Decrease in the concentration of the protein C_{II} below the threshold P_{10} stops the synthesis of the mRNA (*pre*), containing the cistron C_I. This takes place at min 15. The relative concentration of the protein *tof* falls below the threshold P_8. Hence it follows in the next minute an increase of the concentration of the mRNA (P_c), which contains the cistron C_I and corresponds to the scripton L_1. The protein C_I is synthesized on this mRNA. The lysogenic state is completely established by min 26. The only functioning scripton is L_1, the others are turned off.

4.2.4. Modelling of the effect of some mutations of the lambda regulatory system

a) Mutations eliminating phage genome replication

As a result of some mutations, the λ-phage loses its capacity for the autonomous replication of the genome. Examples of such mutations are deletions of the cistrons O and P or of the replicator *ori*. How mutations of this kind can affect the dynamics of the PDRS is considered in the case of the two models with parameter set values No. I (see Tables 1 and 2). Suppose that the model simulates a deletion of the replicator *ori*, i.e. the synthesis of the proteins O and P is possible. The results of the computer calculations are represented schematically in figs. 9, 10. During the given interval of time, when replication is absent ($I(t) = 0$ resp. $a^8_{scr} = 0$), the synthesizing blocks of the PDRS are turned on and off in sequence characteristic of the lytic ($e_{51} = 0$ resp. $x^5_{A1} = 0$) and lysogenic ($e_{51} = 1$ resp. $x^5_{A1} > a^5_{A1}$) regimes. The absolute protein concentrations of the deficient system and the concentration of protein per genome are the same. The kinetic curves for the unstable proteins C_{II}, C_{III}, and N are unimodal and, after reaching maximum, tend to zero with increasing t both at $e_{51} = 0$ and $e_{51} = 1$; the curves for the stable proteins O and P are plateaus. Tending to maximum values, the protein concen-

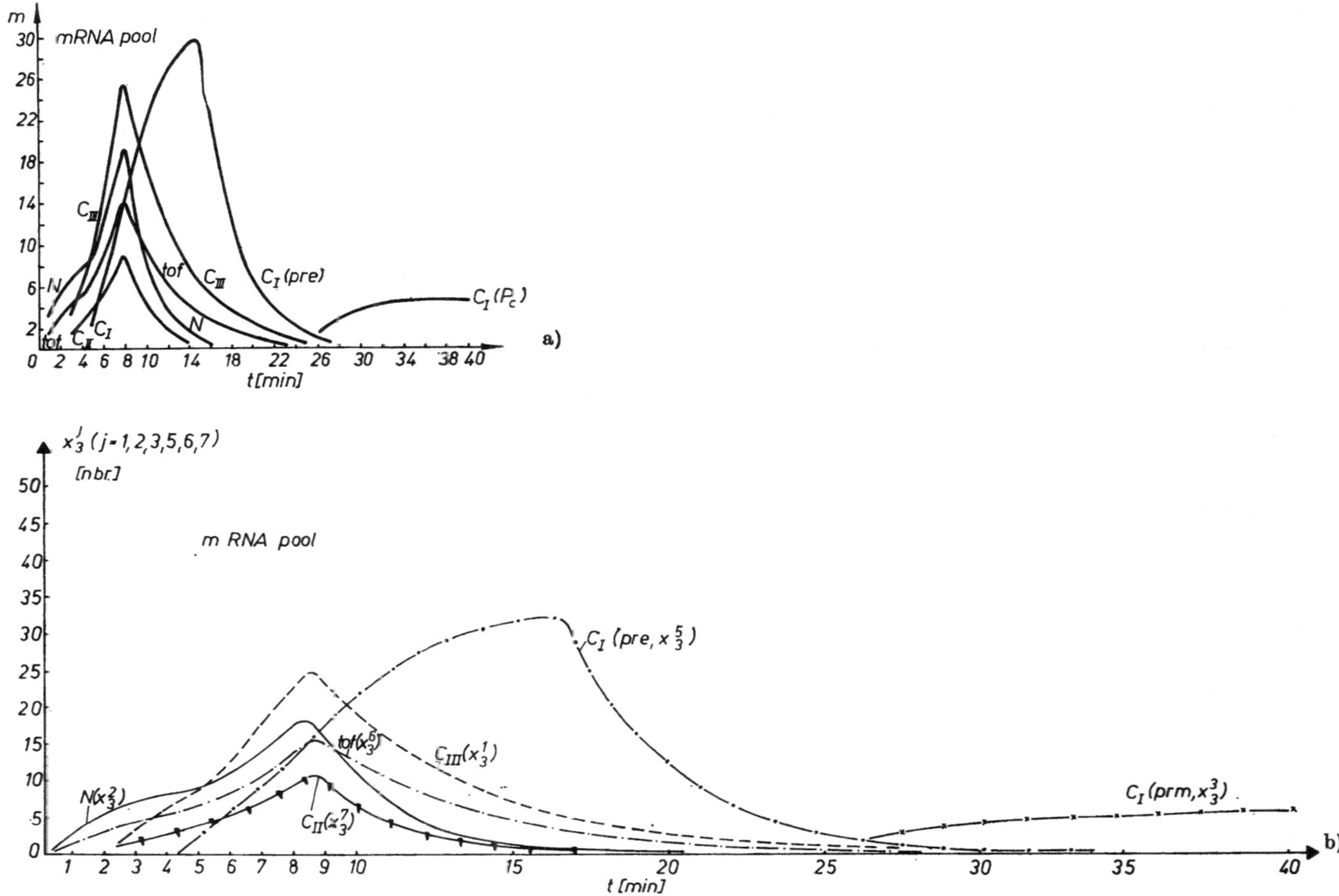

Figs. 6a, b. Lysogenic regime based on parameter set N I: kinetic curves for concentrations of mRNA. Designations are the same as in fig. 3. a) TH, b) SD.

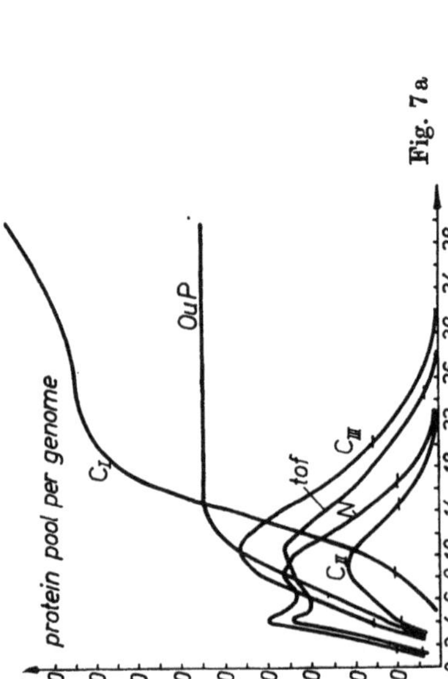

protein pool per genome
OuP
t [min]

Fig. 7a

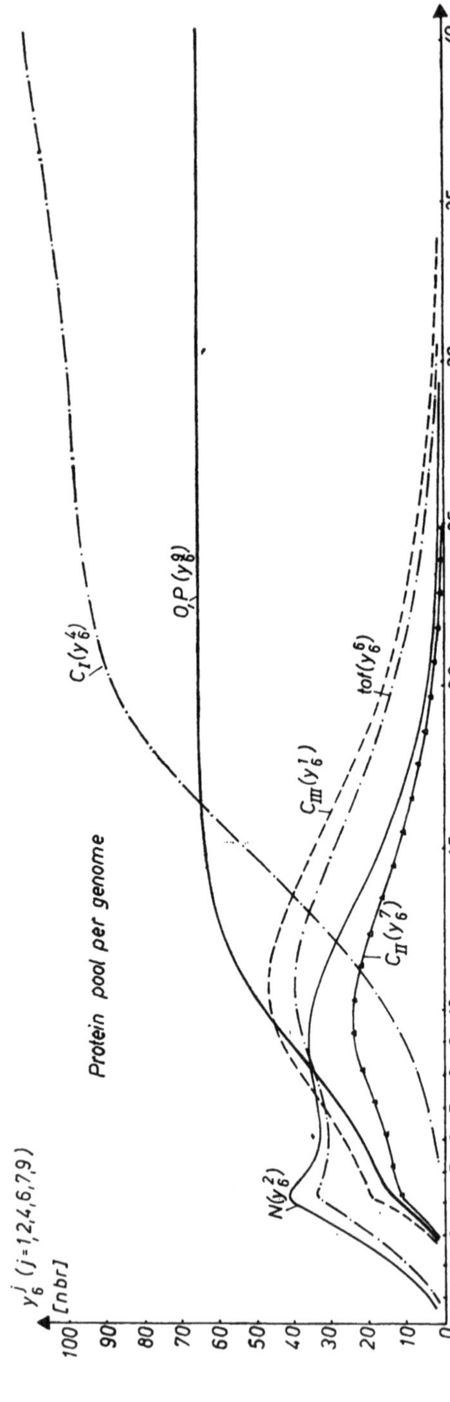

Protein pool per genome
t [min]
[nbr]

Fig. 7b

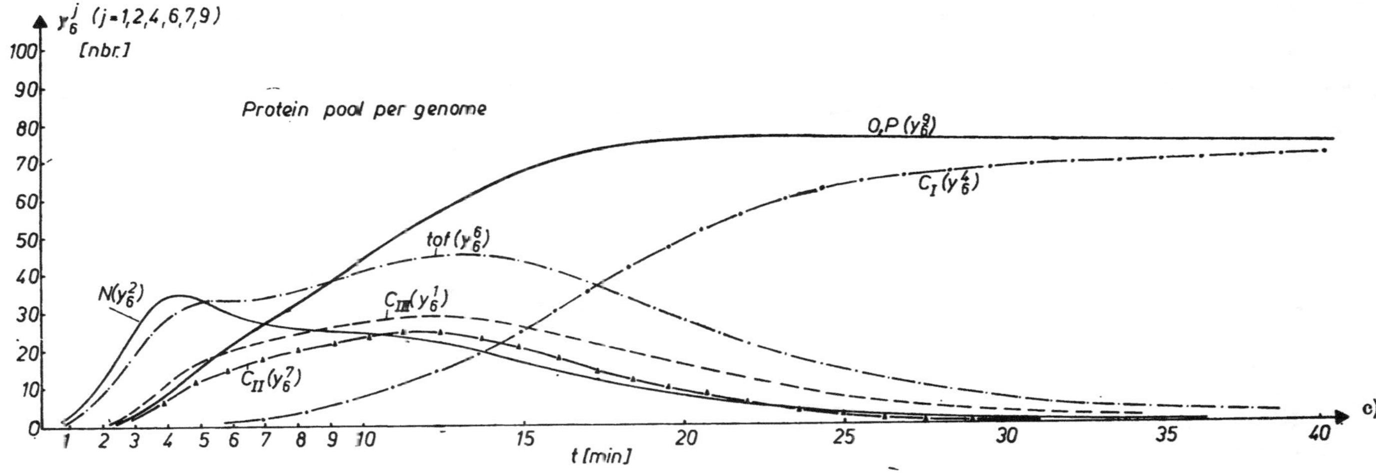

Figs. 7 a, b, c. Lysogenic regime based on parameter set N I: kinetic curves for *relative* concentrations of regulatory proteins. Designations are the same as in fig. 4. a) TH, b) SD, c) SD: lysogenic regime based on parameter set N I but with sigmoid induction and repression control characteristics (parameter $W^j_{Al} = 2$ and $V^j_{Al} = 2$ or $V^j_{Al} = 0$ in the control terms g^j_{Al} see equ. (16) and figs. 3 a, b in contribution 2.

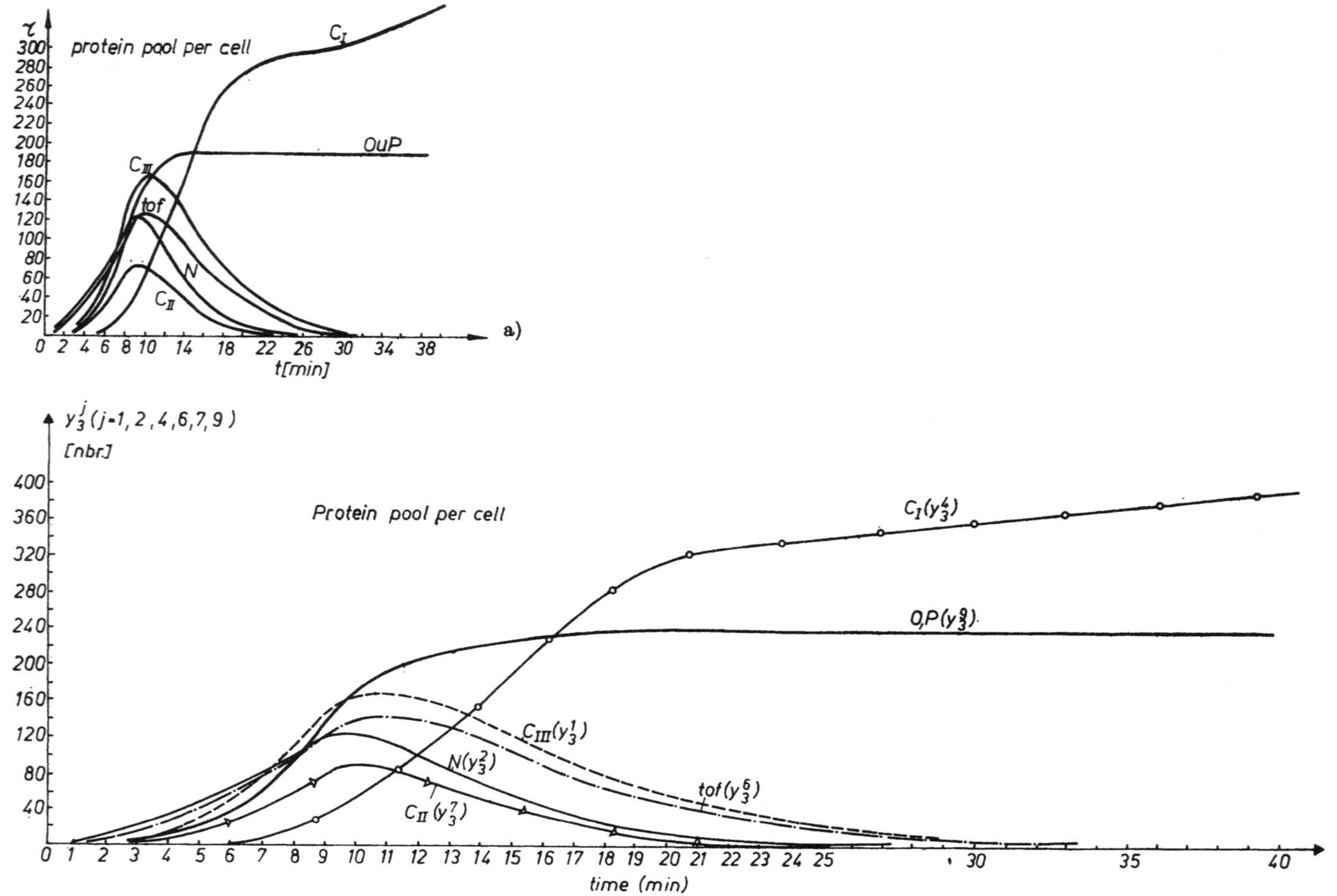

Figs. 8a, b. Lysogenic regime based on parameter set N I: kinetic curves for *absolute* concentrations of regulatory proteins (protein pod per cell). Designations are the same as in fig. 5. a) TH, b) SD.

trations rise more slowly in all the synthesizing blocks of a defective system than in a normal one; however, the relative concentrations rise more slowly than the absolute (see figs. 9, 10). Consequently, many events occur earlier in a defective system: the attainment of the threshold P_9 by the protein *tof*, the decrease in the unstable proteins C_{II}, C_{III}, N and the plateaus of the stable proteins O and P. Generally speaking, a defect in the replication system does not affect, to any significant extent, the dynamics of the undefective part of the system regulating λ-phage development. However, such a defective phage reproduces only in the state of a prophage or a plasmid. In both cases, the replication of the phage genome is the necessary condition for the function of the replication system of the bacterial host cell. If the cistrons O and P are defective, a helper-phage has to be present.

b) Mutation of the cistron *tof*

To gain some insight into the function of the product of the cistron *tof*, it is worthwhile to see how the deletion of *tof* affects the dynamics of the PDRS. Helpful in this connection are the experimental data of Echols (1972) revealing the influence of this mutation of the cistron *tof* on the kinetics of single cistrons of the PDRS. We simulated the deletion of the cistron *tof* with the two models, the relevant computed data are given in figs. 11a, b. The kinetic curves for the defective system under the lytic regime differ very much from the curves for the normal system. The concentrations in these curves increase monotonously during the given interval of time, with the concentration of the stable proteins increasing faster. This is in line with the data of Echols et al. (1972) suggesting that **the role of the protein *tof* is to economize substrate and energy for the production of regulatory proteins.**

In a normal system, once redundant, the synthesis of the regulatory proteins is turned off by the protein *tof*. After the establishment of the lysogenic regime, it becomes the function of the repressor C_I to economize in a similar way. Hence, under the lysogenic regime, the concentration of the unstable proteins in a defective system behaves like the concentration in a normal system. The concentration of the stable proteins O and P form a plateau.

A comparison between the real system's behaviour and the protein pool dynamics calculated by simulation is shown in fig. 12. One may observe qualitatively equal kinetics.

4.3. Computer analysis of the "sensitivity" of the developmental regimes to random parameter variations

4.3.1. Investigations with the TH-model

So far we investigated the dynamics of the PDRS with constant values of parameters. Some of the parameters were experimental data (for instance, the half-life of the protein N, the rate of the synthesis of the protein C_I), other parameters were obtained indirectly. There are several ways of proving that the developmental regimes are invulnerable to random fluctuations. Our proof was based on a random choice of such parameters of computer models as the frequency of RNA polymerase and ribosome binding to initiation sites on templates, thresholds and other parameters. The choice was made from intervals of given length. The next step was to test the functional capacity of the PDRS

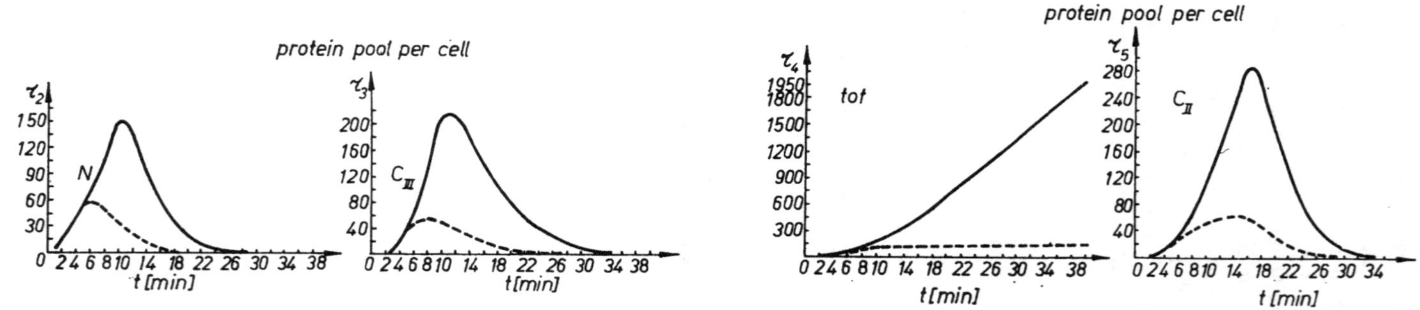
protein pool per cell
τ_2
N
t [min]
τ_3
C_{III}
t [min]
protein pool per cell
τ_4
tot
t [min]
τ_5
C_{II}
t [min]

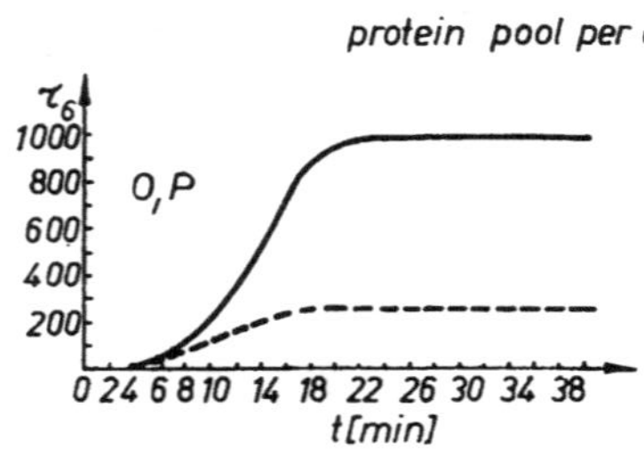
protein pool per cell
τ_6
O, P
t [min]

Fig. 9a

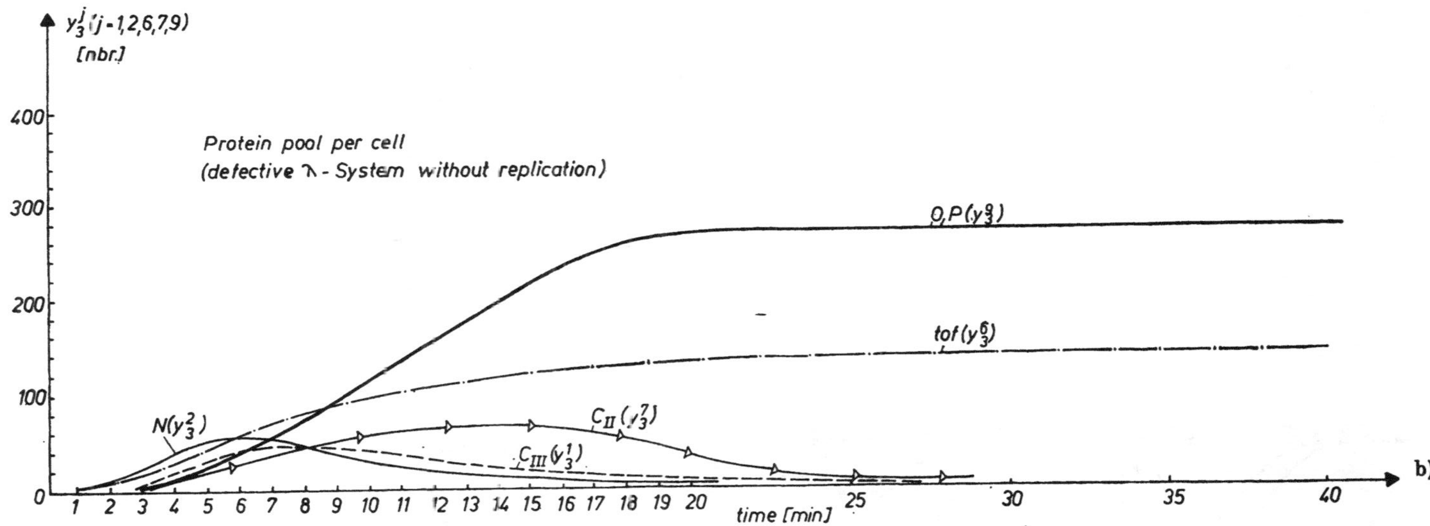

Figs. 9a, b. Lytic regime based on parameter set N I: comparison of the functioning of the normal regulatory system of λ-phage development with the system with deleted replicator *ori*. Solid lines: kinetic curves of the normal system, dotted lines: kinetic curves of the defective system. r_2: protein N concentration; r_3: protein C_{III} concentration; r_4: protein *tof* concentration; r_5: protein C_{II} concentration; r_6: proteins O and P concentrations. a) TH, b) SD.

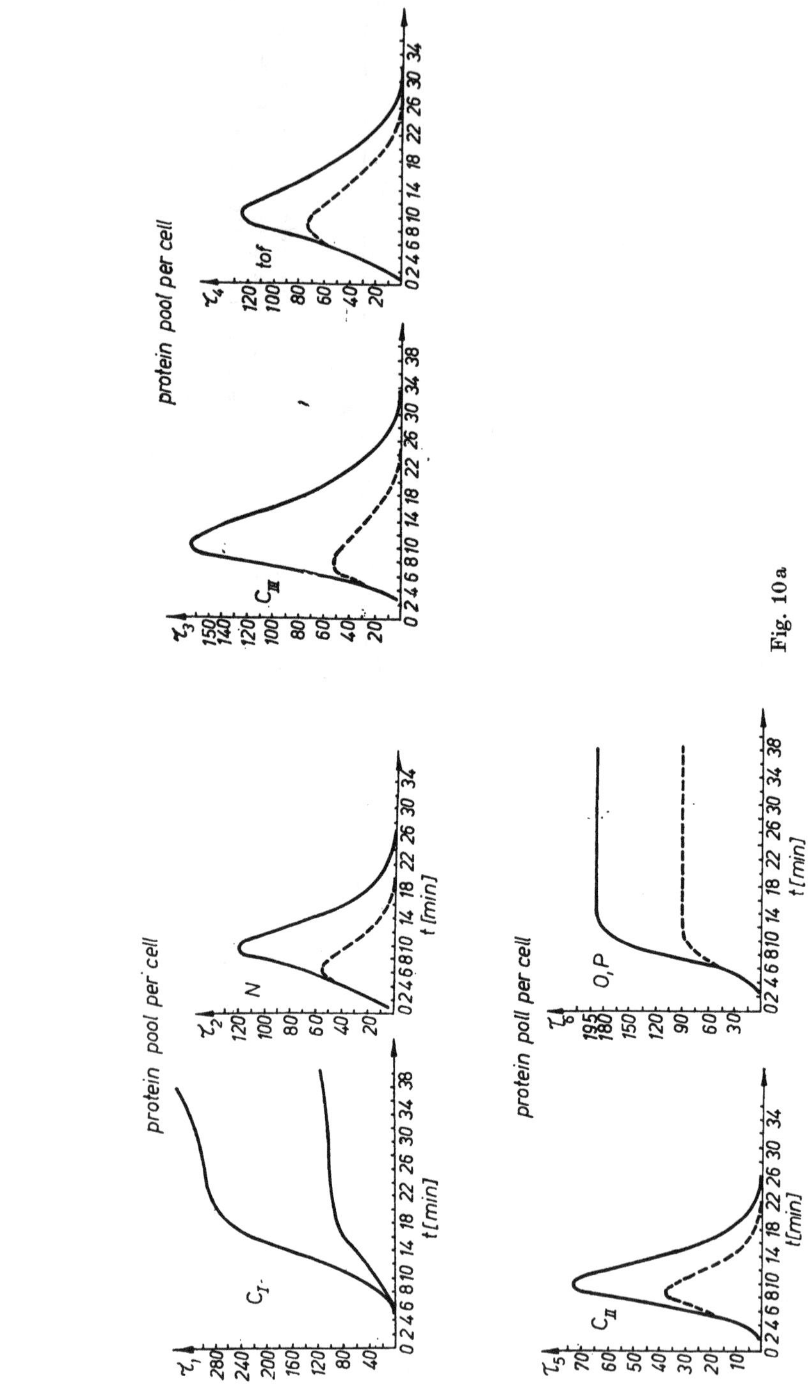

Fig. 10 a

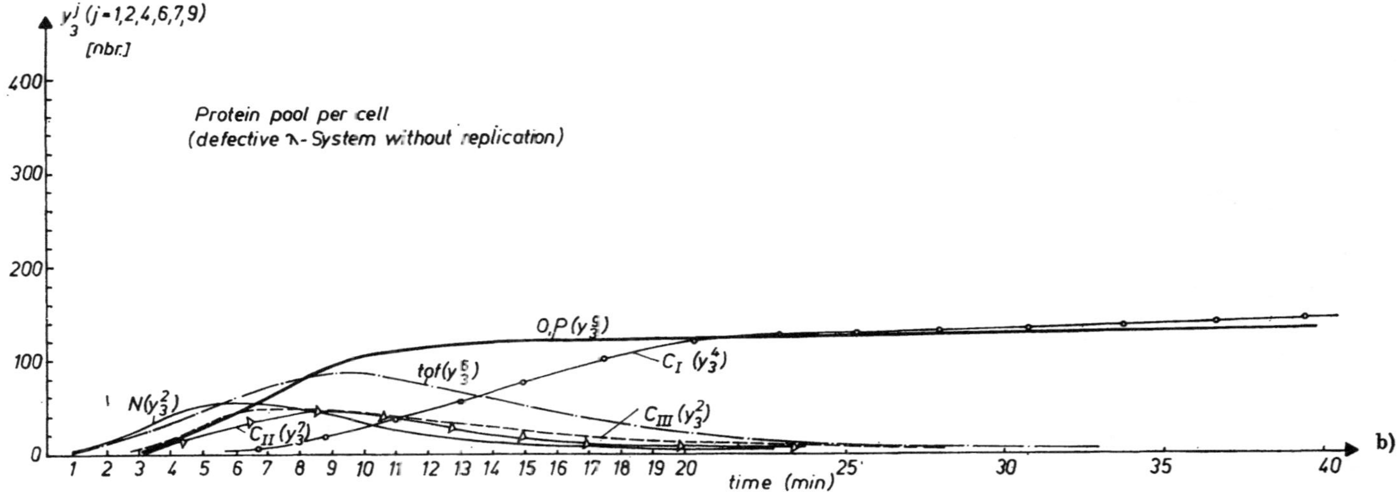

Figs. 10a, b. Lysogenic regime based on parameter set N I: the comparison of functioning of normal regulatory system of λ-phage development with system, with replicator *ori* deletion. r_I: concentration of protein C_I. Other designations are the same as in fig. 9. a) TH, b) SD.

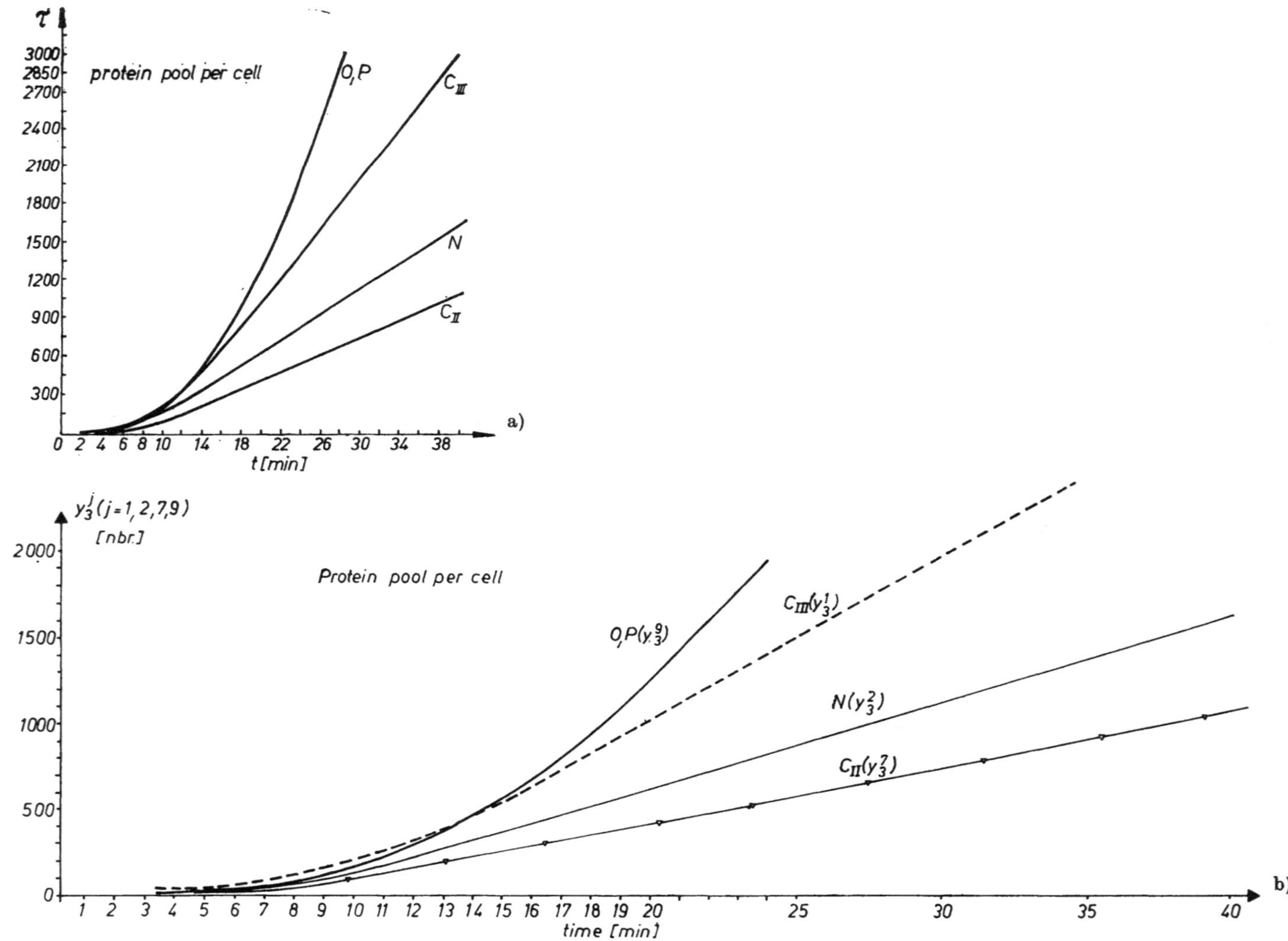

Figs. 11 a, b. Lytic regime based on parameter set N I in regulatory system λ-phage development with deletion of cistron *tof*. Designations are the same as in fig, 5. a) TH, b) SD.

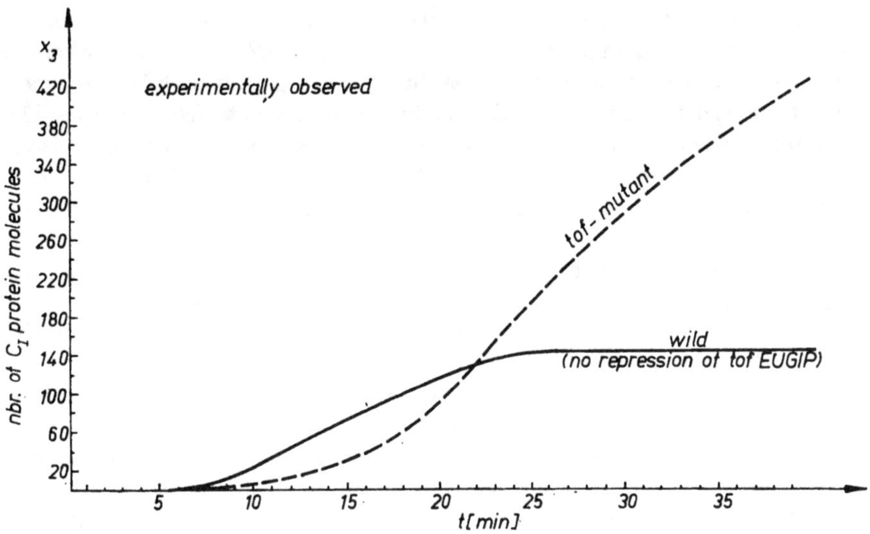

a)

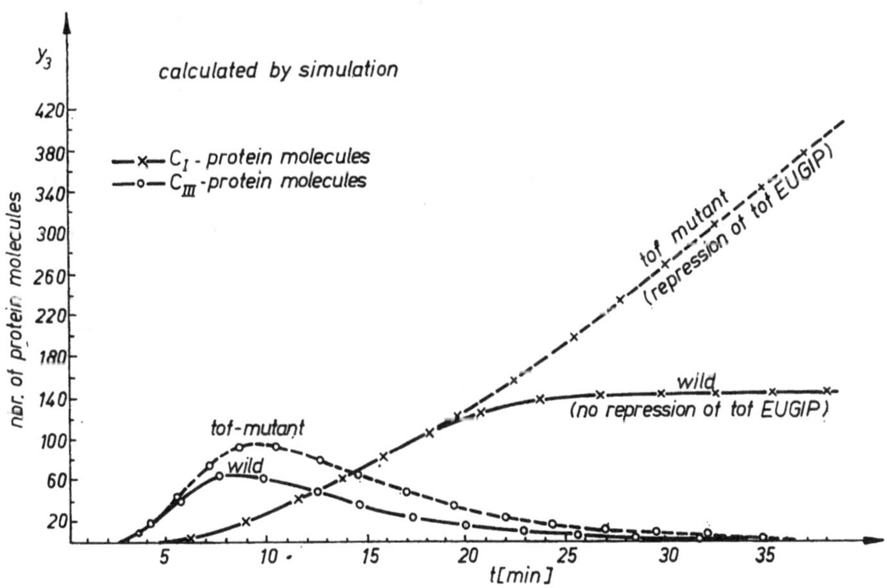

b)

Fig. 12. Kinetics of C_I protein molecules, a) experimentally observed, b) calculated by simulation on the basis of a appropriate parameter set (SD model).

under the two regimes by assigning the chosen set of values to its parameters. In this case, the time interval was 40 min. After repeated tests, computations were made for the frequencies of the lytic regime (a), the lysogenic (b) and both regimes (c). We thus obtained a characteristics of the "sensitivity" of the regimes expressed as the ratio of the frequencies computed to the total number of random choices.

There were two series of computer experiments. In the first series, the values of the parameters were chosen randomly in the $(x_i - 0.2x_i, x_i + 0.2x_i)$ intervals, where x_i is a parameter from the set "good" in the sense that it is compatible with both regimes. The capacity of the PDRS to function under the lytic regime was tested as follows. The attainment of the concentration thresholds of the proteins O, P, Q, and *tof* was decisive. The attainment of the concentration threshold of the proteins O and P evidenced that the concentration of the protein N reached the level earlier. The attainment of these threshold during the given time interval was taken as an indication of the establishment of all the stages of the lytic regime. The capacity of the system to function under the lysogenic regime was tested in a similar manner. The necessary conditions were the attainment of the threshold concentrations of the proteins C_{II}, C_{III}, and C_I and the inaccessability of the respective threshold P_{13} for the protein Q. These conditions are fulfilled by the lysogenic regime under which the protein C_I is synthesized, the replication and the transcription of all the cistrons, except C_I, are repressed.

It should be noted that modelling the control system of the λ-phage under the lytic and lysogenic regimes discounted the block synthesizing the protein Q. Firstly, the output of this block is not connected with the other PDRS blocks and, secondly, an appropriate choice of the values for P_{12} and P_{13} consistently meets the requirements of the regimes for the protein Q. Computer analysis of this block was necessary, because random parameter changes could leave unfulfilled some of the regime requirements.

The programme developed was to compute two systemes of intervals including two initial "good" parameter sets (Table 1, 2). Within each of the two systems of intervals 500 random parameters were chosen. Our final results are given in Table 3.

Table 3. Results of computer simulation of stochastic variants (mutants) with the threshold model

Computer simulation	Nbr. of parameters set	perentage of cases, when system can reproduce itself under		
		Lytic regime	Lysogenic regime	Both regimes
1	I (Tabl. 1)	78	12	12
1	II (Tabl. 2)	94	61	61
2	*	37	1	1

(*) — see page 104

In the second experimental series, the computer obtained a random sample of parameter sets from a much wider interval with boundaries sometimes set by purely physical considerations. The interval of the value for each coefficient a_{1i} was (0.2—60). This seemed reasonable, because not more than 60 initiations of RNA-polymerase per promoter per min are possible. Quite obviously, a polymerase can bind to a promoter only if it is disengaged of the polymerase already on it; this can occur not earlier than 1 sec after its binding (promotor length $\geqq 40$ nucleotides and transcription rate approximately 42 bases per sec). The lower boundary of the interval corresponds to 1 RNA-polymerase initiation in 5 min, i.e., several tenfolds less than the transcription initiation frequency for the tryptophan operon of *Escherichia coli* (BAKER, YANOFSKY, 1972). Granting that the half-life of mRNA is in the order of 1—7 min, a period corresponding to (0.1—0.7) interval is assumed for the coefficients a_{2i}. The interval range for the coefficients b_{1i} is (0.5—30) and for b_{2i} it is (0.07—0.7). The choice of the threshold is made in the (3—1000) interval.

The computer chose at random 1000 parameter sets within these intervals. It is to be noted in 37% of parameter variants the system is capable of functioning under the lytic regime and in 1% of the variants under the lysogenic regime too. What is important is that the system seems to be capable to function in 37% of parameter variants under such wide parameter variations. Thus, the computer experiments confirmed that the regulatory system of phage development is stable, in the broad sense.

4.3.2. Stochasticity in the process of mRNA and protein generation and degradation shown by the DS-model

As a rule the processes of genetic information processing run down stochassically (random binding of polymerases and ribosomes on the template, random inactivation and degradation of mRNA and enzymes), i.e. the control functions F_1^j, F_2^j [equ. (14), (15) in constributation 2], are realized non-deterministically.

The processes of transcription, translation and of mRNA and protein degradation (or inactivation resp.) are considered to be random ones, influenced by the host properties.

The stochasticity of these single processes was realized as described in section 2.2.g. of contribution 2.

It is obvious that by introduction of a stochasticity concerned with the generation and degradation rates of mRNA's and proteins all other variables become stochastic ones too.

The stochasticity may be varied in the system by variation of the dispersion $D^2 = \sigma^2$ and by variation of the duration d of a single random effect (see fig. 13).

On the basis of experimental hints and of results of computer simulation by NEUMANN and KREISCHER (see contribution 2) it seemed reasonable to set $d = 1$; one time step corresponds to the real time needed for generation of 0.5⋯1.0 molecules.

σ was varied between 0 (deterministic case) and 1, and d was varied between 1 and 5.

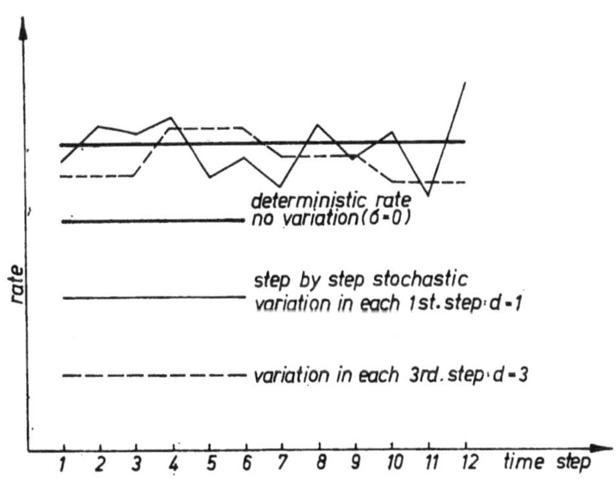

Fig. 13. Illustration of different duration of single random effect (SD model).

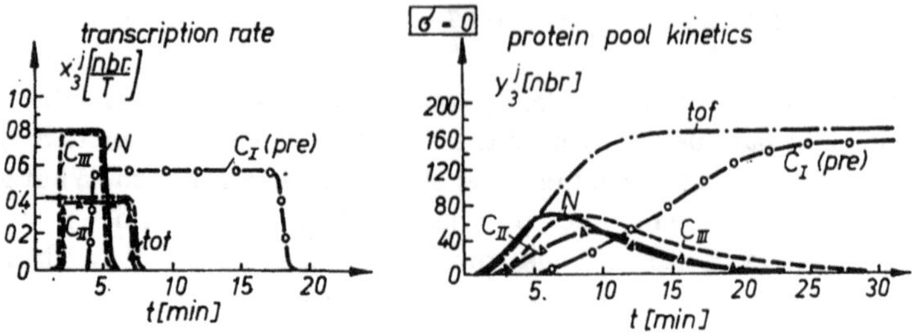

Fig. 14a, b

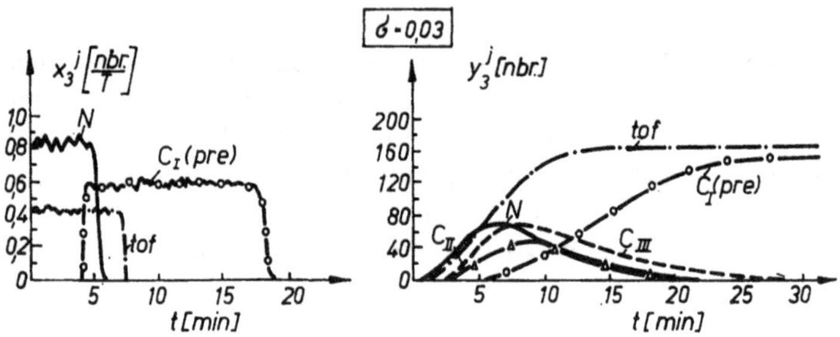

Fig. 15a, b

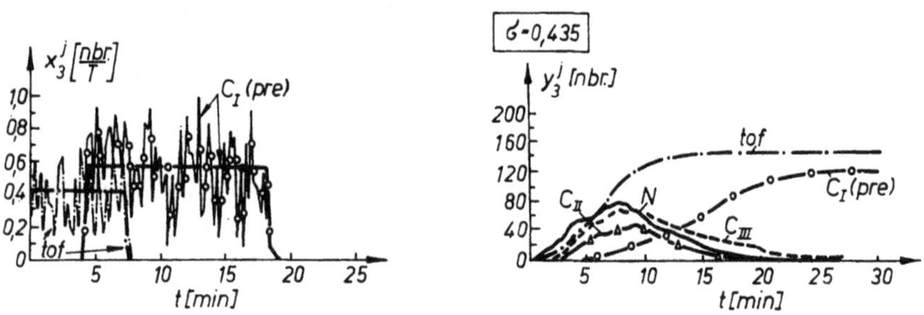

Fig. 16a, b

Fig. 14, 15, 16. Effect of stochastic influences on the sub-processes transcription, translation, mRNA and protein degradation on the early lysogenic development of the λ-system (SD model). The transcription rate and protein pool kinetics depending on σ are shown for illustration (normal distributed random variables with duration $d = 1$).

The results for $d = 1$ are shown in fig. 14, 15, 16 and tables 4, 5. In the figures the time courses of the transcription rates and protein pool concentrations of a single stochastic run chosen at random are demonstrated. In spite of a stochasticity of $\sigma = 0.435$ concerned with the essential genetic signal generation and degradation processes no important qualitative and mostly quantitative deviations from the deterministic case $\sigma = 0$ are observable, i.e., stochastic variability of subprocesses of the complex functional genetic network obviously do not affect the "global stability" of this system, that develops normally towards lysogeny after 5 min.

Table 4. Mean values and standard deviations of some characteristic indices of the global behavior of the λ-system, depending on different σ, got by Monte Carlo simulations with the stochastic difference equation model (see BÖTTNER, 1977); 50 runs, log normally distributed random variables with duration $d = 1$.

$d = 1$	mean value					standard deviation %			
σ	0	0.03	0.30	0.50	1.00	0.03	0.30	0.50	1.00
app. C_I	27.0	26.86	27.5	29.1	28.1	2.91	7.1	8.6	23.6
app. C_{III}	13.0	13.3	12.5	13.9	14.4	4.3	13.9	16.3	22.5
app. N	4.0	4.02	4.32	4.32	4.64	3.5	14.3	17.7	20.7
app. *tof*	5.0	4,68	4.8	4.9	5.1	4.1	15.7	15.0	20.6
disapp. C_{III}	173.0	172.7	175.5	170.6	177.2	1.62	7.0	6.8	13.1
disapp. *tof*	129.0	129.4	129.6	127.5	128.7	1.97	6.4	6.5	11.9
final C_I	143.04	142.5	137.5	134.8	125.8	4.03	16.0	16.8	33.4
final *tof*	154.16	154.4	157.2	165.3	184.6	5.12	16.4	19.6	36.3

app.: appearence time of the specific protein [nbr. of time steps]
disapp.: time of disappearence of the specific protein [nbr. of time steps]
final: final concentration of the specific protein [nbr. of molecules]

Table 5. Mean values and standard deviations of some characteristic indices of the global behavior of the λ-system, depending on different σ and durations d of a single stochastic effect; 50 runs, normally distributed random variables with duration $d = 1$ and $d = 5$.

mean value						standard deviation %				
σ	0	0,03		0,30		0	0,03		0,30	
duration of a single random effect		$d = 1$	$d = 5$	$d = 1$	$d = 5$		$d = 1$	$d = 5$	$d = 1$	$d = 5$
app. C_{III}	13.0	13.1	13.2	13.5	14.0	0	2.6	3.6	6.0	13.9
app. N	4.0	4.0	4.0	4.2	4.4	0	0	0	9.6	17.8
app. *tof*	5.0	5.0	4.9	4.5	4.9	0	2.8	5.6	11.0	13.8
app. C_I	27.0	27.0	27.0	27.4	28.5	0	0	1.3	4.2	9.2
disapp. C_{III}	173.0	172.5	172.8	173.7	172.6	0	0,3	0.7	2.8	6.9
disapp. N	129.0	129.1	129.6	129.9	131.4	0	0.5	0.8	3.5	6.9
final C_I	143.0	143.3	142.8	143.3	140.6	0	0.7	1.6	6.9	16.3
final *tof*	154.2	154.5	154.2	154.8	160.6	0	0.8	2.3	8.5	16.5

app.: appearence time of the specific protein [nbr. of time steps]
disapp.: time of disappearence of the specific protein [nbr. of time steps]
final: find concentration of the specific protein [nbr. of molecules]

More precisely, C_I protein appears, *tof* and C_I protein concentrations are increasing for 10 and 25 min. resp. and converge against a stable value. Furthermore, the moments of maximal concentrations of the C_{II}, C_{III}, and N proteins are the same ones that were observed in the deterministic case ($\sigma = 0$). The confidence intervals for an error of first kind ($\alpha = 5\%$) are shown in figs. 17, 18 for the case of $\sigma = 1$.

By table 4 and fig. 18 the statistical analysis of 50 Monte Carlo runs is represented. Here the duration of a single random effect $d = 1$ is considered.

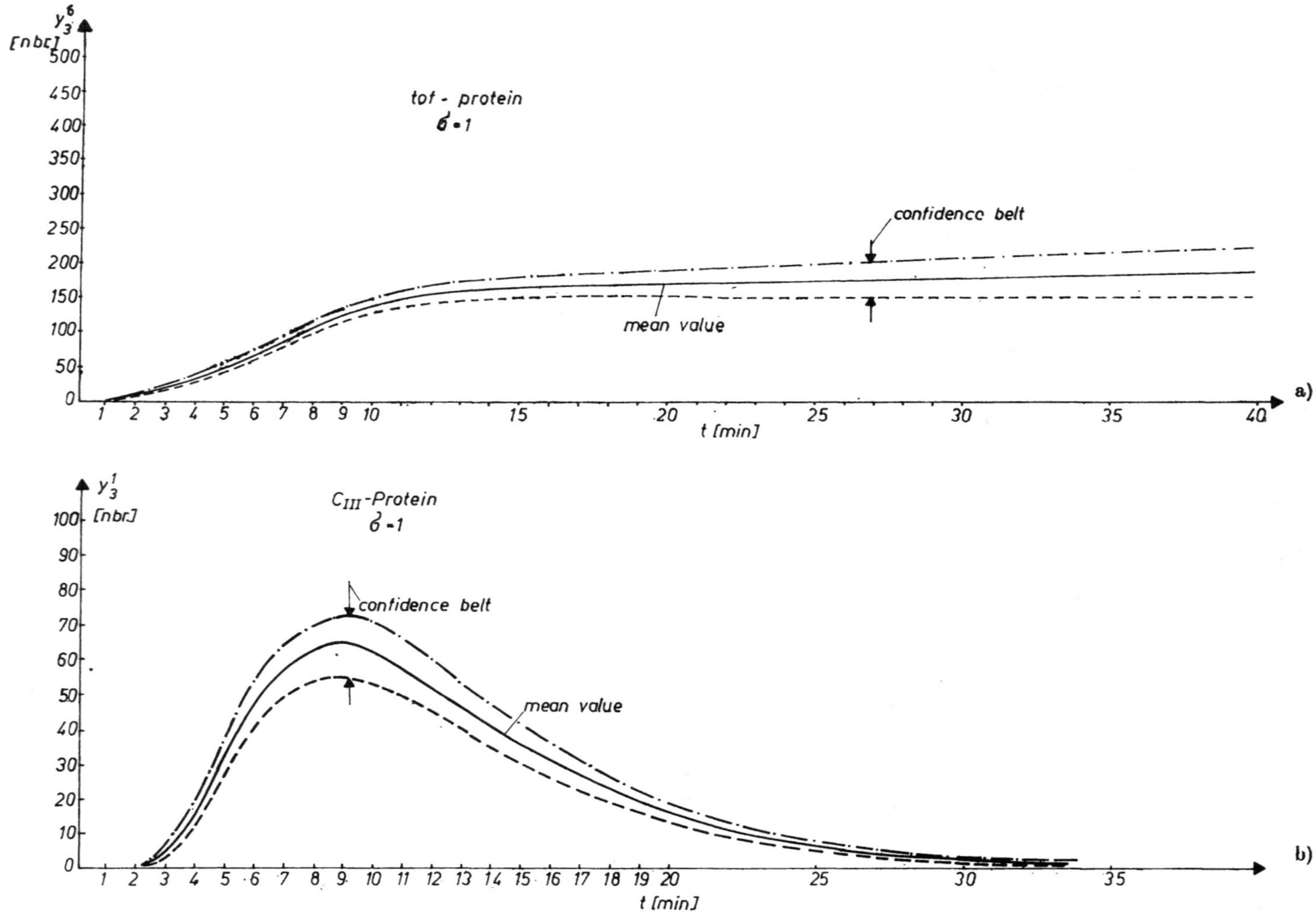

Fig. 17a, b. $\alpha = 5\%$ confidence intervals of the kinetics of C_{III} and *tof* protein for $\sigma = 1$ (result of 50 runs, log normal distributed random variables with duration $d = 1$, SD model).

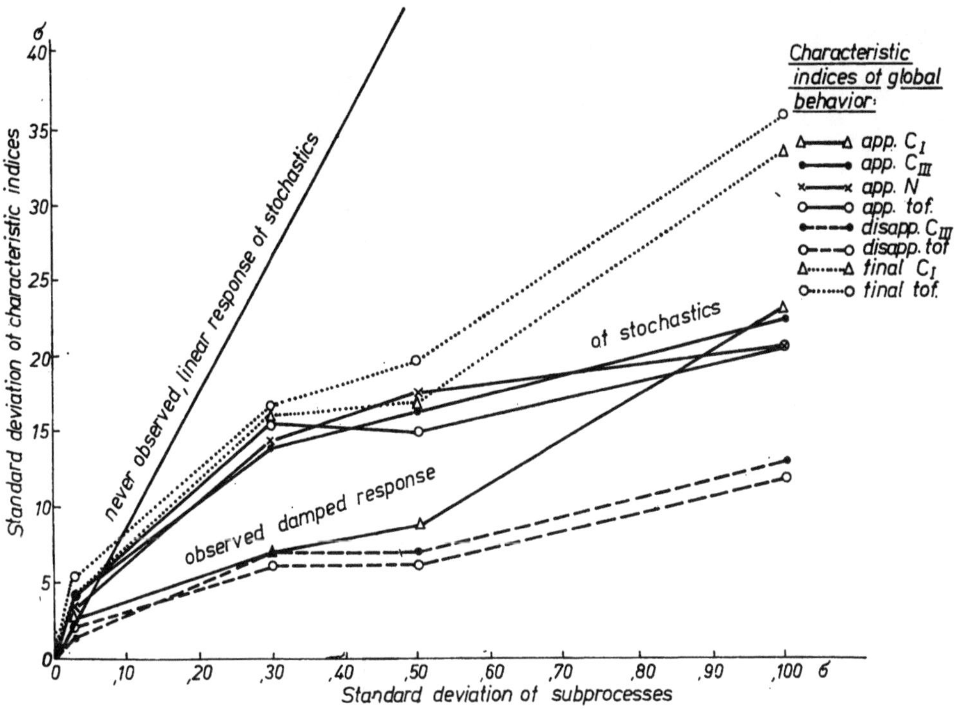

Fig. 18. Standard deviation of the characteristic indices of the global behaviour of the λ-system depending on the degree of random variations of the sub-processes (see table 4, SD model).

The moment of appearence of the proteins C_{III}, N, tof, and C_I, the moment of disappearence of the proteins C_{III} and N, and the final concentrations of the pools of the proteins C_I and tof were chosen as characteristic indices for the global behaviour of the λ-system during its lysogenic development.

The mean values of the global indices are scarcely changed even in the case of strong internal stochasticity, with the exception of final tof for $\sigma = 1.0$. It is remarkable that the stochasticity of the global indices is much lower than the stochastic variation of the elementary processes, i.e., the structural and functional features of the λ-system causes a reduction of internal stochasticity.

The model that represents the real system qualitatively and quantitatively, shows a stable behaviour in spite of unreliably working subprocesses. The stochasticity of the global indices is strongly damped (see fig. 18). The question arises which degree of global stochasticity may be observed if the frequency of the random changes is lowered, i.e., if the duration of a single random effect is increased. In table 5 a comparison between $d = 1$ time steps and $d = 5$ time steps for different σ is given (normally distributed random variables were used).

The mean values remain nearly unchanged in spite of a little more lasting random effects. Opposite to this the standard deviations are approximately doubled by choosing $d = 5$ in contrast to the case $d = 1$. If a stochasticity of this kind is to be assumed in single cases drastic deviations from the normal global behaviour are possible. An extreme variant is shown in fig. 19.

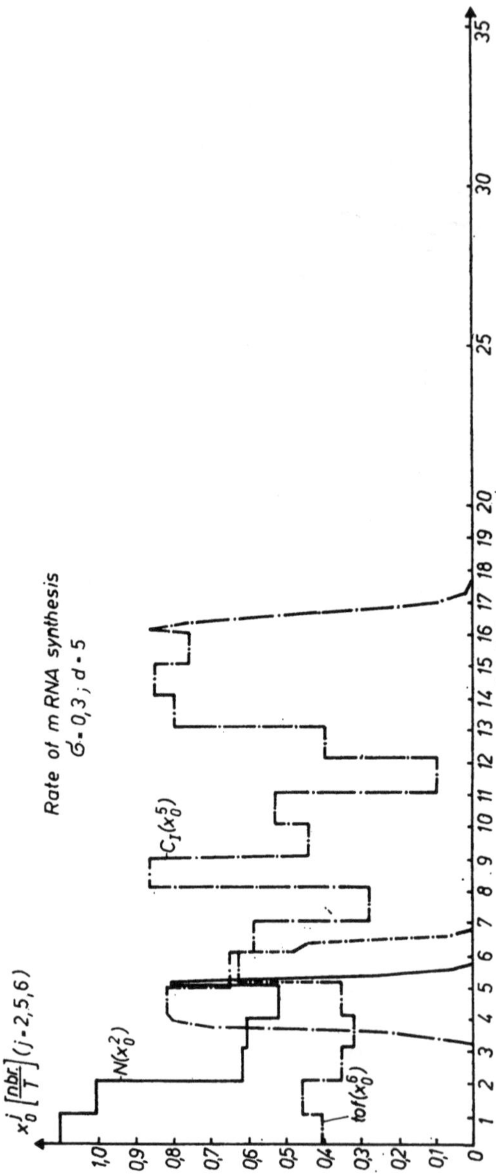

$x_0^j \left[\frac{nbr.}{T}\right] (j = 2,5,6)$
Rate of m RNA synthesis
σ = 0,3 ; d = 5
$N(x_0^2)$
$C_I(x_0^5)$
$tof(x_0^6)$
1,0
0,9
0,8
0,7
0,6
0,5
0,4
0,3
0,2
0,1
0
1 2 3 4 5 6 7 8 9 10 11 12 13 14 15 16 17 18 19 20 25 30 35
t [min]

Fig. 19a

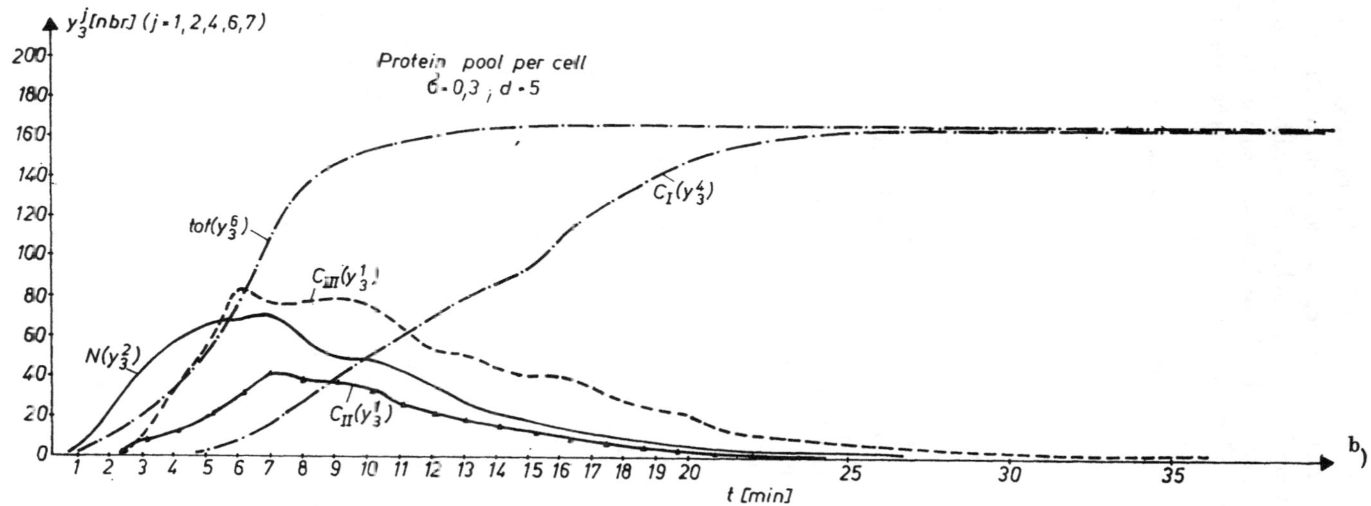

Fig. 19a, b. Transcription rate and protein pool kinetics (SD model) based on a stochasticity of $d = 5$ (normal distributed random variables; see also figs. 14, 15, 16). a) rate of mRNA Synthesis, b) protein pool per cell.

As a consequence of the results of the last 3 tables, if variations are time independent (table 3: mutations of the parameter values for the blocks of the PDRS model; variations of the genetic state) or time depending (tables 4, 5: modifications of rates of subprocesses of the model; variations of the dynamic processes) we must conclude that during evolution (mutation and selection) stable control systems have established; their stability is founded obviously on their complexity and types of feedbacks. Such systems are characterized by low sensibility against chance variations of static and dynamic parameters.

5. Discussion

The model of the regulatory system of the λ-phage is a **portrait**. This means that its purpose is to reflect not only the basis qualitative features of the ontogenetic event that is modelled, but also, at certain parameters, the quantitative dynamics of this event.

Portrait models are of theoretical and practical significance. **Theoretically**, portrait models are advantageous in that they permit to discount some mechanisms, and in doing so, to make more narrow the range of permissible parameter values. When dealing with a system, inadequately studied by experiment, a parallel analysis of its portrait model gives some idea of the inadequacies and suggests diagnostic experiments to fill the gaps in our knowledge. In practice, portrait models make possible predictions of the system's state, so far amply described, depending on its parameters, initial and external conditions, and to foresee how perturbations imposed from outside will affect its function.

The regulatory system of the λ-phage development appears until now the only ontogenetic system sufficiently well studied in a way providing abundant facts concerning the interplay of its diverse basic mechanisms. Models of the development of the λ-phage are far-reaching: they introduce new theoretical dimensions into modelling of living organism in the course of their development and guide our choice of modelling methods. Heuristic principles for the interaction between theory and experiment may be thus explicitly formulated.

Real ontogenetic processes enfolding in the λ-phage at the molecular level have an almost deterministic pattern. Hence, they are a priori sufficently well described by equations of chemical kinetics. The presence of links with small molecule number in the genetic systems of phages and bacteria does not affect, to any appreciable extent at least, the rough ontogenetic dynamics of these systems. Moreover the association constants of the interacting elements are very large (RATNER, 1977).

We lack compelling evidence of how the elements of these systems interact exactly. This makes difficult the application of the laws of chemical kinetics to the analysis of molecular and genetical systems. Even in the case when the details are known, events are expressed by non-linear differential equation quite tedious, if not impossible, to solve. It is readily apparent that ways and means must be found to describe approximately the dynamics of the genetic regulatory system, to translate kinetic events into simple mathematical language.

As applied to the λ-phage here, the method of threshold model and the method of the stochastic difference equation model seem to be promising. By introducing threshold concentrations and logical functions or products of elementary control terms, we described, to some approximation, the interactions between the regulatory proteins (repressors, inducers and others) and the specific sites of recognition (operaters, terminators and others). These interactions are complex and still controversial. In the case of the threshold model changes in the concentrations of each element, at constant values of each set are discribed by linear systems of differential equations that are easy to solve. The solutions are "joined" at the boundaries of the regions differing in the values of the logical variables. The resulting solution gives an overall description of the dynamics of a complex system. It is to be stressed that essentially a similar idea underlies the method of threshold models and POLETAEV's method of limiting factors (POLETAEV, 1973).

In case of the stochastic difference equation model we compute the solution step by step over the whole time range, treating the nonlinearities of control characteristics with numerical calculations.

Comparing the simulation results of both methods reveals in all working regimes of λ-phage (lytic, lysogenic, mutants) sufficient correspondence (shape of the kinetic curves; time moments of starts of synthesis and of maxima; maximal values; stationary values; stability-properties). Small differences of the variables result from rounding errors in calculation of the constants of degradation and from principle differences of the continuous and discrete approach. With this example we have shown, that continuous as well as discrete methods are suitable to describe molecular genetic processes in an consistent manner.

The threshold model is distinguished by short calculating time (use of analytical solution of the **linear** differential equations), whereas the stochastic difference equation model is characterized by great flexibility (structure of a modular program system, allowing the construction of different epigenetic networks with variable control characteristics, including thresholds, by suitable input data, without programming) and applicability (prokaryotes as well as eukaryotes) and tolerable calculating time.

The chief simplification in case of the TH-model is the assumed existence of **thresholds** during the action of regulatory substances on the respective genome sites. In some cases, this assumption is logically deduced from the sigmoid kinetics of the interaction between the regulatory substance and the operator, as established for the protein C_I (WU et al., 1972). Acceptance of this assumption in other cases is an approximation to reality.

To check the influences of sigmoid kinetics on the dynamic properties of the PDRS we used the stochastic difference equation model and calculated the kinetics of the regulatory proteins in this case (see fig. 7c). One can see that the detailed shape of the curves is changed (for instance lack of the typical two peaks in the *tof*- and N-curves; the *tof* peak is higher than the C_{III}-peak, stationary value of O, P proteins is higher than the stationary value of C_I-repressor protein in contrast to the curves with threshold control characteristics) but there is no qualitative change in the dynamic behaviour of the PDRS, i.e. the dynamic properties of the λ-phage development are insensitive against variation in the synthesis-, degradation- and threshold parameters as well as against variation of the shape of control characteristics.

Under this assumption the system is workable, if the constant for the regulatory substance-DNA complex is large enough for the rapid (not longer than 1 min) binding of a small number of regulatory molecules per genome to all the sites specific of the given regulatory protein. This seems to concern the proteins C_I, N, C_{II}, C_{III}, O and P, because these proteins interact with the respective sites promptly after their concentration increases (see contribution 2). In the case of the protein *tof* the constant of its association with the operator O_L' seems to be small and there may be a dissociation in the time of threshold attainment. However, analysis of the model indicated that the "time scatter" does not, within certain limits, interfere with the PDRS function. From the analysis it follows that the regulatory system of λ-phage development is such that phage reproduction under one of the ontogenetic regimes is provided by wide intervals of parameter values. This evidences that the regulatory system of λ-phage development is stable in **the broad sense of the term.** Thanks to this property, the phage reproduces under diverse environmental conditions (temperature, ionic strength and other variable factors) as well as under the effects of many mutations affecting the kinetic characteristics of the system.

Kinetic curves for the concentrations of the regulatory proteins were built on the basis of a quantitative analysis of the threshold model and of the discrete stochastic model of the regulatory system of λ-phage development. This analysis helped to understand how the regulatory proteins C_{II}, C_{III}, O and P may disappear. The C_{II} and C_{III} (or at least one of them) seem to be unstable, while O and P are stable. Were the C_{II} and C_{III} stable, then the synthesis of the protein C_I, which is under the control of the scripton L_3, would not be complete in a single cell cycle, which is in contradiction to experimental data (Reichardt, Kaiser, 1971). Interestingly, recent data indicate that these proteins are, indeed, stable (Reichardt, 1975). Were the proteine O and P unstable, the yield of phage particles under the lytic regime would be low. It seems likely that protein stability is determined by their functional role in phage development. For instance, the instability of the protein N ensures **economy of material and energy necessary for the synthesis of the morphogenetic proteins**; the regulatory proteins exist as long as they are needed.

Comparisons of the cistron *tof* kinetic data derived from the model and from the model simulating a *tof* deletion demonstrated the dual role of this protein. The *tof* protein serves to economize substrates and energy for the production of regulatory proteins, on the one hand, and provides **the irreversibility** of the lytic regime. As inferred here, the dynamics computed for the elements (mRNA fractions and proteins) of the modelled system is compatible with real processes, provided that the parameter values are correctly chosen. This is to say that model reflects quite faithfully the main developmental processes in the λ-phage.

References

Adesnik, M., and C. Levinthal (1970): The synthesis and degradation of lactose operon messenger RNA in *E. coli*. Cold Spring Harbor Symp. Quant. Biol., **35**, 451.

Baker, R., and C. Yanofsky (1972): Transcription initiation frequence and translational yield for the tryptophan operon of *Escherichia coli*. J. Mol. Biol., **69**, I, 89.

Belfort, M., and D. Wulff (1974): The roles of the lambda C_{III} gene and the *Escherichia coli* catabolite gene activation system in the establishment of lysogeny by bacteriophage lambda. Proc. Nat. Acad. Sci. USA, **71**, 3, 479.

Bellmann, K., R. Böttner, A. Knijnenburg and H. Neumann (1974): Computer simulation models of gene expression. In: Symp. Biol. Hung. **18** "Mathematical models in metabolic regulation", 227—255. Budapest 1976.

Bertani, G. (1960): On the interpretation of Boyds "multiplicity effect" in lysogenization. Virology, **12**, I, 132.

Blattner, F. R., and J. E. Dahlberg (1972): RNA synthesis startpoints in bacteriophage λ: are the promoter and operator transcribed? Nature New Biol., **237**, 77, 227.

Blattner, F. R., J. E. Dahlberg, J. K. Boettiger, M. Fiandt and W. Szybalski (1972): Distance from a promoter mutation to an RNA synthesis startpoint on bacteriophage λ DNA. Nature New Biol., **237**, 77, 232.

Blattner, F. R., M. Fiandt, K. K. Hass, P. A. Twose and W. Szybalski (1974): Deletions and insertions in the immunity region of coliphage lambda: revised measurement of the promoter-startpoint distance. Virology, **62**, 458.

Böttner, R. (1974): Mathematisch-kybernetische Modellierung der Regulation der Genexpression. In: Tagungsberichte der Akad. Landw.-wiss. DDR, Berlin, 1976, **132**, 153—174.

BÖTTNER, R. (1976): Mathematisch-kybernetische Modellierung der Regulation der Genexpression. Diss. (A), Humboldt-Univ. Berlin.

BÖTTNER, R. (1977): Simulation models for the dynamic analysis of Molecular genetic control systems. Biom. Jour. **19**, 131—142.

CHADWICK, P., V. PIROTTA and R. STEINBERG (1970): The λ and 434 phage repressors. Cold Spring Harbor Symp. Quant. Biol., **35**, 283.

COUNTURIER, M., C. DAMDLY and R. THOMAS (1973): Control of development in temperate bacteriophages. Molec. Gen. Genet., **120**, 3, 231.

DOVE, W. F. (1968): The genetics of lambdoid phages. Ann. Rev. Genet. 2, 305.

ECHOLS, H. (1972): Temperate phage development. Ann. Rev. Genet., 6, 157.

ECHOLS, H., and L. GREEN (1971): Establishment and maintence of repression by bacteriophage lambda: the role of the CI, CII and CIII proteins. Proc. Nat. Acad. Sci. USA **68**, 9, 2190.

ECHOLS, H., L. GREEN, A. B. OPPENHEIM and A. HONIGMAN (1973): Role of the cro gene in bacteriophage λ development. J. Mol. Biol., **80**, 2, 203.

FOLKMANIS, H., and D. FREIFELDER (1972): Studies of lysogeny in *Escherichia coli* with bacteriophage λ. Physical observation of insertion process. J. Mol. Biol., **65**, I, 63.

GILDERMAN, YU. I., K. N. KUDRINA and I. A. POLETAEV (1970): Models of L-systems (systems with limiting factors). In: "Studies on cybernetics", Publ. House "Sov. Radio", Moscow, 165—210 (russ.).

GREEN, M. N., B. GOTCHEL, J. HENDERSHATT and S. KENNEL (1967): Regulation of bacteriophage lambda DNA replication. Proc. Nat. Acad. Sci. USA, **58**, 6, 2343.

GREENBLATT, J. (1973): Regulation of expression of the *N* gene of bacteriophage lambda. Proc. Nat. Acad. Sci. USA, **70**, 2, 421.

GRODZICKER, T., R. R. ARDITTI and H. EISEN (1972): Establishment of repression by lambdoid phage in catabolite activator protein and adenylate cyclose mutants of *Escherichia coli*. Proc. Nat. Acad. Sci. USA, **69**, 2, 366.

GROS, F., N. STERNBERG, M. BOUQUET and P. KOURILSKY (1971): In: "CIBA Foundation Symposium on Strategy of the Viral genome". Churchill Livingstone, London, 207—228.

HAMPACHEROVA, M., E. KONTECKA and Z. NEUBAUER (1973): Two repressors in bacteriophage lambda. Molec. Gen. Genet. **120**, 2, 133.

HEINEMANN, S. F., and W. G. SPIEGELMAN (1970): Role of the gene N product in phage lambda. Cold Spring Harbor Symp. Quant. Biol., **35**, 315.

HERSKOWITZ, I., and E. SIGNER (1970): Control of transcription from *r* strand of bacteriophage lambda. Cold Spring Harbor Symp. Quant Biol., **35**, 355.

KORN, D., and H. THOMAS (1971): Control of plasmid replication in *E. coli*: correlation of the membrane site of DNA replication with the bacterial segregation unit. Proc. Nat. Acad. Sci. USA, **68**, 9, 2047.

KOURILSKY, P., M. F. BOURGUIGNON, M. BOUQUET and F. GROS (1970): Early transcription controls after induction of prophage λ. Cold Spring Harbor Symp. Quant. Biol., **35**, 305.

KUMAR, S., E. CALEF and W. SZYBALSKI (1970): Regulation of the transcription of *Escherichia coli* phage λ by its early genes *N* and *tof*. Cold Spring Harbor Symp. Quant. Biol., **35**, 331.

MANIATIS, T., and M. PTASHNE (1976): A DNA operator-repressor system. Scientific American, **231**, I, 64.

MARK, K., and W. SZYBALSKI (1973): Repressor and rex product of coliphage lambda: lack of collaboration and joint controls. Molec. Gen. Genet., **123**, I, 1.

NAONO, S., and K. TOKUYAMA (1970): On the mechanism of λ-DNA transcription in vitro. Cold Spring, Harbor Symp. Quant. Biol., **35**, 375.

NIJKAMP, H. J. J., K. BOVRE and W. SZYBALSKI (1970): Control of righward transcription in coliphage λ. J. Mol. Biol., **54**, 3, 599.

NIJKAMP, H. J. J., W. SZYBALSKI and E. CALEF (1971): Antirepressor controls the transcription of repressor operon of lambda prophage. In: "Informative Molecules in Biological Systems" (L. G. H. LEDOUX, Ed.), North-Holland Publ. Co., Amsterdam, 241—248.

OGAWA, T., and J. TOMIZAWA (1967): Abortive lysogenization of bacteriophage lambda b2 and residual immunity of non-lysogenic segregants. J. Mol. Biol., **33**, 3, 967.

OPPENHEIM, A. B., Z. NEUBAUER and E. CALEF (1970): The antirepressor: a new element in the regulative of protein synthesis. Nature, **226**, 5240, 31.

PTASHNE, M., and N. HOPKINS (1968): The operators controlled by the λ phage repressor. Proc. Nat. Acad. Sci. USA, **60**, 4, 1282.

RATNER, V. A. (1974): Genetic system controlling λ phage ontogenesis. Uspehi sovr. biol., **77**, I, 4. (Russ.).

RATNER, V. A. (1975): Molecular genetic regulatory systems. Publ. House "Nauka", Novosibirsk.

RATNER, V. A. and R. N. TCHURAEV (1978): Simplest genetic systems controlling ontogenesis: organization principles and models of their function. In: "Progress in Theoretical Biology" (R. ROSEN, Ed.), Acad. Press, New York.

REICHARDT, L. F. (1975): Control of bacteriophage lambda repressor synthesis after phage infection: the role of the *N*, C_{II}, C_{III} and cro products. J. Mol. Biol., **93**, 2, 267.

REICHARDT, L. F., and A. D. KAISER (1971): Control of λ repressor synthesis. Proc. Nat. Acad. Sci. USA, **68**, 9, 2185.

ROBERTS, J. W. (1970): The ϱ factor: termination and anti-termination in lambda. Cold Spring Habor Symp. Quant. Bibl., **35**, 121.

ROLFE, B., J. SCHELL, A. HEIP, K. ONODERA and E. SCHELL-FREDERICK (1973): A colicin — tolerant mutant of *Escherichia coli* with reduced levels of cyclic AMP and a strong blas towards λ lysogeny. Molec. Gen. Genet., **120**, I, 1.

RIGGS, A. D., S. BOURGEOIS and M. COHN (1970): The lac-repressor-operator interaction. III. Kinetic Studies. J. Mol. Biol., **53**, 3, 401.

SCHWARTZ, M. (1970): On the function of the *N* cistron in phage lambda. Virology, **40**, I, 23.

SPIEGELMAN, W. G., S. F. HEINEMANN, P. BRACHET, L. PEREIRA DA SILVA and H. EISEN (1970): Regulation of the synthesis of phage lambda repressor. Cold Spring Harbor Symp. Quant. Biol., **35**, 325.

STEVENS, W., S. ADHYA and W. SZYBALSKI (1971): Origin and bidirectional orientation of DNA replication in coliphage lambda. In: "The bacteriophage lambda" (A. D. HERSHEY, Ed.), Cold Spring Harbor Laboratories, Cold Spring Harbor, 515—533.

SZYBALSKI, W. (1970): Genetic and molecular map of coliphage λ. In: "Handbook of biochemistry", Chem. Rubb. Co., Cleveland, 135—138.

SZYBALSKI, W. (1971): Control of transcription and replication in coliphage lambda. Karl-August-Forster-Lectures. Mainz, 6.

SZYBALSKI, W., K. BØVRE, M. FIANDT, S. HAYES, Z. HRADECNA, S. KUMAR, H. LOZERON, H. J. J. NIJKAMP and W. F. STEVENS (1970): Transcriptional units and their controls in *Escherichia coli* phage λ: operons and scriptons. Cold Spring Harbor Symp. Quant. Biol., **35**, 341.

TAKEDA, Y., K. MATSUBARA and K. OGATA (1975): Regulation of early gene expression in bacteriophage λ: effect of *tof* mutation on strand specific transcription. Virology, **65**, 2, 374.

TCHURAEV, R. N., and V. A. RATNER (1975): Simulation of the dynamics of λ-phage developmental regulatory system. In: "Studies on mathematical genetics" (V. A. RATNER, Ed.), Inst. Cyt. Genet. Press. Novosibirsk, 5—66 (Russ.).

WAKE, R. G., A. D. KAISER and R. B. INMAN (1972): Isolation and structure of phage λ head-mutant DNA. J. Mol. Biol., **64**, 3, 519.

WU, A. W., S. GHOSH, H. ECHOLS and W. G. SPIEGELMAN (1972): Repression by the C_I protein of phage λ: in vitro inhibition of RNA synthesis. J. Mol. Biol., **67**, 3, 407.

WOOD, K. G., and M. KONRAD (1974): Stimulation of RNA chain initiation by the *N* gene protein of bacteriophage λ. Mol. Gen. Genet., **134**, 4, 359.

5. Mathematico-cybernetical model of prokaryotic transcription systems

H. NEUMANN, U. KREISCHER

Contents

1. Introduction

It is one of the declared aims of biochemical and biophysical research to unravel the regulation principles of genetic information processing. Closely connected with this aim is the elucidation of the structure and function of transcription system. However, often enough experimental research is limited for object related, technical, technological reasons, or just for lack of time. As a result, a huge variety of individual data is obtained. Mathematico-cybernetic models offer a possibility to overcome these limitations. They may prove valuable tools to extrapolate single experimental findings to a wider context. A model allows one to test almost an infinite number of variants of structure-function-correlations. By comparing experimental with model-derived results it is possible to verify or to falsify hypotheses about such correlations. The degree of intrica-

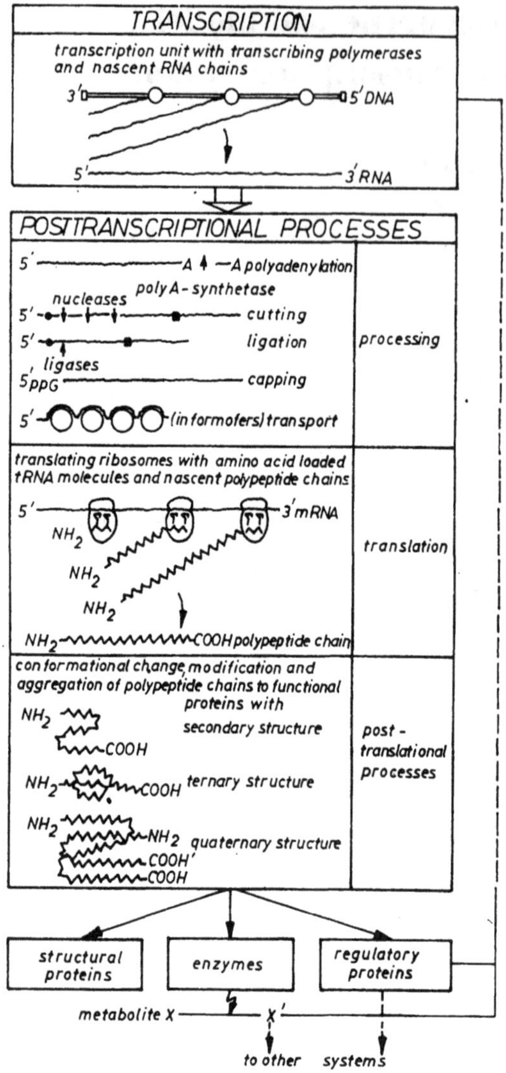

Fig. 1. Scheme of the system of gene expression. According to the organizational level of the organism the single steps of gene expression are marked more or less distinctly.

cy[1] and complexity[2] of the transcription system correlates essentially with the organizational level of the organism, which may be of prokaryotic or eukaryotic nature. In view of the high degree of intricacy and complexity of eukaryotic systems, experimental research has been unable as yet to provide facts and data in a measure which would be necessary for evolving a fundamental model, especially one of mathematico-cybernetical nature. Consequently, the purpose of modelling should be to provide a principle information of simple and well-studied systems, which will enable one to deal with the more

[1] The intricacy of a system is the number of different elements of the system.

[2] The complexity of a system is the kind and the number of relations existing between the elements of the system.

intricated and complicated systems. Hence it is advisable to study first the transcription system of prokaryotic organisms for the fundamental structure and functional pattern. For first considerations, a "training model" was developed on the basis of the classical operon (NEUMANN, 1973; NEUMANN, KREISCHER, 1975).

The aim of further work on genetic regulation in prokaryotes (NEUMANN, 1976), to be described here, is to give a detailed theoretical outline of possible organizational forms of the prokaryotic transcription unit and of the regulation principles involved. For a sufficient representation of the transcription system it is necessary to map the posttranscriptional processes of the system of gene expression, because proteins coded by a transcription system may carry out directly or indirectly a regulatory function on this or other transcription systems (Fig. 1). For this reason the model is generally speaking a model of the system of gene expression, too.

The mathematico-cybernetical model is to analyse

a) principle types of regulation and their effects on the external behaviour (RNA synthesis) of a transcription system, to allow predictions about
b) the kind of action mechanism of a transcription system of certain organizational form and
c) the interaction of its elements.

On this basis the model helps

1. to analyse networks of prokaryotic transcription systems,
2. to study (at least partially) the eukaryotic transcription system and also
3. to describe other biological processes, which are on principle of the same nature (e.g. translation, replication).

Furthermore, in the case of an adequate dynamics between the model and the experimental research

4. a transition from qualitative to quantitative systems analysis might be performed by proceeding specification of the model parameters.

2. Model approach

When using the model approach, the subject (man) interacts with both the original (real object), in this case with the prokaryotic transcription system, and the model. The overall process proceeds within the bounds of the model system consisting of subject, original and model.

Now, starting-point of the model approach is:

1. a given task of the subject, which was fixed in the preceding part of this article, and
2. certain essential conditions and properties of the original, which are manifested by defining a prototype in combination with the elaboration of an abstract-ideal model.

The relation existing between the prototype, the representative of the original in its essential properties, and the mathematico-cybernetic model, which is to be built, represents an analogy regarding the structure, the function or the behaviour. Projection of the prototype to the model level consequently calls for an exact system definition to be made in the mathematico-cybernetical sense. The approach taken was derived from the theory of abstract automatons. This procedure is based on the following assumptions:

1. the transcription system itself is conceived as an automaton, because as a cybernetical system it holds information signals from the environment, stores and processes this

information, and gives information signals to the environment arising as a result of this processing.
2. the single processes of transcription and transcription regulation follow successivly in a determined order, although the processes themselves bear a stochastical character.

On the other hand, it is possible

3. to represent both functional and structural properties of the transcription system and
4. to analyse exactly the genesis of the dynamic system behaviour by means of the state patterns of an automaton.

By reason of the intricacy and complexity of the transcription system reflected by the prototype, which is to be projected by the automaton, a mathematico-analytical representation of the functions describing the behaviour of the automaton is hardly possible. Moreover, an analytic representation would imply an idealization which is not justified. Essential properties of the system to be analysed might not be reflected by such a representation, which is more suitable for a phenomenological description. For that reason these functions are realized by an algorithm. Generally, an algorithm is characterized by a set of operations. In each case the sequence of operations depends on the result of the preceding operation. Therefore the behaviour of the automaton defined from the prototype and consequently also the behaviour of the original is simulated by means of an algorithm. Since random processes (e.g. aggregation and termination events) are represented, the algorithm and thus, the automaton, is undetermined. The mathematico-cybernetical model is the concrete realization of the algorithm on the theoretical basis of the defined automaton. The model was implemented on the computer BESM6 using FORTRAN as the programming language.

3. The prototype

The elements of the transcription system are the transcription unit occurring in most various organizational forms and the molecules interacting with the transcription unit directly or indirectly. The structure of the transcription system is determined by the set of all relations between these elements. Because firstly the transcription process itself and secondly the control of transcription are subject to a large number of influences, it is necessary to make an abstraction with regard to certain properties of the real biological system. Consequently, a prototype is to be defined, that is, a system which does not differ from the real system in certain properties regarded as "essential", and is different from other, "non-essential" properties.

Thus, proceeding from experimental findings in the literature and including several other conceivable organizational forms of the transcription system the prototype is defined as follows:

3.1. Transcription Unit

The transcription unit is a DNA sequence to be transcribed (note 1) and is characterized by the following general constructive principle (Fig. 2):

1. Regulation sequence consisting of an
   - operator region (note 2) $o_1, \ldots, o_{\max}$ ($\max \in \{1, \ldots, j''\}$), which may be absent too, and a
   - promoter region (note 3) $p_1, \ldots, p_{\max}$ ($\max \in \{1, \ldots, i+1\}$);

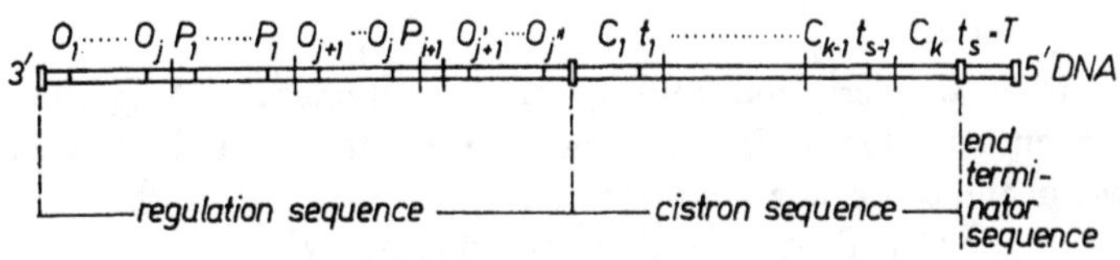

Fig. 2. Abstract-ideal model of possible organizational forms of the transcription unit.

p_l	promoter	$1 \leq l \leq i+1$	
o_m	operator	$1 \leq m \leq j''$	
c_n	cistron	$1 \leq n \leq k$	$l, m, n, r \in N$
t_r	terminator	$1 \leq r \leq s$	

$\sum_{l=1}^{j+1} p_l =$ promoter region $\qquad \sum_{m=1}^{j''} o_m =$ operator region

2. Cistron sequence $c_1, \ldots, c_k$ with
3. Intermediate terminators $t_1, \ldots, t_{s-1}$, which may separate cistrons and/or cistron groups from each other;
4. End terminator sequence $T(= t_s)$.

3.2. Control mechanisms

The transcription of a DNA sequence may be subject to the following control mechanisms (Fig. 3):

a) operator-localized

- negative control accomplished by operator-repressor-inducer interactions in which the transcription is prevented in the presence of the repressor and in the absence of the inducer,
- positive control accomplished by operator-activator-antiactivator interactions in which the transcription is ensured in the presence of the activator and in the absence of the antiactivator

b) promoter-localized positive control accomplished by

- promoter-polymerase (core or holoenzyme)
- promoter-CAP/cAMP-(polymerase)
- polymerase-polymerase subunit (sigma factor) interactions

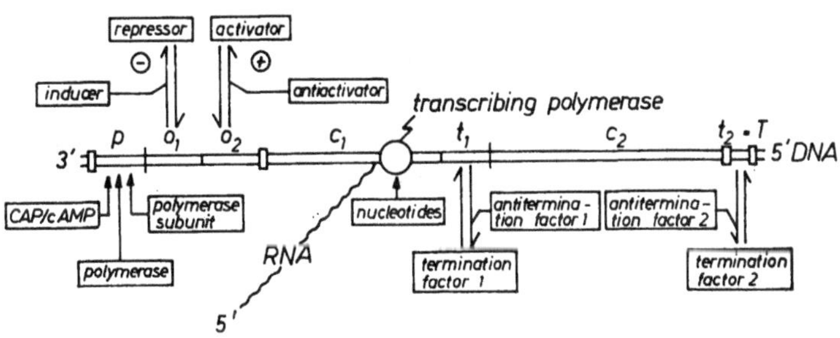

Fig. 3. Some mechanisms of the transcription control.

p promoter, $\quad c_1, c_2$ cistrons
o_1, o_2 operators, $\quad t_1, t_2$ terminators
operator-localized negative (⊖) and positive (⊕) control

c) elongation control by

- changes in the transcription velocity of the polymerase along the transcription unit after mutation loci and/or intermediate terminators and/or specific nucleotide sequences in the codogenic DNA strand,
- the concentration of ribonucleoside triphosphates

d) terminator-localized control

- accomplished by termination and antitermination factors, where a terminator-located polymerase is released in the presence of the termination factor and in the absence of the antitermination factor, runs over the terminator in the presence of the antitermination factor and remains at the terminator in the absence of the two factors and, as a special case,
- elicited by mutations with termination character in the codogenic DNA strand.

3.3. Posttranscriptional processes

For a sufficient representation of the transcription system it is necessary to consider the posttranscriptional processes. These processes are formulated only generally, that is, without giving a detailed representation of structural elements and regulation mechanisms of the underlying systems:

1. Synthesis of the polypeptide chains (translation)
2. Aggregation of the polypeptide chains (formation of the active form of proteins built complexly)
3. Inactivation of the proteins
4. Degradation of translated mRNA chains

3.4. Protein interactions

Proteins coded by the cistrons of a transcription system and synthesized as a result of the posttranscriptional processes are able to mediate directly or indirectly between the output (mRNA) of this transcription system and the input (regulatory molecules)

1. of the same transcription system (feedback) and/or
2. of one or several other transcription systems (coupling), where
   - open chains of cascade-connected transcription systems
   - closed chains of cascade-connected transcription systems
   - input and output ramification of parallelly connected transcription systems

   are to be discriminated.

Combinations of any number of these types are termed networks of transcription systems and the kind of the coupling between the transcription systems coupling strategy.

3.5. Premises

Furthermore the following premises are valid:

a) The polymerase core enzyme is regarded as a whole. Hence the subunits of the core enzyme are presumed to have the proper stoichiometrical proportions.

b) The ions Mg^{2+}, Co^{2+} or Mn^{2+}, respectively, needed from the polymerase are assume to exist in sufficient quantity.
c) The polymerase is to transcribe the different nucleotides with a constant mean velocity.
d) Three by three of the four ribonucleoside triphosphates ATP, UTP, CTP, and GTP needed for the RNA synthesis are to be available in each case at least to such a degree as the fourth that could be assumed.
e) All regulatory molecules of a given transcription system are supposed in the active form.
f) Completeness and optimal productivity of the translation system are presumed.
g) The conditions for an enzyme reaction are to be such that they are sufficient for a given substrate turnover number.

Note 1

The transcribability results from the following steps of the transcription process:

1. **Induction** of the transcription system accomplished by operator-localized control mechanisms.
2. **Initiation** of the transcription closed by formation of the first phosphodiester bond after primary binding of the RNA polymerase to a specific DNA sequence (promoter) and a series of further processes (according to ZILLIG et al. 1970, GOLDTHWAIT et al. 1970, RÜGER 1972):
   - reversible primary binding

     $E + DNA \rightleftharpoons E\text{-}DNA$
   - reversible stabilized binding

     $E\text{-}DNA + \sigma \rightleftharpoons E\sigma\text{-}DNA$
   - activated complex, temperature-dependent regarding reversibility

     $E\sigma\text{-}DNA \xrightleftharpoons{>20°C} E^*\sigma\text{-}DNA^*$
   - initiation complex

     $E^*\sigma\text{-}DNA^* + pppPu \rightharpoonup \underset{pppPu}{E^*\sigma\text{-}DNA^*}$
   - initiation

     $\underset{pppPu}{E^*\sigma\text{-}_{\top}DNA^*} + ppPy + n(ppN + pppN') \rightharpoonup \underset{pppPupPy(pNpN')_n}{E^*{}_{\top}DNA^*} + \sigma + (n+1)\,pp_i.$
3. **Elongation** (actual phase of the RNA synthesis).
4. **Termination** of the transcription by release of the polymerase and RNA product at a specific DNA sequence (terminator).

Note 2

In the usual way the definition of an operator sequence $o_1, \ldots, o_j$ located in front of the promoter region is reasonable only, if it consists of only one operator ($j = 1$) and if there exists a steric hindrance between the operator-specific repressor and the polymerase with regard to their binding to the adequate DNA sequences ("overlapping"). The "overlapping" variants of elements of the promoter-localized and operator-localized control (Fig. 4) are treated in the same way, because they do not differ in the regulatory effect. It must be demanded that the affinity of the repressor to the operator is stronger than the affinity of the polymerase to the promoter, or the primary binding of the polymerase with the promoter must be of stronger lability. Otherwise, with the optimal repressor and polymerase concentrations the transcription system concerned might not be repressed by the repressor. When the polymerase is in the phase of the formation of the "stable complex" the repressor cannot push away the polymerase any longer.

Furthermore, the elements $o_1, \ldots, o_j$ ($j \geqq 1$) introduced as part of the operator region may be functionally established by events occurring prior to the transcription process itself. They are regarded as "overlapping" elements.

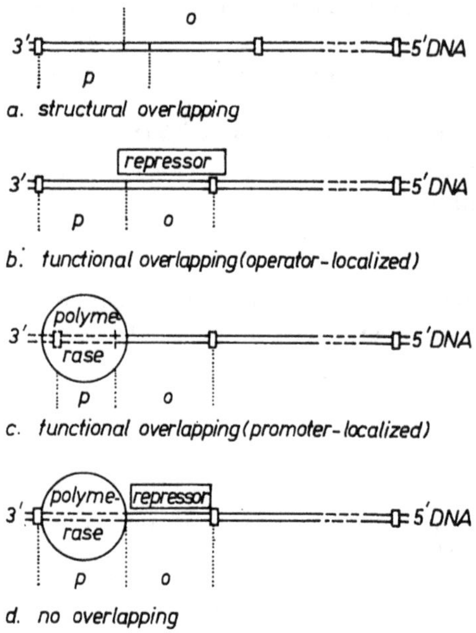

Fig. 4. Overlapping variants of elements of the promoter-localized and operator-localized control.

p promoter
o operator

Note 3

As the findings obtained above all on the phages T4 and T5 (SCHÄFER et al. 1972, 1973) have shown, the promoter region may have a more complicated structure than defined by the "classical" promoter and consists of a longer DNA sequence than it is necessary for the binding of a polymerase molecule. Before the RNA synthesis is initiated, the polymerase has to recognize and to pass, respectively, three DNA sequences ("preinitiator" model):

- DNA sequence for the primary binding of a polymerase ("entry site")
- non-transcribed DNA sequence held several "entry site"-started polymerases ("storage stretch" or "preinitiator")

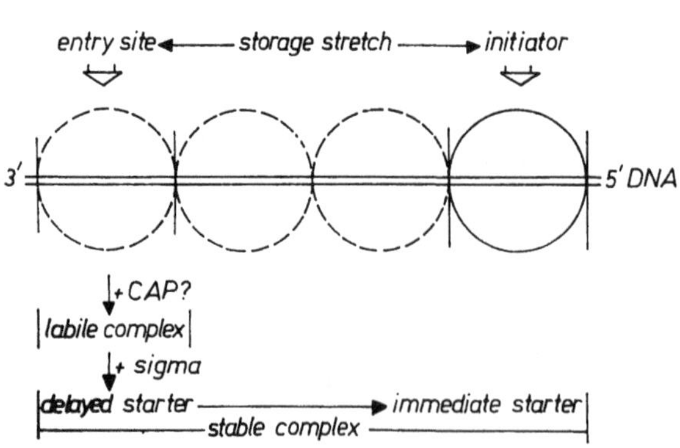

Fig. 5. Scheme of the initiation of transcription at a promoter region of the "preinitiator" type (summarized according to SCHÄFER et al., 1972, 1973).

• DNA sequence called "initiator", at which the RNA synthesis starts actually with pppPu

Furthermore it was postulated regarding the formation of the initiation complex (Fig. 5):

• Formation of a "labile complex" (possibly on the influence of protein factors as e.g. CAP). In this phase the recognition of the "entry site" may take place followed by a rapid transition to
• a "stable complex" of first phase ("delayed starter") in the presence of the sigma subunit. Then a slow transition to
• a "stable complex" of second phase ("immediate starter") takes place, where the "initiator" must be recognized.
• Binding of a purinribonucleoside triphosphate to the "immediate starter"-complex starting the RNA synthesis.

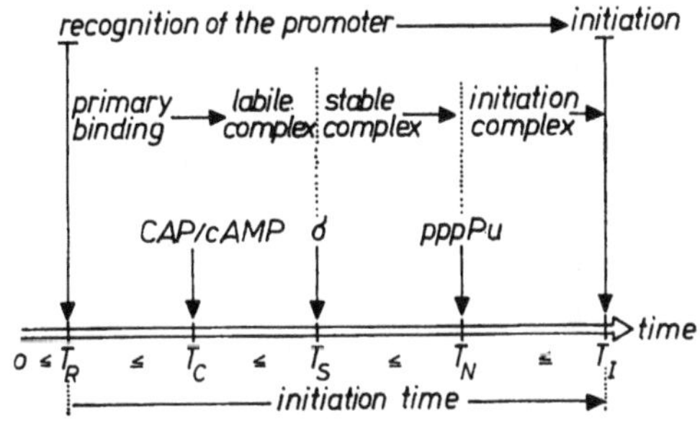

Fig. 6. Temporal sequence of the events to the initiation of a polymerase molecule.

The events leading to the initiation of a polymerase molecule within a certain period of time, the initiation time (Fig. 6), must not take place exclusively on the part of the promoter region having the primary binding with the polymerase. Each of these binding sites of a promoter region and in the case of the "preinitiator" model with the addition of each partial sequence of the "storage stretch" covered by a stored polymerase molecule is defined as promoter $p_l (1 \leq l \leq i + 1)$. Until the initiation complex is formed the polymerase molecule can be translocated to the promoter, which is the next one in each case. Hence a set of different initiation variants INVA results arising from the organizational forms of the promoter region:

INVA 1.1: At **every** promoter $p_1, \ldots, p_i$ **one** polymerase is able to initiate ["multiple" promoters (first type) and, as a special case, "classical" promoter with $i = 1$, Fig. 7a].

INVA 1.2: At **every** promoter $p_1, \ldots, p_i$ $(i > 1)$ **one** polymerase is able to perform the steps towards the formation of the stable complex (binding of the sigma subunit), but can initiate only at p_i ["multiple" promoters (second type), Fig. 7b].

INVA 1.3: At **every** promoter $p_1, \ldots, p_i$ $(i > 1)$ **one** polymerase is able to stay until the labile complex is formed (possibly influenced by the binding of the CAP/cAMP complex), but can perform the next steps towards the initiation only at p_i ["multiple" promoters (third type), Fig. 7c].

INVA 2.1: **Only** at promoter p_1 is a polymerase able to perform the steps towards the formation of the stable complex, passes p_2 up to $p_{\max-1}$ $(\max \in \{i, i + 1\}, i > 1)$, initiating at $p_{\max}$ ["preinitiator" model (first type), Fig. 7d].

INVA 2.2: **Only** at promoter p_1 is a polymerase able to stay until the labile complex is formed, passes p_2 up to $p_{\max-1}$ $(\max \in \{i, i + 1\}, i > 1)$, performing the next steps towards the initiation at $p_{\max}$ ["preinitiator" model (second type), Fig. 7e].

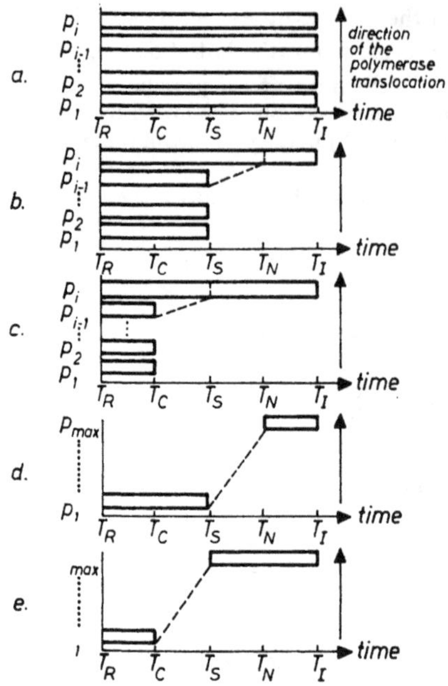

Fig. 7. Initiation variants of polymerase molecules at the promoter regions of different types (cp. Fig. 6).

4. The model

The automaton aut_p defined from the prototype can be divided into five component automata: aut-induction, aut^0-initiation, aut'-elongation, aut^+-termination and aut^*-protein pool formation. The division of the automaton aut_p into the five component automatons results from

- the sequential order of the stages of the transcription process (cf. part 3, note 1) and
- the position of the posttranscriptional processes to the transcription system itself.

The input signals X, X^0, X', X^+, X^* for the individual component automatons are calculated from

1. the actual input X_p (of the automaton aut_p), a set of vectors for the components of wich stand the concentration measures of the regulatory molecules involved, and
2. the actual states of certain other component automatons.

The state vector Z_p of the automaton aut_p is generated by the states Z, Z^0, Z', Z^+, Z^* of the component automatons, where Z is the induction pattern, Z^0 the initiation pattern, Z' the elongation pattern, Z^+ the termination pattern and Z^* the protein pool pattern. The functions δ, δ^0, δ', δ^+, δ^* of the component automatons realizing the state transfer and, hence, δ_p of the automaton aut_p are implemented by an algorithm $A_p =_{df} \{A, A^0, A', A^+, A^*\}$ (Fig. 8). Since random processes are represented, the algorithm A_p and thus, the automaton aut_p, is undetermined.

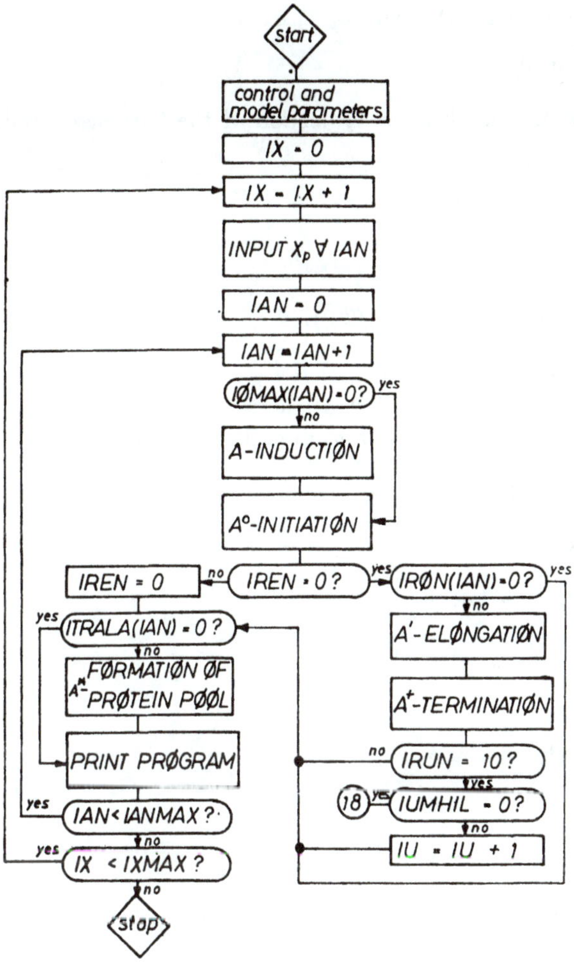

Fig. 8. Block diagram of the algorithm A_p.

The automaton aut_p is defined as follows:

$$\text{aut}_p = (X_p, Y_p, Z_p, A_p, \delta_p, \mu_p, Z_a) \text{ with}$$

$$X_p =_{\text{df}} \left\{\vec{x}_p : \vec{x}_p = (\{C_{\text{ef}}\}, \{C_{\text{aef}}\}, C_{\text{pol}}, C_{\text{sub}}, C_{\text{CAP}}, C_{\text{AMP}}, C_{\text{nuc}}, \{C_{\text{tf}}\}, \{C_{\text{atf}}\})\right\},$$

where the components of the vector $\vec{x}_p$ are concentration measures of the effectors (repressors, activators), the antieffectors (inducers, antiactivators), the polymerase (core or holoenzyme), the polymerase subunit (sigma factor), the catabolite activation protein, the 3′, 5′-cyclic AMP, the nucleoside triphosphates, the termination factors and the antitermination factors, respectively.

$$Y_p =_{\text{df}} Z^*$$

$$Z_p =_{\text{df}} \{\vec{z}_p : \vec{z}_p = (\vec{z}, \vec{z}^0, \vec{z}', \vec{z}^+, \vec{z}^*)\},$$

where $\vec{z} \in Z$, $\vec{z}^0 \in Z^0$, $\vec{z}' \in Z'$, $\vec{z}^+ \in Z^+$, $\vec{z}^* \in Z^*$

Z_a $=_{df}$ non-empty set of initial states with $Z_a \subseteqq Z_p$

A_p $=_{df}$ algorithm realizing δ_p

The text state function δ_p describes the projection of the Cartesian product $Z_p \times X_p$ to the set Z_p by means of A_p:

$$\delta_p : A_p \circ (Z_p \times X_p) \rightarrow Z_p,$$

where for each time T is:

$$\vec{z}_p(T+1) = \delta_p\big(A_p \circ (\vec{z}_p(T), \vec{x}_p(T))\big).$$

$\vec{z}_p(T+1)$ is not unequivocally determined, that is, it is only known that $\vec{z}_p(T+1)$ being realized with a probability $p(\vec{z}_p(T+1))$ descends from a certain subset Z_p' : $\vec{z}_p(T+1) \in Z_p' \subset Z_p$. The output function μ_p describes a simple transformation:

$$\mu_p : Z_p \rightarrow Y_p = Z^*,$$

where for each time T is:

$$\vec{y}(T) = \mu_p(\vec{z}_p(T+1)).$$

4.1. Induction

The automaten **aut-induction** is defined as follows:

aut $= (X, Z, A, \delta)$ with

$X =_{df} \{(X_1, Z^0, Z')\}$ and $X_1 =_{df} \{\vec{x}_1 : \vec{x}_1 = (\{C_{ef}\}, \{C_{aef}\})\}$

$Z =_{df} \{\vec{z} : \vec{z} = (E(o_m) \vee o_m : 1 \leq m \leq j'')\}$ **induction pattern,** where

$$E(o_m) = \begin{cases} 1 & \text{repression} \\ 0 & \text{induction} \end{cases} \quad \text{of the transcription system}$$

A $=_{df}$ algorithm realizing δ and δ: $A \circ (Z \times X) \rightarrow Z$
Substantial operations of the algorithm A include:

1. The positive control is described by the implication $p \rightarrow q$, and the negative control by the non-implication $\sim (p \rightarrow q)$:

		$E(o_m)$	
p	q	$p \rightarrow q$	$\sim(p \rightarrow q)$
1	1	1	0
1	0	0	1
0	1	1	0
0	0	1	0

$$p = \begin{cases} 1 & \text{if} \quad IP = f(C_{ef}(o_m, T) > RN, \\ 0 & \text{otherwise} \end{cases} \qquad q = \begin{cases} 1 & \text{if} \quad IP = f(C_{aef}(o_m, T) > RN \\ 0 & \text{otherwise} \end{cases}$$

RN is a random number from the interval [0, 1). T is the time. IP is the concentration-dependent interaction probability between the regulatory molecules and the molecules and the adequate binding partners. IP is from the interval (0, 1], and generally, it is:

$$IP = f(C(T)) = n + m \cdot e^{-z(\ln C(T) - \ln C')^k},$$

where n, m, z, k are parameters, and C' is the saturation level of the regulatory molecule concerned.

2. The binding stability of an operator-effector complex is assumed as a relative value $B(o_m) \in [0, 1]$. If a such complex exists, then:

$$E(o_m) := \begin{cases} \sim E(o_m) & \text{if } B(o_m) \leqq RN, \text{ i.e. decay of the complex} \\ E(o_m) & \text{if } B(o_m) > RN, \text{ i.e. no decay of the complex} \end{cases}$$

4.2. Initiation

The automaton **aut°-initiation** is defined as follows:

$\text{aut}^0 = (X^0, Z^0, A^0, \delta^0)$ with

$X^0 =_{\text{df}} \{(X_2, X_3, Z, Z')\}$ and $X_2 =_{\text{df}} \{\vec{x}_2 : \vec{x}_2 = (C_{\text{pol}}, C_{\text{sub}}, C_{\text{CAP}}, C_{\text{AMP}})\}$

$X_3 =_{\text{df}} \{\vec{x}_3 : \vec{x}_3 = (C_{\text{nuc}})\}$

$Z^0 =_{\text{df}} \{\vec{z}^0 : \vec{z}^0 = (F(p_l) \ \forall \ p_l : 1 \leqq l \leqq i)\}$ **initiation pattern**, where $F(p_l) \in \{1, \ldots, 11\}$ state of the promoter p_l in the initiation process of a polymerase molecule:

1: no binding of a polymerase at p_l, because p_l is covered by a p_{l-1}-started polymerase
2: no binding of a polymerase at p_l, because the polymerase concentration is too low and/or $E(o_m) = 1$ for at least one "overlapped" operator o_m
3: primary binding of the polymerase at p_l
4: CAP/cAMP level being too low delays the formation of the labile complex
5: formation of the labile complex
6: sigma factor level being too low delays the formation of the stable complex
7: formation of the stable complex
8: nucleotide level being too low delays the formation of the initiation complex
9: formation of the initiation complex
10: initiation and the polymerase has not yet left p_l
11: translocation of the polymerase from p_l to p_{l+1}

$A^0 =_{\text{df}}$ algorithm realizing δ^0 and $\delta^0 : A^0 \circ (Z^0 \times X^0) \to Z^0$
Substantial operations of the algorithm A^0 include:

1. The initiation time T_I (cf. Fig. 6) is composed of four intervels to assume:

$$T_I = T_1 + T_2 + T_3 + T_4,$$

where $T_1 = T_C - T_R,\quad T_2 = T_S - T_C,\quad T_3 = T_N - T_S,\quad T_4 = T_I - T_N.$

2. The initiation time T_I is extended by

$$\frac{T_I}{IP} - T_I,$$

if one or several of the concentrations of the polymerase, polymerase subunit (sigma factor), CAP, cAMP and nucleotides are not at saturation level C', that is, if

$$C < C'$$

and the interaction probability $IP \in (0, 1]$ is

$$IP = f(C(T)) < 1,$$

where $C \in \{C_{\text{pol}}, C_{\text{sub}}, C_{\text{CAP}}, C_{\text{AMP}}, C_{\text{nuc}}\}$ and T time.

3. The initiation variants (cf. part 3, note 3 and Fig. 7) are assumed as control parameters of the algorithm.

4.3. Elongation

The automaton **aut′-elongation** is defined as follows:

aut′ $= (X', Z', A', \delta')$ with

$X' =_{\text{df}} \{(X_3, Z, Z^0, Z^+)\}$

$Z' =_{\text{df}} \{\vec{z}' : \vec{z}' = (G(\text{pol}_h) \forall \text{pol}_h : 1 \leq h \leq k \text{ and } R(\text{pol}_h) = 0)\}$ **elongation pattern** (occupancy of the transcription unit with transcribing polymerases), where $G(\text{pol}_h) \in \{0, 1, \ldots, l\}$ position of the polymerase pol_h on the transcription unit having a length of l nucleotide pairs, and
pol_1 is the first-initiated polymerase,
pol_k is the last-initiated polymerase,

$$R(\text{pol}_h) = \begin{cases} 1 & \text{if } \text{pol}_h \text{ was released at one of the terminators or mutation sites} \\ 0 & \text{otherwise} \end{cases}$$

$A' =_{\text{df}}$ algorithm realizing δ' and $\delta' : A' \circ (Z' \times X') \to Z'$
Substantial operations of the algorithm A' include:

1. The elongation velocity of the RNA chain synthesized by the polymerase pol_h is determined by the degree of the probability of concentration-dependent nucleotide assembly

$IP = f(C_{\text{nuc}}(T)) \in (0, 1]$ with

$$G(\text{pol}_h) := \begin{cases} G(\text{pol}_h) + 1, & \text{if } IP > RN \\ G(\text{pol}_h), & \text{if } IP \leq RN. \end{cases}$$

RN is a random number from the interval $[0, 1)$.
Simultaneously, $G(\text{pol}_h)$ is a measure for the length of the RNA chain synthesized by the polymerase pol_h.

2. Changes in the transcription velocity of a polymerase molecule along the transcription unit having a length of l nucleotide pairs are simulated by means of the parameter $TV(d)$ with $TV(d) \in [0, 1]$ and $d \in \{0, 1, \ldots, l\}$, where $TV(d) = 1$ means the maximal transcription velocity assumed. Let:

$$G(\text{pol}_h) := \begin{cases} G(\text{pol}_h) + 1, & \text{if } TV(d) > RN \\ G(\text{pol}_h), & \text{if } TV(d) \leq RN \end{cases} \quad \text{with} \quad d = G(\text{pol}_h) + 1.$$

4.4. Termination

The automaton **aut⁺-termination** is defined as follows:

aut⁺ $= (X^+, Z^+, A^+, \delta^+)$ with

$X^+ =_{df} \{(X_4, Z')\}$ and $X_4 =_{df} \{\vec{x}_4 : \vec{x}_4 = (\{C_{tf}\}, \{C_{atf}\})\}$

$Z^+ =_{df} \{\vec{z}^+ : \vec{z}^+ = (\text{pol}_h \forall h \in \{1, \ldots, k\} : G(\text{pol}_h) = P(t_r) \vee Q(m_i)$ and $H(t_r) \vee I(m_i) = 1$, $\forall t_r : 1 \leq r \leq s \vee \forall m_i : 1 \leq i \leq n)\}$ **termination pattern** [polymerases pol_h (and with these the RNA chains having a length of $G(\text{pol}_h)$ nucleotides) released from the terminators t_r or the mutation sites m_i], where $P(t_r), Q(m_i) \in \{1, \ldots, l\}$ position of the terminator t_r or mutation site m_i on the transcription unit having a length of l nucleotide pairs,

$$H(t_r), I(m_i) = \begin{cases} 1 & \text{termination} \\ 0 & \text{no termination} \end{cases}$$ of the transcription for the polymerase pol_h at the terminator t_r or mutation site m_i.

$A^+ =_{df}$ algorithm realizing δ^+ and $\delta^+ : A^+ \circ (Z^+ \times X^+) \to Z^+$
Substantial operations of the algorithm A^+ include:

1. The terminator-localized control accomplished by the interactions between the terminator, the polymerase, the termination and antitermination factor is described by the non-implication $\sim (p - q)$:

		$H(t_r)$	
p	q	$\sim (p \to q)$	
1	1	0	a)
1	0	1	
0	1	0	a)
0	0	0	b)

$$p = \begin{cases} 1 & \text{if } IP = f(C_{tf}(t_r, T) > RN, \\ 0 & \text{otherwise} \end{cases} \qquad q = \begin{cases} 1 & \text{if } IP = f(C_{atf}(t_r, T) > RN, \\ 0 & \text{otherwise} \end{cases}$$

where $RN \in [0, 1)$ random number, $IP \in (0, 1]$ interaction probability and T time.
If $G(\text{pol}_h) = P(t_r)$ for a polymerase pol_h, then

$$H(t_r) = \begin{cases} 1 & \text{termination} \\ 0 & \text{no termination with a) } \text{pol}_h \text{ runs over } t_r \\ & \qquad\qquad\qquad\quad \text{b) } \text{pol}_h \text{ remains at } t_r \end{cases}$$

2. If $G(\text{pol}_h) = Q(m_i)$ for a polymerase pol_h, then

$$I(m_i) = \begin{cases} 1 & \text{termination,} & \text{if } PR(m_i) \leq RN \\ 0 & \text{no termination (pol}_h \text{ runs over } m_i), & \text{if } PR(m_i) > RN. \end{cases}$$

RN is a random number from the interval [0, 1).
$PR(m_i) \in [0, 1]$ is an assumed probability of the polymerase release at the mutation site m_i.

4.5. Protein pool formation

The automaton **aut*-protein pool formation** is defined as follows:

aut* = $(X^*, Z^*, A^*, \delta^*)$ with

$X^* =_{\mathrm{df}} \{(Z', Z^+)\}$

$Z^* =_{\mathrm{df}} \{\vec{z}^* : \vec{z}^* = (J(\mathrm{pro}_{c_n}) \forall c_n : 1 \leq n \leq k)\}$ **protein pool pattern,** where $J(\mathrm{pro}_{c_n}) \in \{0, 1, \ldots\}$ number of those active protein molecules pro_{c_n} that or the polypeptide chains of which are coded by the cistron c_n of the respective transcription system

$A^* =_{\mathrm{df}}$ algorithm realizing δ^* and $\delta^* : A^* \circ (Z^* \times X^*) \to Z^*$

Substantical operations of the algorithm A^* include:

1. Synthesis of the polypeptide chains poly_{c_n} (translation) on the RNA copy of a cistron c_n synthesized by the polymerase pol_h is calculated as follows:

$$T_x = T_b(\mathrm{pol}_h) + \big((e(c_n) - b(c_m))/3\big) \cdot V + TD(c_n) \cdot TF(c_n, \mathrm{pol}_h),$$

where

T_x — calculated moment of the synthesis of a complete polypeptide chain;
$T_b(\mathrm{pol}_h)$ — moment of the onset of translation of the RNA copy of the cistron c_n synthesized by the polymerase pol_h;
$e(c_n)$ — end of the RNA copy of the cistron c_n (in nucleotides);
$b(c_n)$ — beginning of the RNA copy of the cistron c_n (in nucleotides) with

$$m = \begin{cases} 1, \text{ at polycistronic translation} \\ 0, \text{ at monocistronic translation;} \end{cases}$$

V — translation velocity (amino acids per time);
$TD(c_n)$ — time interval between two translating ribosomes on the RNA copy of the cistron c_n;
$TF(c_n, \mathrm{pol}_h) \in \{1, \ldots, k\}$ translation frequency of the RNA copy of the cistron c_n synthesized by the polymerase pol_h, where k is the maximal number of possible polypeptide chain copies; with each synthesized polypeptide chain $TF(c_n, \mathrm{pol}_h)$ is increased to $TF(c_n, \mathrm{pol}_h) + 1$ before k is reached.

Hence it follows for the pool of the polypeptide chains poly_{c_n} at the time T:

$$K(\mathrm{poly}_{c_n}) := \begin{cases} K(\mathrm{poly}_{c_n}) + 1, & \text{if } T_x = T \\ K(\mathrm{poly}_{c_n}), & \text{if } T_x > T. \end{cases}$$

2. With the last ribosome translating the RNA copy of the cistron c_n synthesized by the polymerase pol_h a stepwise degradation of the RNA takes place, which means:

$$G(\mathrm{pol}_h) = G(\mathrm{pol}_h) - 1 \quad \text{at each time interval } T_k.$$

3. The formation of a protein pro_{c_n} the active form of which consists of (one or) several polypeptide chain subunits is realized as follows:

$$J(\mathrm{pro}_{c_n}) := \begin{cases} J(\mathrm{pro}_{c_n}) + 1, & \text{if } L = K(\mathrm{poly}_{c_n}) \\ J(\mathrm{pro}_{c_n}), & \text{if } L > K(\mathrm{poly}_{c_n}), \end{cases}$$

where $L = NS(\mathrm{pro}_{c_n}) \cdot \big(J(\mathrm{pro}_{c_n}) + 1\big)$,
$NS(\mathrm{pro}_{c_n})$ number of the polypeptide chain subunits of the protein pro_{c_n}.

4. The mean period between the synthesis of a protein molecule pro_{c_n} coded by the cistron c_n and the degradation or inactivation of which is assumed with the parameter $T_h(pro_{c_n})$. If at the time T the period $T_h(pro_{c_n})$ is run down for i protein molecules pro_{c_n}, then $J(pro_{c_n})$ is decreased to $J(pro_{c_n}) - i$.

Note:

The digital operating principles of an automaton is based on a temporally discontinuous processing of discrete signals. For that reason an abstraction from the continuous time to the discrete time must take place. The time interval $T_{j+1} - T_j$ $(0 \leq j < m)$ between the recent moment T_j and the subsequent moment T_{j+1} is called a "time unit" T. If all "time units" are of equal length $l(T_k) = l(T_{k+1}) \; \forall \; k: 1 \leq k < m$, then these intervals are equidistant. Consequently, within the space of the period T_m a sequence of in each case m discrete signals of a certain kind can be given. The length of the "time unit" T can be chosen you like.

In this case within the space of a period assumed the polymerase is able to transcribe a certain number of nucleotides. Let this period be a "time unit" T. A "time unit" represents, on the one hand, a real time but, on the other hand, a spatial measure, a number of nucleotides. With this "normalization" by a fixed measure, the "time unit" T being defined like this, all the temporal and spatial parameters are in conformity with the discontinuous processing of discrete signals by the defined automaton aut_p.

Example:

The length of the "time unit" T is fixed with 10 nucleotides or 0.25 seconds, i. e., within the space of a "time unit" T 10 nucleotides can be transcribed; this is adequate to a transcription velocity of 40 nucleotides per second. If the period of the experiment P_{exp} is to be five minutes and the length of the transcription unit L_{tu} 6000 nucleotide pairs, then:

$$P_{exp} = 1200 \text{ "time units"}, \; L_{tu} = 600 \text{ "time units"}.$$

5. General representation of simulation results

The model allows an exceedingly detailed analysis of the external (or overall) behaviour of a given transcription system. The external behaviour will be reflected by the rate of RNA synthesis; it is result of a certain internal system behaviour. This internal behaviour is described by the state vector Z_p of the automaton aut_p which is the theoretical basis of the model. By means of a strongly simplified transcription system this state of affairs will be illustrated by two simulation experiments. Moreover, for technical reasons it is not possible to give a detailed, stepwise representation of the state patterns given out by the print program of the model in the following paragraphs.

5.1. Transcription system assumed

The following transcription system is assumed:

1. The promoter region is to consist of two promoters. In the presence of the sigma subunit, CAP and cAMP, the polymerase molecules are able to initiate completely at each of the two promoters. The length of a promoter is to correspond to the diameter of the polymerase and to amount to 40 nucleotide pairs.
2. The operator region is to consist of only one operator with a length of 20 nucleotide pairs. The system is subject to a negative control mechanism accomplished by opera-

tor-repressor-inducer interactions. An "overlapping" between the operator and promoter region is not to exist.

3. The cistron sequence is to consist of two cistrons and to have a length of 220 nucleotide pairs.
4. The cistrons are to be separated from each other by an intermediate terminator localized between the nucleotide pair 100 and 110. The end terminator is to be added directly at the cistron sequence (nucleotide pair 260—270). The release of the transcribing polymerases is to take place in the presence of the intermediate and end termination factors, respectively.
5. The transcription velocity of a polymerase molecule is to be constant along the entire transcription unit and to amount to 40 nucleotides per second. The length of a "time unit" T is assumed to be 0.25 s, which is in conformity with a spatial measure of 10 nucleotides and nucleotide pairs, respectively.
6. The initiation time of the polymerase is to amount to 1.75 s, where the primary binding is to take up 0.25 s, the formation of the labile, stable and initiation complex in each case 0.5 s.

For both simulation experiments (se) an experiment lasting for 12.5 seconds ($T_{\max} = 50$) is assumed. The experiments are performed with the following inputs (X_p of the automaton aut$_p$):

se 1: The concentrations of all regulatory molecules (including the nucleoside triphosphates) are to be in each case above the saturation level, that is, $C(T) = 100$ for every $T \in \{1, \ldots, 50\}$. However, the intermediate termination factor is to be absent.

se 2: Let for the input component

- inducer: $C(T) = \begin{cases} 0 & \text{for every } T \in \{13, \ldots, 31\} \\ 100 & \text{otherwise} \end{cases}$

- cAMP: $C(T) = \begin{cases} 0 & \text{for every } T \in \{15, \ldots, 21\} \\ 100 & \text{otherwise} \end{cases}$

- intermediate termination factor:

$$C(T) = 25 \text{ for every } T \in \{1, \ldots, 50\}.$$

All other input components are to be as in se 1.

The transcription system defined responds to the inputs in each case with a determined internal behaviour reflected by certain state patterns. The states $\vec{z}_p$ ($\vec{z}_p \in Z_p$ of the automaton aut$_p$) producing these patterns were listed stepwise in Tab. 1 and Tab. 2, respectively. There are:

column A: induction pattern Z
column B: initiation pattern Z^0
column C: elongation pattern Z'

(and in this case termination pattern Z^+ indirectly)

Note:

The protein pool pattern Z^* does not appear, because for this system the automaton aut*-protein pool formation was not defined.

Table 1. State pattern of the transcription system simulated in se 1.

"time unit" T	real time [s]	A $E(o_1)$	B $F(p_1)$	$F(p_2)$	C $G(\text{pol}_h)$, pol_h			
1	0.25	0	3	3				
2	0.50	0	5	5				
3	0.75	0	5	5				
4	1.00	0	7	7				
5	1.25	0	7	7				
6	1.50	0	9	9				
7	1.75	0	11	10	0 1			
8	2.00	0	11	10	10 1			
9	2.25	0	11	10	20 1			
10	2.50	0	11	10	30 1			
11	2.75	0	11	10	0 2	40 1		
12	3.00	0	3	10	10 2	50 1		
13	3.25	0	5	10	20 2	60 1		
14	3.50	0	5	10	30 2	70 1		
15	3.75	0	7	10	40 2	80 1		
16	4.00	0	7	3	50 2	90 1		
17	4.25	0	9	5	60 2	100 1		
18	4.50	0	11	5	70 2	110 1		
19	4.75	0	11	7	80 2	120 1		
20	5.00	0	11	7	90 2	130 1		
21	5.25	0	11	9	100 2	140 1		
22	5.50	0	11	10	0 3	110 2	150 1	
23	5.75	0	11	10	10 3	120 2	160 1	
24	6.00	0	11	10	20 3	130 2	170 1	
25	6.25	0	11	10	30 3	140 2	180 1	
26	6.50	0	11	10	0 4	40 3	150 2	190 1
27	6.75	0	3	10	10 4	50 3	160 2	200 1
28	7.00	0	5	10	20 4	60 3	170 2	210 1
								→ Fig. 9 B
29	7.25	0	5	10	30 4	70 3	180 2	220 1
30	7.50	0	7	10	40 4	80 3	190 2	230 1
31	7.75	0	7	3	50 4	90 3	200 2	240 1
32	8.00	0	9	5	60 4	100 3	210 2	250 1
33	8.25	0	11	5	70 4	110 3	220 2	260 1
34	8.50	0	11	7	80 4	120 3	230 2	
35	8.75	0	11	7	90 4	130 3	240 2	
36	9.00	0	11	9	100 4	140 3	250 2	
37	9.25	0	11	10	0 5	110 4	150 3	260 2
38	9.50	0	11	10	10 5	120 4	160 3	
39	9.75	0	11	10	20 5	130 4	170 3	
40	10.00	0	11	10	30 5	140 4	180 3	
41	10.25	0	11	10	0 6	40 5	150 4	190 3
42	10.50	0	3	10	10 6	50 5	160 4	200 3
43	10.75	0	5	10	20 6	60 5	170 4	210 3
44	11.00	0	5	10	30 6	70 5	180 4	220 3
45	11.25	0	7	10	40 6	80 5	190 4	230 3
46	11.50	0	7	3	50 6	90 5	200 4	240 3
47	11.75	0	9	5	60 6	100 5	210 4	250 3
48	12.00	0	11	5	70 6	110 5	220 4	260 3
49	12.25	0	11	7	80 6	120 5	230 4	
50	12.50	0	11	7	90 6	130 5	240 4	
								→ Fig. 9 C

Table 2. State pattern of the transcription system simulated in se 2.

"time unit"	real time	A	B		C			
T	$[s]$	$E(o_1)$	$F(p_1)$	$F(p_2)$	$G(\text{pol}_h)$, pol_h			
1	0.25	0	3	3				
2	0.50	0	5	5				
3	0.75	0	5	5				
4	1.00	0	7	7				
5	1.25	0	7	7				
6	1.50	0	9	9				
7	1.75	0	11	10	0 1			
8	2.00	0	11	10	10 1			
9	2.25	0	11	10	20 1			
10	2.50	0	11	10	30 1			
11	2.75	0	11	10	0 2	40 1		
12	3.00	0	3	10	10 2	50 1		
13	3.25	1	5	10	20 2	60 1		
14	3.50	1	5	10	30 2	70 1		
15	3.75	1	7	10	40 2	80 1		
16	4.00	1	7	3	50 2	90 1		
17	4.25	1	9	4	60 2	100 1		
18	4.50	1	11	4	70 2	110 1		
19	4.75	1	11	4	80 2	120 1		
20	5.00	1	11	4	90 2	130 1		
21	5.25	1	11	4	100 2	140 1		
22	5.50	1	11	5	150 1			
23	5.75	1	11	5	160 1			
24	6.00	1	11	7	170 1			
25	6.25	1	11	7	180 1			
26	6.50	1	11	9	190 1			
27	6.75	1	11	10	0 3	200 1		
28	7.00	1	11	10	0 3	210 1		
								→ Fig. 9B
29	7.25	1	11	10	0 3	220 1		
30	7.50	1	11	10	0 3	230 1		
31	7.75	1	11	10	0 3	240 1		
32	8.00	0	11	10	10 3	250 1		
33	8.25	0	11	10	20 3	260 1		
34	8.50	0	11	10	30 3			
35	8.75	0	11	10	0 4	40 3		
36	9.00	0	3	10	10 4	50 3		
37	9.25	0	5	10	20 4	60 3		
38	9.50	0	5	10	30 4	70 3		
39	9.75	0	7	10	40 4	80 3		
40	10.00	0	7	3	50 4	90 3		
41	10.25	0	9	5	60 4	100 3		
42	10.50	0	11	5	70 4	110 3		
43	10.75	0	11	7	80 4	120 3		
44	11.00	0	11	7	90 4	130 3		
45	11.25	0	11	9	100 4	140 3		
46	11.50	0	11	10	0 5	110 4	150 3	
47	11.75	0	11	10	10 5	120 4	160 3	
48	12.00	0	11	10	20 5	130 4	170 3	
49	12.25	0	11	10	30 5	140 4	180 3	
50	12.50	0	11	10	0 6	40 5	150 4	190 3
								→ Fig. 9C

5.2. Results

For both experiments the external behaviour of the defined transcription system resulting from a certain internal behaviour is represented in Fig. 9a.

Fig 9b and c show the internal behaviour of the transcription system transferred from the Tab. 1 and 2 to the moment $T = 28$ and $T = 50$, respectively.

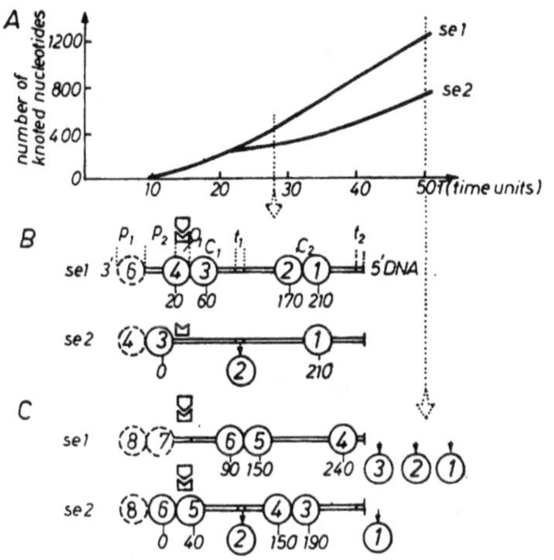

Fig. 9. External (A) and internal (B, C) behavior of a transcription system in two simulation experiments (se 1, 2) with different inputs (see legend to Fig. 23 of contribution 6).

On the conditions of the simulation experiment se 1 and after the lapse of the period of the experiment, the transcription system reaches a capacity of RNA synthesis of 1240 knoted nucleotides. Six polymerases have been involved, with three complete RNA chains being synthesized. On the conditions of the se 2, five polymerases knoted 740 nucleotides. The synthesis of only one complete RNA chain was found to happen; this deficit of synthesis arised from:

1. the release of the polymerase pol_2 at the intermediate terminator,
2. the delay of the initiation of the polymerase pol_3 by reason of the catabolite repression (cAMP),
3. the blocking of the preinitiated polymerase pol_3, because with the temporary absence of the inducer an operator-localized repression of the transcription system has taken place.

The positions of the transcribing polymerases (elongation pattern) at the moment $T = 50$ in the experiment se 2 are in conformity with those in the se 1 at the moment $T = 41$. This temporal difference of nine "time units" is the result of the effects named in the points 2 and 3, where the first one lasted for five "time units" ($T = 17, \ldots, 21$) and the second one for four "time units" ($T = 28, \ldots, 31$). However, in consequence of the different termination patterns the capacity of RNA synthesis at both moments is not adequate.

5.3. Conclusions

From these simulation experiments and their results it can be concluded:

1. The external behaviour of a transcription system assumed is explicable from the internal system behaviour.
2. At any moment of a simulation experiment it can be established, whether or not and if so, which control factors and mechanisms have effects on the transcription process and to which extent.
3. If in each case two of the components, the structure of a transcription system assumed, external behaviour (RNA synthesis) and input, are known and constant, then it can be concluded on the third one.
4. Single experimental findings can be extrapolated to a wider context.
5. The model parameters can be directly interpreted in their biological significance.

6. Results and discussion of simulation studies

6.1. Theoretical aspects of the transcription control

I. In the following we want to analyse transcription systems with different initiation variants INVA arising from the structural principles of the promoter region (cf. part 3, note 3) for their capacity of RNA synthesis.

INVA 1.1: At every promoter $p_1, \ldots, p_i$ one polymerase is able to initiate.
The results of the experiments with $i = 1, 2, 3, 4$, and 5 are represented in Fig. 10. The RNA synthesis capacity of the transcription system with a promoter region consisting of only one promoter ($i = 1$, "classical" promoter) achieved after a definite period of time is the reference point for the others. It is obvious that systems with "multiple" promoters of INVA 1.1 are much more effective in synthesizing RNA than those with

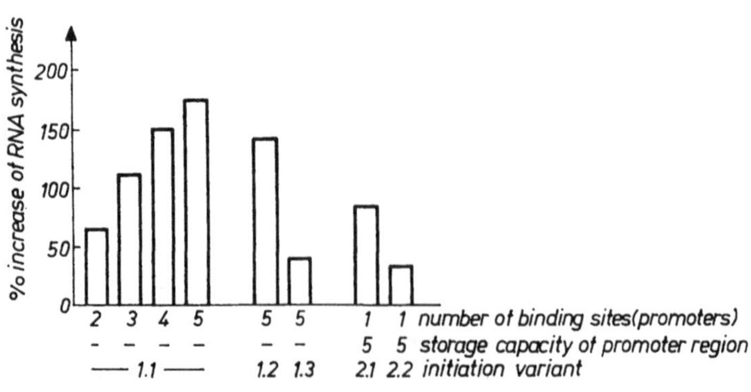

Fig. 10. Percent increase of the RNA synthesis of transcription systems with different initiation variants as against the RNA synthesis of a transcription system with "classical" promoter.

a "classical" promoter. But there is no linear relationship between the size of the promoter region and the rate of synthesis. For instance, the system with five promoters remains far below the five-fold synthesis rate of the system with one promoter; only an increase by about 175 p.c. is reached.

INVA 1.2: At every promoter $p_1, \ldots, p_i$ ($i > 1$) one polymerase is able to perform the steps towards the formation of the stable complex, but can initiate only at p_i.

INVA 1.3: At every promoter $p_1, \ldots, p_i$ ($i > 1$) one polymerase is able to stay until the labile complex is formed, but can perform the next steps towards the initiation only at p_i.

For both variants, experiments with $i = 5$ were carried out, which revealed a clear difference in the rate of RNA synthesis (Fig. 10) attributable to the different initiation conditions. The synthesis rates of the two systems are in part distinctly below that of comparable system with INVA 1.1 and $i = 5$.

INVA 2.1: Only at promoter p_1 is a polymerase able to perform the steps towards the formation of the stable complex, passes p_2 up to $p_{\max-1}$ ($\max \in \{i, i+1\}$, $i > 1$), initiating at $p_{\max}$.

INVA 2.2: Only at promoter p_1 is a polymerase able to stay until the labile complex is formed, passes p_2 up to $p_{\max-1}$ ($\max \in \{i, i+1\}$, $i > 1$), performing the next steps towards the initiation at $p_{\max}$.

These two variants are characteristic for transcription systems having a promoter region of the type of the "preinitiator" model (SCHÄFER et al., 1972, 1973), where p_1 is the "entry site", $p_1, \ldots, p_{\max}$ "storage stretch" and $p_{\max}$ "initiator".

The results of the experiments with the two initiation variants and a "storage stretch" capacity of five polymerase molecules ($\max = i = 5$) show (Fig. 10) that such transcription systems are more effective than the comparable system with the "classical" promoter (INVA 1.1, $i = 1$) because the latter one also has only one polymerase binding site. The enhanced synthesis rates are related to the concept of the "preinitiator" model which distributes the time necessary for initiation of a polymerase molecule from one promoter to several sequences ("entry site", "storage stretch" and "initiator").

As a criterion for evaluating the effectivity of one initiation variant may serve the rate of RNA synthesis achieved by the transcription system under optimal input conditions. Depending on the given initiation variant, a constant synthesis rate is established (Fig. 11). From the slope of the linear portion of the curve of a RNA synthesis kinetics we can derive an effectivity measure (EM); thus we obtain (related to INVA 1.1, $i = 1$):

INVA 1.1,	$i = 1$ "classical" promoter	EM = 1.0
INVA 1.1,	$i = 5$ „multiple" promoters (first type)	EM = 2.7
INVA 1.2,	$i = 5$ "multiple" promoters (second type)	EM = 2.4
INVA 1.3,	$i = 5$ "multiple" promoters (third type)	EM = 1.4
INVA 2.1,	max = 5 "preinitiator" model (first type)	EM = 2.0
INVA 2.2,	max = 5 "preinitiator" model (second type)	EM = 1.4

Summarizing, we can say:

1. A transcription system having a promoter region of the type of the "preinitiator" model is more effective than one having a "classical" promoter with regard to the rate

of RNA synthesis, but is inferior to one having "multiple" promoters (multiplex polymerase binding sites).
2. Therefore, an evolutionary trend from "classical" promoter → "preinitiator" model → "multiple" promoters would be conceivable.
3. Binding of the sigma subunit for stabilization of the polymerase-DNA complex seems to proceed only on this portion of the promoter region, with which the respective polymerase established the primary binding.

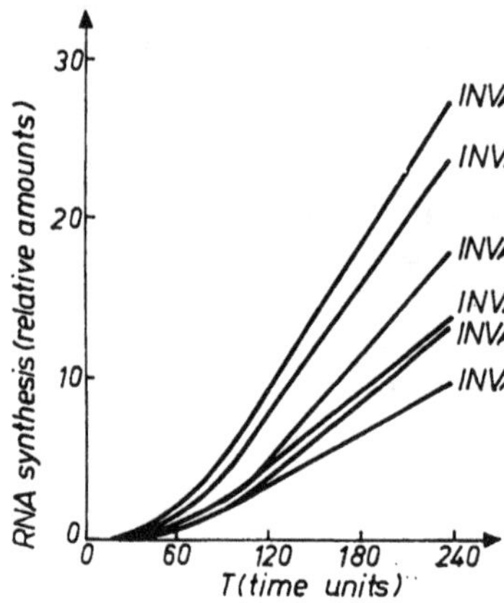

Fig. 11. Kinetics of the RNA synthesis of transcription systems with different initiation variants.

II. The operator region, the second essential structural element of the regulation sequence of a transcription unit, is characterized by a sequence of operators $o_1, \ldots, o_{j'}$. Each of these operators may come into interactions with protein effectors (repressors and activators, respectively), which may be weakened or abolished by antieffectors (inducers and antiactivators, respectively). These interactions form control mechanisms of transscription (of negative or positive nature) by influencing the initiation or migration of polymerase; both the stability of binding of an effector with the operator and the size of the effector bulk resulting from the number of binding sites (operators) of an operator region should have an influence on the transcription frequency. In order to answer this question for systems with negative control mechanisms, experiments have been performed, the results of which are summarized in Fig. 12. Systems with an operator region $o_{j+1}, \ldots, o_{j'}$ put up in front of the promoter region (cf. Fig. 2) are being considered, where $j = 0$ and $j' = 1, 2, 3, 4, 7$, and 10. The degree of binding stability $B(o_m)$ of a repressor molecule to the operator o_m $(1 \leq m \leq j')$, where $B(o_m) \in [0, 1]$ can be determined by

1, the half-life of the repressor-operator complex ("macrolability") and/or
2. temporary, mostly very brief instabilities of the repressor-operator complex within the half-life ("microlability") and/or
3. the concentration-dependent influence of a specific inducer on the repressor-operator interactions.

As shown in Fig. 12, for systems with an operator region $o_1, \ldots, o_{j'}$, starting from about $j' = 4$, no distinctly different rates of RNA synthesis can be observed. This findings permits conclusions in a twofold respect:

1. If transcription systems are controlled via several operators by repressors of the same kind, then these single controls via the operator regions $o_1, \ldots, o_{j'}$, starting from about $j' = 4$, have no lasting effects on RNA synthesis.
2. If transcription systems are controlled via several operators by repressors of different kinds, then these multiple controls via the operator regions $o_1, \ldots, o_{j'}$, starting from about $j' = 4$, appear to be quite reasonable because in addition to the all-or-none-principle the transcription frequency is influenced only by the binding stabilities of the various repressors to the corresponding operators, but not by the degree of multiplex control resulting from the number of the various repressors.

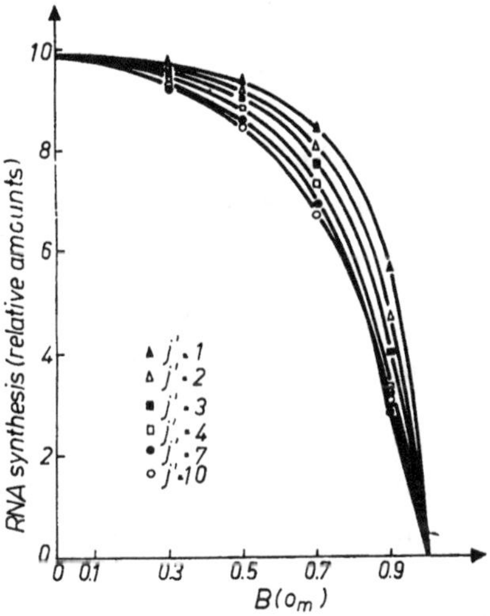

Fig. 12. Effect of binding stability of the repressor ($B(o_m)$, $1 \leq m \leq j'$) to the operators o_m of the operator regions $o_1, \ldots, o_{j'}$ with $j' = 1, 2, 3, 4, 7$, and 10 on the RNA synthesis.

III. Certainly, the polymerases transcribe along the transcription unit not always with a constant maximum velocity; this may be related to specific nucleotide sequences in the DNA and/or progressing functional inefficiency of the polymerases itself. Thus we obtain for the polymerases along the respective DNA sections a certain average transcription velocity and, for every single polymerase molecule, a purely random transcription velocity scattered about the corresponding mean value. For this purpose, systems with the transcription velocity profiles along the transcription unit as represented in Fig. 13a, are simulated. As is obvious from the elongation patterns (Fig. 13b) there occurred in the distal section of the transcription unit average transcription distances which are approximately proportional to the given average transription velocities. Whereas in system I the polymerases transcribe at an uniform maximum velocity, 200 nucleotide pairs spaced apart, the transcription distances for system II are reduced on the average to 152, for III, to 96, and for IV, to 61 nucleotide pairs.

In addition to controlled events at the intermediate terminators and events elicited by mutations in the DNA strand, a decrease of the transcription velocity from 3'- to 5'-end of a transcription unit possibly may contribute to the phenomenon of polarity. The stronger the decrease of transcription velocity along the distal DNA sections the

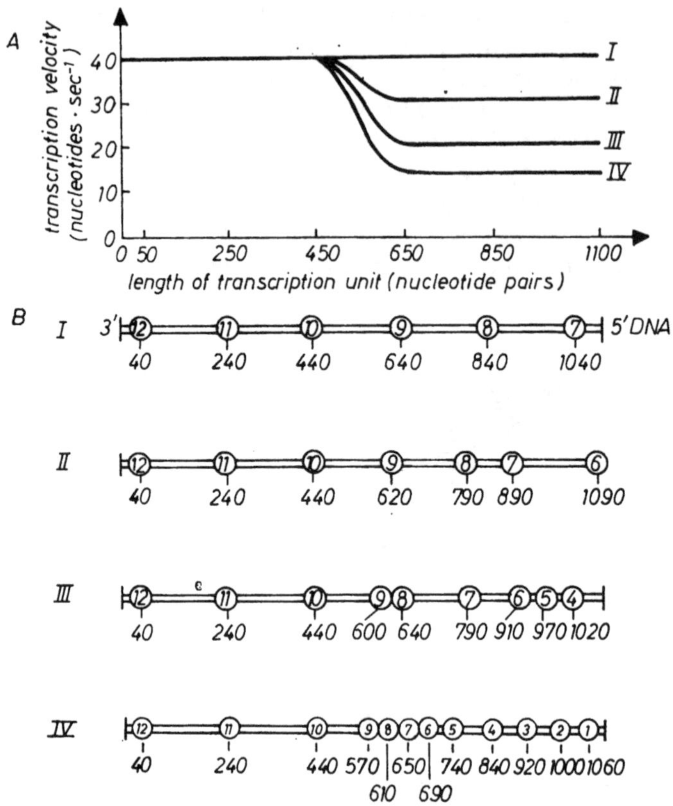

Fig. 13. Elongation patterns at a moment of simulation (B) for the transcription velocity functions I, II, III, IV depending on the length of transcription unit (A). Legend see Fig. 9.

greater the relative proportion of RNA copies of proximal DNA sections over those of distal ones.

IV. Termination is not only the last stage of the transcription process but is of considerable importance for the transcription control too. According to experimental studies (SCHÄFER, ZILLIG, 1973) the system of the bacteriophage T4 offers a good possibility for simulating blocking and termination events for a real object in a concrete manner. SCHÄFER, ZILLIG studied the effect of ionic strength on the termination of transcription of *T*-phage DNA and the action of protein factors rho (ϱ) and kappa ($\varkappa$). The transcription of T4-DNA by *E. coli* polymerase at low ionic strength is stopped at specific low-ionic-strength blocking sites without release of the RNA product. In the presence of the factors ϱ and $\varkappa$ the transcription is arrested at other blocking sites in each case. But the RNA product is not released in these cases either. At high ionic strength correct termination with release of both polymerase and RNA product is accomplished at high-ionic-strength terminator. Hence, the factors ϱ and $\varkappa$ would have to be looked on as inhibitory factors of elongation rather than termination factors. **In vivo**, perhaps, they require additional factors to realize correct termination, or to remove the block for complete transcription of the DNA up to the high-ionic-strength terminator.

In Fig. 15, there are plotted the results of corresponding simulation experiments (A) in comparison with the data experimentally obtained by SCHÄFER, ZILLIG (B). Since the simulation experiments were performed on the basis of the model postulated in Fig. 14 for the organization of the T4-DNA, the proposed model should not be rejected.

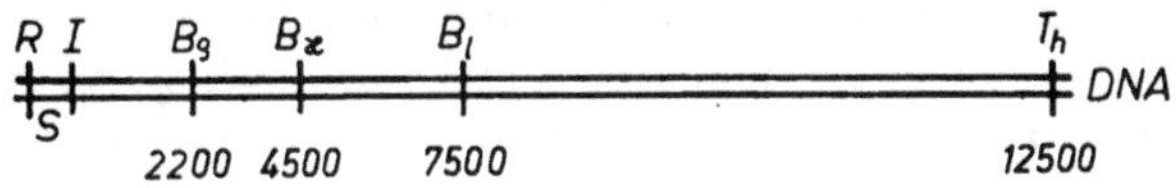

Fig. 14. Average positions (in nucleotide pairs) of different blocking and termination sites of one transcription unit on T4-DNA (according to SCHÄFER/ZILLIG 1973).

R entry site
S storage stretch
I initiator
B_ϱ ϱ-blocking site
$B_\varkappa$ $\varkappa$-blocking site
B_l low-ionic-strength blocking site
T_h high-ionic-strength terminator

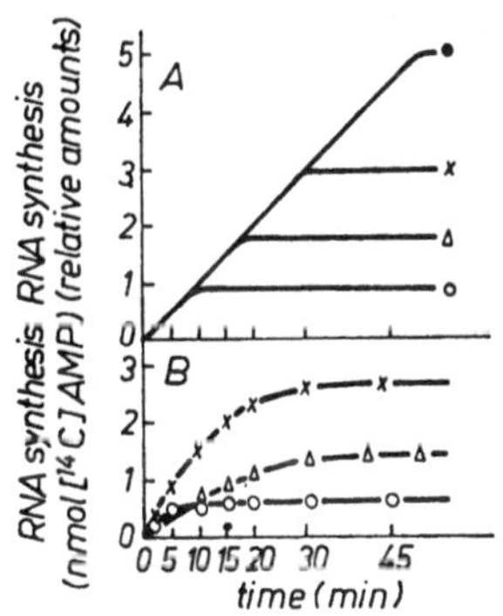

Fig. 15. Comparison of low-ionic-strength kinetics of RNA synthesis on T4-DNA in the absence of factor (×),
with factor ϱ (○),
with factor $\varkappa$ (△)
A. results from simulation experiments high-ionic-strength kinetics (•)
B. experimental curves (SCHÄFER/ZILLIG, 1973).

6.2. Feedback mechanisms of the transcription regulation

Feedback mechanisms are characterized by the fact that proteins control directly or indirectly the expression of their own structural genes (cistrons). Thus, the transcription system concerned are embedded into closed control circuits via feedbacks which may be of negative or positive nature.

1. The repressible tryptophan (trp) system is subject to a negative feedback mechanism on the operator side (Fig. 16).
   In Fig. 17 showing the simulated time course of the tryptophan synthetase (TSase) and RNA concentration the feedback principle is clearly seen. If the TSase and,

hence, the tryptophan concentration exceed a certain value, the RNA synthesis is stopped. The tryptophan acting as a corepressor increasingly activates aporepressor molecules which, as repressors by binding to the operator trpO, prevent the transcription of the trp-operon by means of further polymerases. The TSase concentration then remains constant over a certain period of time because synthesis and degradation of active TSase molecules are at equilibrium. When the TSase and thus the tryptophan concentration declines below a certain level, RNA synthesis is started again. With decreasing tryptophan concentration in this phase progressively less

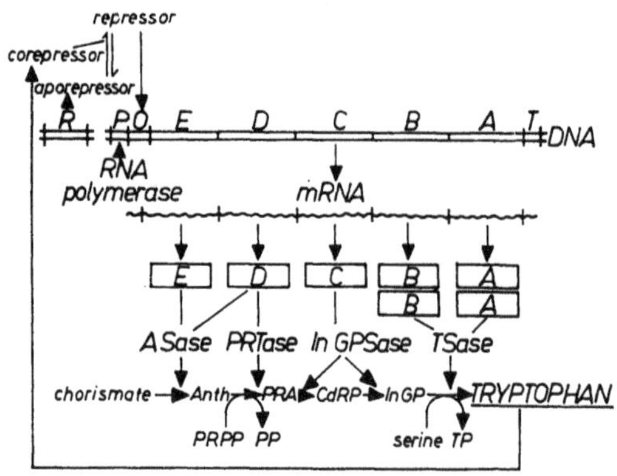

Fig. 16. The tryptophan system in *E. coli.*

▭	Polypeptide
ASase	Anthranilate synthetase
PRTase	Phosphoribosyl-anthranilate transferase
InGPSase	Indole glycerolphosphate synthetase
TSase	Tryptophan synthetase
Anth	Anthranilate
PRA	N-5′-Phosphoribosyl-anthranilate
CdRP	1-o-Carboxy phenylamino-1′-deoxyribulose-5′-phosphate
InGP	Indole-3-glycerol-phosphate
PRPP	5-Phosphoribosyl-pyrophosphate
PP	Pyrophosphoric acid
TP	Triosephosphate

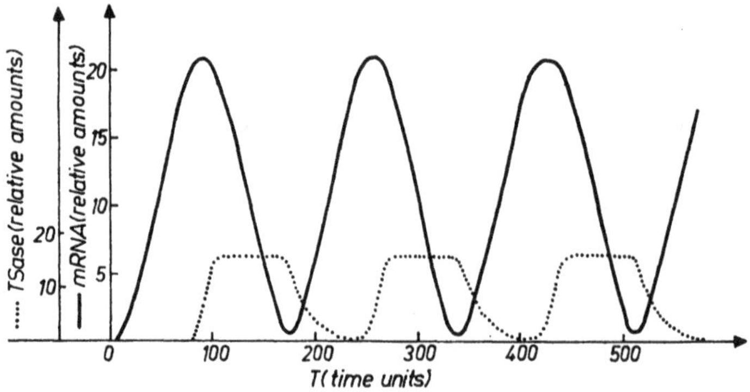

Fig. 17. Time course of the tryptophan system after complete removal of tryptophan at zero time.

aporepressor molecules are being activated, so that the trp-operon can be transcribed again, because the latter are no longer operator-blocked. Thus, the feedback mechanism causes an interplay between RNA synthesis, enzyme and tryptophan synthesis, respectively.

Certainly, the generated concentration curves for the trp-system are not necessarily of biological relevance because relatively short half-lives were given for active enzyme molecules. **In vivo**, the TSase concentration, not dropping completely to zero, should constantly oscillate undampedly around a mean value. The RNA peaks would then appear more in a pulse-like fashion at certain intervals.

2. Systems subject on the operator site to a negative **and** positive regulation mechanism, like the arabinose (ara) system, or to a purely positive regulation mechanism are essentially inducible systems (Fig. 18). Such a system codes for enzymes necessary for the catabolism of the substance induced the system. The results of an experiment for simulation of the ara-system are shown in Fig. 19. By addition of arabinose at

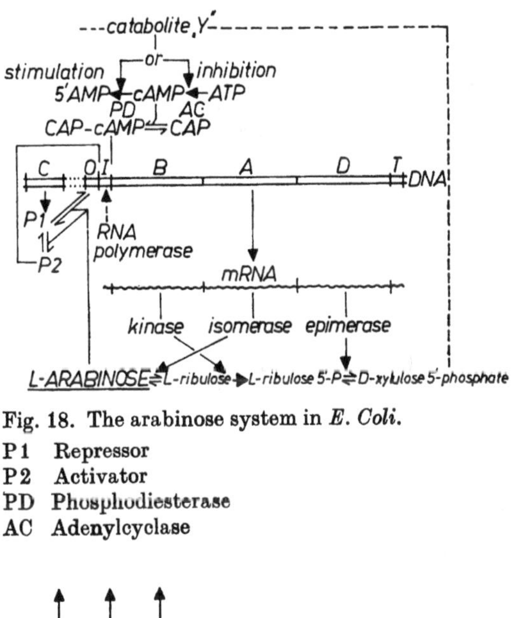

Fig. 18. The arabinose system in *E. Coli.*

P1 Repressor
P2 Activator
PD Phosphodiesterase
AC Adenylcyclase

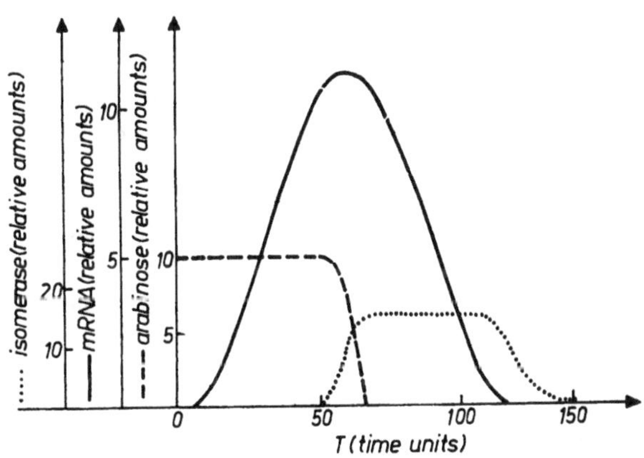

Fig. 19. Time course of the arabinose system after addition of arabinose at zero time.

zero time the product of the regulator gene araC is activated and binds as activator P2 indirectly or directly to the operator sequence araO, so that the ara-operon is released for transcription by polymerases starting at the promoter araI. With a certain delay then starts the synthesis of the isomerase which converts the arabinose into ribulose. The arabinose concentration, which falls with rising isomerase concentration, causes the araC product to remain in an increasing measure as repressor P1. Thus, the ara-operon is repressed via araO from a certain P1 concentration onwards. On repeated addition of arabinose the described feedback mechanism would become active anew. The catabolite repression mechanism to which the ara-system (like all the systems coding for the enzymes of the sugar metabolism) is also subject, did not become effective under the conditions of this simulation experiment because arabinlse was added in low amounts and not permanently.

3. The lactose (lac) system being an inducible system is regulated on the operator side through a negative mechanism (Fig. 20). It is also subject to the positively regulating influence of the CAP/cAMP complex. The mechanism of catabolite repression

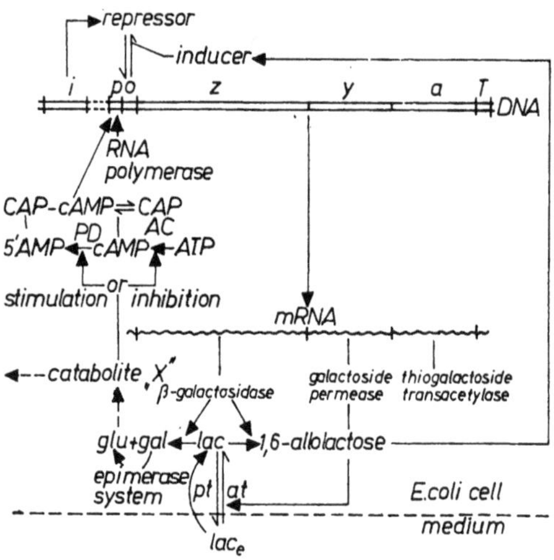

Fig. 20. The lactose system in *E. coli*.

pt passive transport of the lactose in the cell

at active transport by the galactoside permease

β-galactosidase with two functions

1. production of the 1,6-allolactose from the lactose → metabolic induction
2. cleavage of the lactose in glucose and galactose → catabolite repression

implemented through this complex effects a fine regulation of the lac-system, contrary to the control on the operator side which provides a rougher regulation because of the all-or-none-principle of the operator-repressor-inducer interactions. The modus operandi of this mechanism is demonstrated by the simulation results shown in Fig. 21.

The inducer lactose inactivates repressor molecules, polymerases can initiate, and RNA synthesis starts. With the appearance of β-galactosidase lactose is being cleaved, glucose concentration rises and the cAMP level falls. Since with falling cAMP concentration the initiation times of the polymerases become increasingly higher, the latter

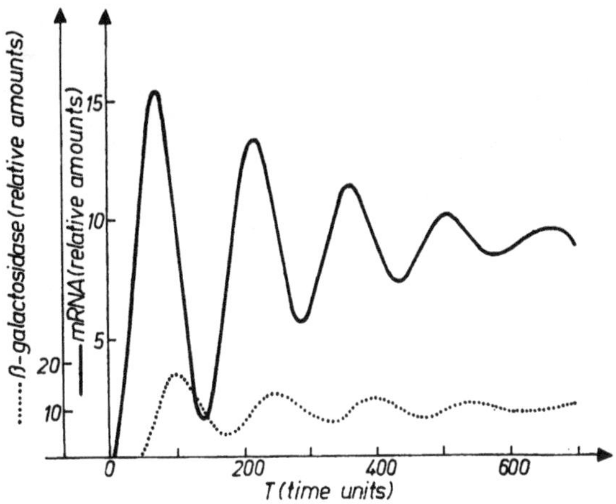

Fig. 21. Time course of the lactose system after addition of lactose in excess at zero time

initiate ever more rarely, causing strong inhibition of RNA synthesis. The transcription frequency is reincreased only if the cAMP level rises as a result of decrease of β-galactosidase and hence glucose concentration. Since lactose is to flow permanently and, consequently, no repression of transcription, on the operator side, can occur, β-galactosidase and mRNA concentration start dampedly oscillating. The two show a phase shift of 90° relative one another, that is, where the mRNA concentration has its maximum, there is the steepest rise of the β-galactosidase concentration. With time the concentrations become adjusted to a certain level, which is caused by the mechanism of catabolite repression of the regulation pathway
β-galactosidase $\rightarrow$ glucose $\rightarrow$ cAMP $\rightarrow$ cAMP/CAP-promoter-(polymerase) complex $\rightarrow$ mRNA $\rightarrow$ β-galactosidase.

6.3. Coupled transcription systems

The expression of a complete genetic program is achieved through the coordinated regulation of a network of transcription systems. The freely selected model of two coupled transcription systems shown in Fig. 22 was simulated as simple example for description and analysis of such networks. Fig. 23a illustrates the time course of the defined overall system. Through addition of the inducer the repressor molecules are inactivated, the operator blockage of the transcription unit 1 is removed, polymerases can initiate, and mRNA synthesis starts. With a certain delay then starts the synthesis of the c_1-coded protein C_1 which, acting as an activator, interacts with the operator of transcription unit 2. Thus, at a certain C_1 concentration the mRNA synthesis for transcription system (TS) 2 is initiated which is then accompanied by the occurrence of the s_2-coded protein S_2. S_2 molecules, on the one hand, degrade the inducer which leads to repression of TS 1 because with increasing S_2 concentration and, hence, decreasing inducer concentration repressor molecules are ever more rarely inactivated. On the other hand, S_2 also represses the expression of its own structural gene by weakening the activator C_1 with regard to its binding stability to the operator. The switching on and off of the two transcription

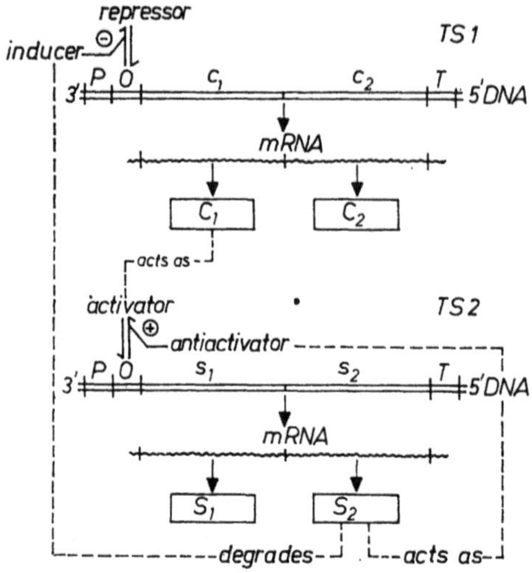

Fig. 22. Abstract-ideal model of two coupled transcription systems (TS).

p "classical" promoter
o operator
T end terminator
c_1, c_2, s_1, s_2 cistrons, which code fort he proteins C_1, C_2, S_1, S_2
⊖ negative and ⊕ positive transcription control

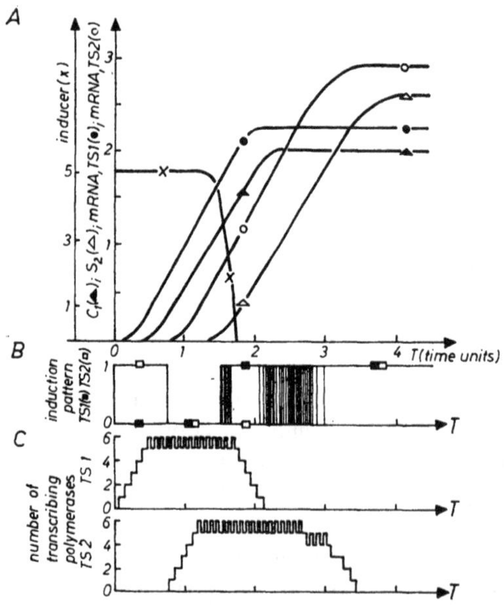

Fig. 23. Behaviour of two coupled transcription systems (TS) — given in Fig. 22 — after addition of inducer at zero time; concentrations of inducer, C_1, S_2 and mRNA in relative amounts.

induction pattern: TS repressed (1), TS induced (0)

systems is illustrated by the induction patterns in Fig. 23b. While TS2 is switched on abruptly, because by reason of the temporary absence of the antiactivator already the first C_1-operator interaction has initially a permanent effect, above all TS2 and also TS1 is switched only after some time of "flickering" between "on" and "off". This is due to the stochastically occurring interactions of the respective molecules, depending on their given concentrations. In the activation phase, 20 polymerases for TS1 and 26 polymerases for TS2, as is obvious from Fig. 23c, come to initiation.

Therefore, the overall system given in Fig, 22 comprises two mutually connected regulation circuits:

1. coupling of the two TS:
   inducer $\rightarrow$ mRNA(TS1) $\rightarrow C_1$ (activator) $\rightarrow$ mRNA(TS2) $\rightarrow S_2 \rightarrow$ inducer
2. feedback TS2:
   activator $\rightarrow$ mRNA(TS2) $\rightarrow S_2$ (antiactivator) $\rightarrow$ activator

by which a time-shifted, coordinated expression of the c_2- and s_1-coded proteins, which may be elements of a certain genetic program, are obtained in response to a certain stimulus from the environment (inducer).

References

GOLDTHWAIT, D. A., D. A. ANTHONY and C. W. WU (1970): Studies with the RNA polymerase. In: RNA Polymerase and Transcription (SILVESTRI, L., ed.), 10 (1st Lepetit Colloquium), North-Holland, Amsterdam.

NEUMANN, H. (1973): Simulationsmodell zur Beschreibung einer prokaryotischen Transkriptionseinheit, Diplomarbeit, Humboldt-Universität, Berlin.

NEUMANN, H. (1976): Mathematisch-kybernetisches Modell prokaryotischer Transkriptionssysteme, Dissertation, AdW der DDR, Berlin.

NEUMANN, H., and U. KREISCHER (1975): Simulationsmodell zur Beschreibung einer procaryotischen Transcriptionseinheit. Biom. Jour. **17**, 373–390.

RÜGER, W. (1972). Die Transcription der genetischen Information und ihre Regulation durch Proteinfaktoren. Angew. Chemie **84**, 961–972.

SCHÄFER, R., and W. ZILLIG (1973): The effects of ionoc strength in termination of transcription of DNAs from bacteriophages T_4, T_5, and T_7 by DNA dependent RNA polymerase from *E. coli* and the nature of termination by factor ρ. Eur. Journ. Biochem. **33**, 215–226.

SCHÄFER, R., W. ZILLIG and K. ZECHEL (1972): On the function of DNA-dependent RNA polymerase. In: Polymerization in Biological Systems, Ciba Foundation Symposium 7, published by ASP (Elsevier, Excerpta Medica, North-Holland) Amsterdam, 41.

SCHÄFER, R., W. ZILLIG and K. ZECHEL (1973): A model for the initiation of transcription by DNA-dependent RNA polymerase from *E. coli*. Eur. Journ. Biochem. **33**, 207–214.

ZILLIG, W., K. ZECHEL, D. RABUSSAY, M. SCHACHNER, V. S. SETHI, P. PALM, A. HEIL and W. SEIFERT (1970): On the role of different subunits of DNA-dependent RNA polymerase from *E. coli* in the transcription process. In: Cold Spring Harbor Symp. Quant. Biol. **35**, 47–58.

6. SIMTRA — a programming system for modelling of molecular genetic processes

U. KREISCHER, H. NEUMANN

Contents

1. Introduction

Often the experimental research in the field of molecular-biology is limited for object-related, technical, and technological reasons, or just for pure lack of time. In these cases only subsystems can be investigated and as result a large variety of individual data is obtained without realizing the global relations. Simulation models basing on mathematical models and being imaged on a computer offer the possibility to overcome these limiations. They describe the essential properties of the original system and permit an experimental research work to be carried out several times and with a clear aim in such a way, that the wanted information concerning the original system can be obtained with the help of these models (see KARPLUS, 1976 and SCHWARZE, 1977). Getting to relevant results about the causal relations between the microscopic structure of the molecular genetic objects and their macroscopic properties it is necessary to consider the real objects on the molecular imaging level (see BELLMANN, 1976). In view of the intricacy and complexity of the molecular genetic systems the realization of the simulation results in a considerable modelling expense. For the purpose of a reduction of this expense the systems were examined with regard to repeated structures with the former being abstracted and implemented in the sense of relatively independent subsystems (see KREISCHER and KNIJNENBURG, 1976). In doing so the aspect of a multiple application of the submodels and their connection with more complex models was expecially investigated. In particular the last demand resulted in the necessity to make suitable means available to the user for the purpose of a simple handling of the submodels. For this reason terms of language were defined and allocated to the corresponding submodels, so that the user would be able to realize his molecular genetic simulation models directly on the computer employing these terms of language (model instructions) mentioned above. The terms of language refer to the formulation of the model, the control of the processing, and the output. They form in their totality the SIMTRA-language and represent that component of the programming system SIMTRA (**SIM**ulation of **TRA**nscription systems)

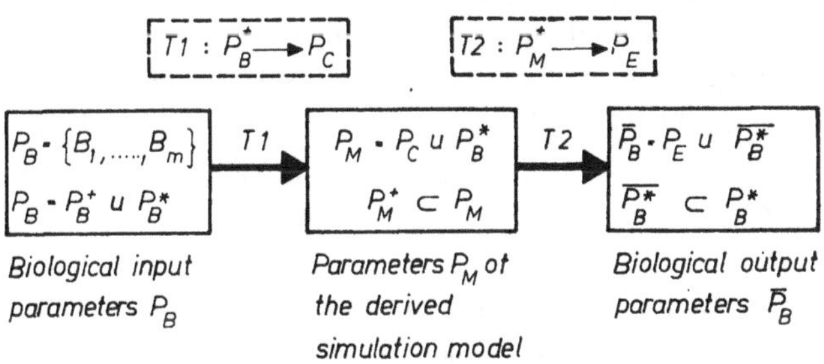

Fig. 1. Parameter transformations in the programming system SIMTRA.

with which the user will work (see KREISCHER, 1978). By means of this language the user describes his molecular genetic simulation model with the parameters of the model $P_B = \{B_1, \ldots, B_m\}$ being exclusively of biological nature. A subset $P_B^+ \subset P_B$ of these biological parameters is brought into a set of controlling parameters P_c of the derived simulation model[1] by means of transformation routines. The set P_B^* of biological parameters which is complementary to P_B^+ turns out to be a direct constituent of the model parameter P_M of the derived simulation model. The latter does possess the same degree of complexity as the simulation model; however, it is much more complicated than the simulation model. The resulting data $\overline{P}_B$ are again of biological nature after all. They consist of the set P_E of efficiency parameters, i.e. parameters enabling statements on the efficiency of the molecular genetic system, and of a set $\overline{P_B^*}$ of biological parameters $\left(\overline{P_B^*} \subset P_B^* \subset P_B\right)$. The efficiency parameters will be calculated from the model parameters $P_M^+ \subset P_M$ of the simulation model by means of transformation routines.

The transformations T1 and T2 will be carried out automatically by means of the programming system SIMTRA, so that the user of SIMTRA is working exclusively with molecular genetic parameters.

[1] Here the derived simulation model means that simulation model is implemented for processing on the computer.

1.1. Efficiency and aims of the programming system SIMTRA

SIMTRA is a programming system for the simulation of processes of gene expression with the emphasis lying on the description of prokaryotic transcription systems including their regulation. The system comprises the present state of the knowledge of the real object.

The programming system SIMTRA

— presents terms of language, which make it possible
  - to describe the structure of transcription systems
  - to represent the sequential molecular genetic processes of the transcription together with their corresponding control mechanisms
  - to describe posttranscriptional processes of the system of gene expression
  - to simulate coupled transcription systems (together with their feedback mechanisms);
— yields the possibility to set up models for the real objects to be investigated without any knowledge in the field of simulation and programming. The steps of modelling and programming will be carried out simultaneously by means of SIMTRA;
— represents the preliminary step of a programming system for the simulation of the global system of the gene expression.

Formulating his molecular genetic simulation models by means of the programming system SIMTRA the user can utilize several advantages in the field of modelling and programming:

— the programming of a molecular genetic problem corresponds in the main part to the establishment of a flow chart of the sequential molecular genetic processes
— the model parameters of the simulation model are exclusively of biological nature;
— modifications both of the simulation model and of the model parameters can be carried out with a minimum programming expence;
— the programming system based on a time sequencing control cares for a correctly passing process. In this case SIMTRA calculates automatically for every molecular genetic model an optimal simulation time interval;
— the functions being used for the representation of concentration distributions can be both of empirial and theoretical, discrete or steady nature;
— the possibilities for storage of model states (and the model itself too) and for the continuation of the simulation are available;
— the output of the results is carried out in a standardized form demanding no additional programming expense. It comprises characteristic statistical data of the simulated problem;
— a user's output enables the modification of the standard output and the printing of essential values;
— the statements of the programming system SIMTRA can be coded unformattedly on punched cards;
— knowledge of the operating system BAMOS of the BESM6 computer[1] is not necessary for programming, because the user can call the programming system SIMTRA

[1] The Soviet computer BESM6 is a one-address computer with a high computing speed (nearly 1 million operations per second) and with a main storage capacity of 32 up to 128 k words (1 BESM-word corresponds to 50 bit). Seven selector channels are available for connection of peripheral storages (drums and disks) and I/0-devices.

by means of a standard card set (by which the system connection is realized) in a simple manner;
- a comprehensive treatment of errors facilitates the correction of an user's program formulated in SIMTRA. An automatic error correction is carried out to a limited extent. The possibility of this correction is based on a "useful" redundancy of the SIMTRA-language and the field-specific background of the programming system.

Because the image of the transcription system is carried out on a molecular level, SIMTRA is distinguished by the following essential properties:
- the action of the controlling mechanisms of the system accounts for the behaviour of a given transcription system;
- at any random moment of a simulation experiment it is possible to find out, whether controlling mechanisms intervene in the transcription process and if the answer is in the affirmative which of them do act to what extent;
- if two of three model components of a molecular genetic system — structure of a given transcription system, global behaviour and input — are known at the same time, the third one can be directly ascertained;
- global connections can be derived from a sum of single experimental findings;
- the ambiguity of conclusions is greatly reduced.

1.2. Possibilities for action

The programming system SIMTRA can be used both theoretically and practically in the field of molecular genetic research work for several different possibilities of application.

The programming system SIMTRA
- admits a description of the transcription together with its regulation in the same manner as this in the case in the cell (in vivo or in vitro). Adjusting sufficiently the model or the parameters of the model to a given real object laboratory attempts can be replaced by simulation experiments and predictions concerning the result of such attempts can be made. Moreover, simulation experiments can be used for the explanation of such problems, which cannot be treated practically for object-related, technical, and technological reasons, or just for pure lack of time;
- permits transition from a qualitative to a quantitative representation provided that the corresponding dynamics between model and experiment is fulfilled by advancing concretization of the model parameters;
- is suited for testing the influence of chemical reagents, drugs or cell internal factors on the RNA synthesis, if these substances are effective on the transcription level;
- offers possibilities for the investigation, whether eucaryotic transcription systems and regulation mechanisms can be sufficiently described at least on principle or partly concrete by prokaryotic transcription systems;
- can be used for the solution of problems of differential geneexpression and hormon action under the condition of corresponding adjustment.

2. The prokaryotic transcription system

The programming system SIMTRA is based on a mathematico-cybernetical model, developed by Neumann and Kreischer (1982), which describes the prokaryotic transcription system in its essential properties as follows:

I. The transcription unit is a DNA sequence to be transcribed and is characterized by the following general constructive principle:

1. Regulation sequence consisting of an
   - operator region $o_1, \ldots, o_{\max}$ ($\max \in \{1, \ldots, j''\}$), which may be absent too, and a
   - promoter region $p_1, \ldots, p_{\max}$ ($\max \in \{1, \ldots, i + 1\}$)
2. Cistron sequence $c_1, \ldots, c_k$ with
3. Intermediate terminators $\tilde{t}_1, \ldots, \tilde{t}_{s-1}$, which may separate cistrons and/or cistron groups from each other
4. End terminator sequence $\tilde{T} (= \tilde{t}_s)$

II. The transcription of a DNA sequence may be subject to the following control mechanisms

1. operator-localized
   - negative control accomplished by operator-repressor-inducer interactions in which the transcription is prevented in the presence of the repressor and in the absence of the inducer,
   - positive control accomplished by operator-activator-antiactivator interactions in which the transcription is ensured in the presence of the activator and in the absence of the antiactivator

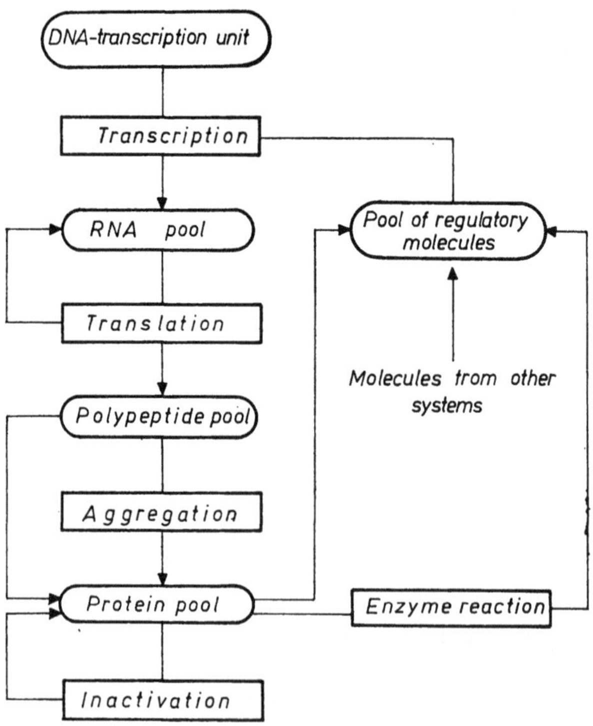

Fig. 2. Formalized pathway for expression of a transcription unit.

2. promoter-localized positive control accomplished by
   - promoter-polymerase (core or holoenzyme)
   - promoter-CAP/cAMP-(polymerase)
   - polymerase-polymerase subunit (sigma factor) interactions
3. elongation control by
   - changes in the transcription velocity of the polymerase along the transcription unit after mutation loci and/or intermediate terminators and/or specific nucleotide sequences in the codogenic DNA strand,
   - The concentration of ribonucleoside triphosphates
4. terminator-localized control
   - accomplished by termination and antitermination factors, where a terminator-located polymerase is released in the presence of the termination factor and in the absence of the antitermination factor, runs over the terminator in the presence of the antitermination factor and remains at the terminator in the absence of the two factors

   and, as a special case,
   - elicited by mutations with termination character in the codogenic DNA strand

III. For a sufficient prepresentation of the transcription system it is necessary to consider the posttranscriptional processes. These processes are formulated only general-

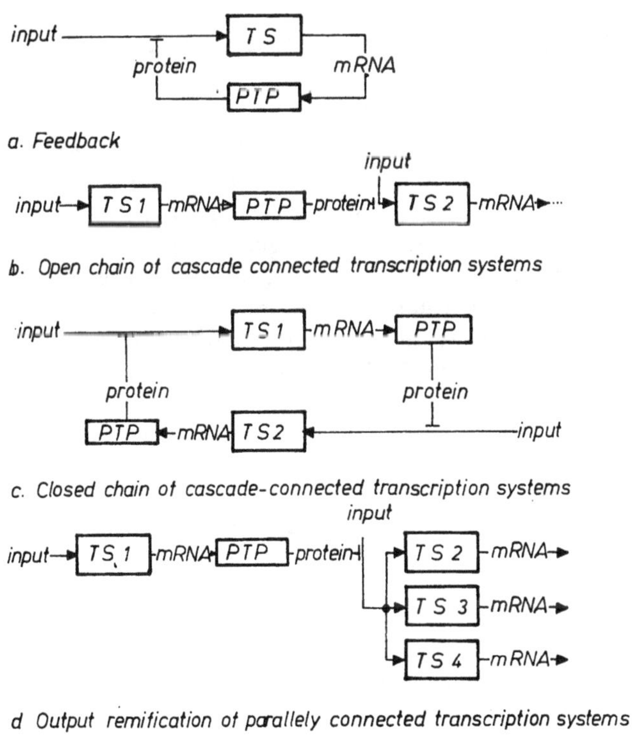

Fig. 3. Coupling strategies between transcription systems.
TS Transcription System
PTP Post Transcriptional Processes

ly, that is, without giving a detailed representation of structural elements and regulation mechanisms of the underlying systems:

1. Synthesis of the polypeptide chains (translation)
2. Aggregation of the polypeptide chains (formation of the active form of proteins built complexly)
3. Inactivation of the proteins
4. Degradation of translated mRNA chains

IV. Proteins coded by the cistrons of a transcription system and synthesized as a result of the posttranscriptional processes are able to mediate directly or indirectly between the output (mRNA) of this transcription system and the input (regulatory molecules)

1. of the same transcription system (feedback) and /or
2. of one or several other transcription systems (coupling), where
   - open chains of cascade-connected transcription systems
   - closed chains of cascade-connected transcription systems
   - input and output ramification of parallely connected transcription systems

   are to be discriminated.

Combinations of any number of these types are called networks of transcription systems and the kind of coupling between the transcription systems coupling strategy.

3. Principles of realization of the programming system SIMTRA

3.1. Generation of an executable program

The implementation of a special simulation language demands the necessity of a special translator. A program formulated in SIMTRA is not directly transformed into a machine program, but the transformation is carried out in two steps. In a first step the SIMTRA-program is transformed into an intermediate language implemented in the operating system of the BESM6 computer by means of the special translator, the so-called SIMTRA-Compiler. The second step comprises the transformation of the intermediate language into the load modules by means of the FORTRAN-Compiler. These load modules can be put together by the LOADER of the BESM6 computer with the modules of the SIMTRA — run-time system being available in the core image library to an executable program. The fig. 4 illustrates these facts and shows, that both the multi-stepped transformations of a molecular genetic model described just now and the properties of the computer remain by far invisible for the user.

The three essential constituents of the programming system SIMTRA are:

— **SIMTRA-Compiler**

The SIMTRA-Compiler, implemented in ALGOL, was designed as multi-pass compiler. The three passes of the compiler possess the following basic functions:

pass 1 — Establishment of the symbol tables
pass 2 — Generation of statements in FORTRAN
pass 3 — Combining the statements generated to complete FORTRAN-programs

The SIMTRA-Compiler contains also the routines for realization of the transformation T1 (see section 1).

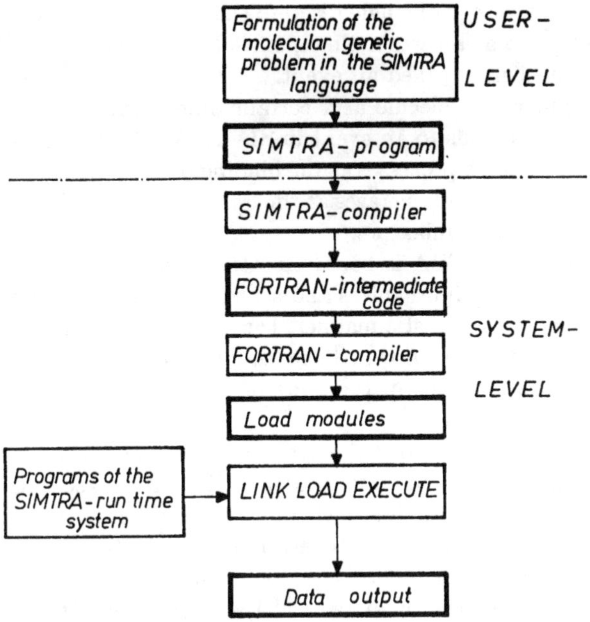

Fig. 4. Execution of a molecular genetic problem, formulated by a user, by means of the programming system SIMTRA.

– **SIMTRA – run-time system**

  The subroutines of the SIMTRA – run-time system, the programming of which were carried out in FORTRAN, are needed for the processing of the SIMTRA-programs. They refer to the time flow control, the realization of elementary processes (see FRANK, 1977) and housekeeping functions (e.g., computation of mathematical functions, generation of random variables, calculation of statistical values, arithmetical operations etc.).

  The SIMTRA – run-time system contains also the routines for realization of the transformation T2 (see section 1).

– **SIMTRA-Compiler – interface module**

  The interface module programmed in ASSEMBLER provides information from the operating system of the BESM6 computer for the SIMTRA-Compiler and calls the latter for processing.

The programming of SIMTRA was carried out in FORTRAN and ALGOL, because high-level languages are much more independent from the speciality of hardware than machine languages and because these languages are implemented on a wide variety of different computers.

3.2. About time flow control

SIMTRA is a programming system with a time-discrete simulation flow, i.e. the time flow takes place in single steps (discrete time intervals). This form of time flow is useful at detailed modelling of systems, in which the relations between the elements are achieved through the exchange of single discrete elements. It can then be supposed, that an exchange takes place at a certain time point. The state of the system changes only at such a time point. This is called an event.

The activity of a polymerase molecule at a certain time point or the change of the power of protein pools corresponds to an event in SIMTRA. The activities of the polymerases contain for instance such processes as the binding of the polymerases at the promoter region, the blocking of the polymerases at the operator region of the release of the polymerase at the intermediate terminators.

The programming system SIMTRA is based on a time sequencing control (period or interval orientated method), which organizes and schedules events so, that a simulated system moves through time in a realistic manner. During this an internal clock, the so-called system clock, controls the time flow. At the beginning of the simulation this clock shows zero, later it shows the number of the units of the simulation time, which went away since the beginning[1].

Regarding the time sequencing control the system clock is advanced in fixed equal steps of time, the so-called simulation time intervals ΔT. After executing all calculations at T_i with $0 \leqq i \leqq n$, $n \in N$, and $T_0 = 0$ the system clock is set upon the time point $T_{i+1} = T + \Delta T$. After that the calculations falling into the time interval (T_i, T_{i+1}) have to be carried out, and so on. It is characteristic of this procedure that at any time point T_i all conditions for the realization of events have to be tested. It is supposed that events occur only at time point T_i and one generally ignores their existing distribution over the interval ΔT (see fig. 5). The correct selection of the time interval ΔT is therefore of great importance for accuracy of the modelling of the original, for a false ΔT

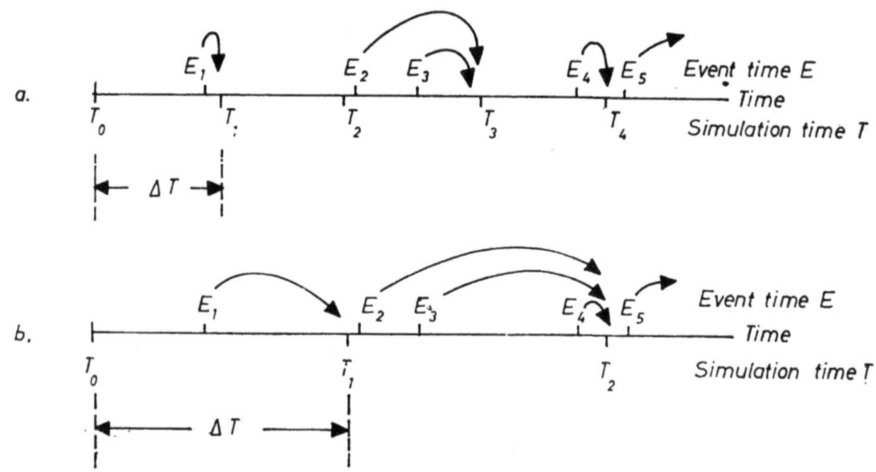

Fig. 5. Time-oriented simulation

a. with simulation time interval ΔT

b. with simulation time interval $2 * \Delta T$

Concerning the time-oriented simulation the simulation time takes on subsequently values $T_0, T_1, \ldots, T_n$ corresponding to the simulation time interval ΔT chosen, which need not agree absolutely with the time points $E_1, E_2, \ldots, Em$ of the events.

---

[1] Here the term simulation (or system or real or clock) time means the time being measured in the simulated system rather than the time needed for the execution of the simulation run.

can lead to logically impossible model states during the simulation. If the time interval ΔT is too large for instance certain events in the simulation model will take place at the same time which in reality occur at different time points. In this case the simulation model distinguishes in its "actions" from the original.

The correct selection of the time interval ΔT is especially complicated at coupled transcription systems, because the molecular genetic processes can occur time shifted and/or different speed in them, i.e. each system can have its own different time scale. The interaction between the systems without deviation from the original can only be described, if a "synchronization" of the processes takes place. Thereby the time interval has to be selected so, that a localization of the events in the space-time-continuum is possible for each transcription system. Therefore the molecular genetic structure of the DNA, especially the discrete order of the nucleotides, has to be taken into account for the calculation of the simulation time interval ΔT.

Owing to these restrictions ascertainment of an optimal time interval becomes very complicated for the user. Therefore the determination of the time interval is carried out by the programming system SIMTRA itself. Above all the molecular genetic parameters of the TIME-statements are needed for it. That way the user influences the value of ΔT. The automatic calculation of ΔT guarantees that the portion of computation time which possibly at very small ΔT does not change the system state, which means some kind of idling time, can be minimized.

3.3. About storage allocation

In the programming system SIMTRA there are standard declarations for some statements of the system for storage allocation. In doing so both the allowed number of elements of each type of statement and the organization of storage location was fixed

Table 1. SIMTRA-statements with fixed storage allocation.

type of statements	statement	maximum number of the restricted statements
definition	FUNCTION	92
	LIST	100
statement	POOL	100
	RNGENERATOR	15
	SCALAR	20
	VARIABLE	100
block	TIME	5
	INDUCE	5
statement	INITIATE	5
	ELONGATE	5
	TERMINATE	5
	TRANSLATE	5
	AGGREGATE	5
	INACTIVATE	5
	COUPLE	5
output	SYSTEMOUTPUT	5
statement	SELECTOUTPUT	5

standardly within that storage area intended for one element. Even if the corresponding statement does not occur in the simulation program, the storage location is occupied. These declarations fixed by standard cannot be influenced by the user (see table 1). These are not explicit restrictions for the statements not mentioned in this table. But their number is implicitly restricted by the logic of the simulation model and the size of the program area, in which the executable SIMTRA-program is stored. This kind of memory space administration is of advantage for the user, who need not take care about the memory space requirements of his molecular genetic model, i.e. need not deal with the internal storage organization of the computer.

Table 1 shows, that at present the modelling of five transcription systems (including the corresponding posttranscriptional processes), which can all be coupled with each other, is simultaneously possible.

Regarding the usage of FUNCTION-statements (one FUNCTION-statement describes one respective element of the molecule group of the considered molecular genetic system) it has to be taken into account, that the 92[1] standardly fixed functions are

Table 2. Distribution of FUNCTION-statements onto the molecule groups defined in SIMTRA.

molecule group	maximum number of the corresponding FUNCTION-statements
polymerase	2
polymerase subunit	2
CAP	1
cAMP	1
nucleotide	1
effector* (repressor, activator)	20
anti-effector* (inducer, anti-activator)	20
termination factor*	20
antitermination factor*	20

distributed as follows acc. to table 2. The elements of the molecule groups can be dynamically distributed within the bounds fixed by table 2 onto the considered transcription systems. For the distribution of the molecule groups marked with a asterisk '*' onto the transcription systems the following mathematical relation is given:

$$\sum_{i=1}^{S} n_{ij} \leqq N_j^{\max}.$$

Hereby mean:

S — maximum number of the considered transcription systems

n_{ij} — number of the elements of the molecule group j corresponding to the transcription system i

$N_j^{\max}$ — maximum number of elements of the molecule group j.

[1] Corresponding to the possible number of modelled transcription systems there are five FUNCTION-statements for the description of the transcription velocity of the polymerase along the transcription unit.

4. SIMTRA-program constitution

4.1. Notation and terminology

In this article an extended BACKAUS-NAUR-notation (see MAURER, 1969) is used for the description of global syntactic relations of SIMTRA. But the detailed description of the syntactic structure of SIMTRA-statements is carried out with the help of a unitary representation scheme, introduced in chapter 5.

The extended BACKAUS-NAUR-form, applied here, contains the following metalanguage-symbols:

::= — definition symbol
⟨ ⟩ — limiter of a metalanguage-variable
| — alternative-symbol
[ ] — limiter of optional usable elements
... — repetition-symbol

These symbols are not parts of the SIMTRA-language.

Interpretation:

Metalanguage-variables are used for the denotation of syntactic constructions. They are represented by strings (letter sequences and alphanumerical characters) so that one can acknowledge the semantics of the constructions out of the names of variables. The metalanguage-variables for the description of the SIMTRA-language are enclosed in angular brackets ⟨ and ⟩. Thereby it is possible to distinguish between metalanguage-variables and the basic symbols of the SIMTRA-language.

A metalanguage-variable standing on the left of the definition symbol ::= can be replaced by a symbol standing on the right of the symbol sequence.

The symbol | is the alternative sign. Symbol sequences in square brackets [and] can but need not occur at the corresponding positions.

The three points ... allow to omit some of the alternatives in the notation, if the rules of forming the alternatives can be perceived without doubt.

4.2. Syntactic structure of the SIMTRA-statements

In the programming system SIMTRA one program consists of a sequence of SIMTRA-statements. Corresponding to their mode of action one distinguishes four different statement types:

⟨SIMTRA-statement⟩ ::= ⟨regime statement⟩ | ⟨definition statement⟩|
⟨block statement⟩ | ⟨output statement⟩

These statement types possess different syntactic structures. But each SIMTRA-statement has a key word; its semantics characterizes the action of the statement.

Syntactic structure of the regime statements

The regime statements consist only of key words, that is their syntactic structure is given by:

⟨regime statement⟩ ::= ⟨key word⟩

The semantics of the statements depends on the appropriate key word.

Syntactic structure of the block statements

⟨block statement⟩ ::= ⟨key word⟩ ⟨parameter list⟩
⟨parameter list⟩ ::= ⟨list of necessary parameters⟩
[,⟨list of additional parameters⟩]
⟨list of necessary parameters⟩ ::= ⟨necessary parameter⟩
[,⟨list of necessary parameters⟩]
⟨list of additional parameters⟩ ::= ⟨additional parameter⟩
[,⟨list of additional parameters⟩]
⟨necessary parameter⟩ ::= ⟨constant⟩ | ⟨reference symbol⟩ | ⟨short key⟩
⟨additional parameter⟩ ::= ⟨constant⟩ | ⟨reference symbol⟩

The parameters of the block statements are divided in necessary and additional parameters. If necessary parameters are absent in the parameter list of a block statement, this statement becomes ineffectual. The additional information can but need not be made. They define additional object-specific facts and activate subroutines, which generally are not called during processing.

In general the following specifications are applied to the parameters of the parameter list:

- The parameters are positional parameters, that is, the sequence of the parameters within the parameter list is significant.
- The meaning of each parameter is fixed by the corresponding statement itself, that is, it depends on the appropriate key word.
- The number of parameters in SIMTRA-statements differs.

Syntactic structure of the definition statements

⟨definition statement⟩ ::= ⟨identifier⟩ ⟨keyword⟩ ⟨parameter list⟩
[⟨statement specification⟩]

Whether a definition statement has a specification or not depends on the corresponding key word of this statement. The same dependence is valid for the syntactic structure and the semantics of the statement specification. Furthermore, one can fix the following conventions about the syntactic constructions standing above:

⟨identifier⟩ ::= ⟨symbolic name⟩ | ⟨identification number⟩
⟨parameter list⟩ ::= ⟨list of necessary parameters⟩
[,⟨additional parameter⟩]
⟨list of necessary parameters⟩ ::= ⟨necessary parameter⟩
[,⟨list of necessary parameters⟩]
⟨necessary parameter⟩ ::= ⟨constant⟩ | ⟨reference symbol⟩ | ⟨short key⟩
⟨additional parameter⟩ ::= ⟨constant⟩ | ⟨reference symbol⟩

The above definitions about sequence, meaning and number of the parameters are also applied here to the parameters of the parameter list.

Syntactic structure of the output statements

⟨output statement⟩ ::= ⟨key word⟩ ⟨parameter list⟩
⟨parameter list⟩ ::= ⟨list of necessary parameters⟩
[,⟨additional parameter⟩]

⟨list of necessary parameters⟩ ::= ⟨necessary parameter⟩
[,⟨list of necessary parameters⟩]
⟨necessary parameter⟩ ::= ⟨constant⟩ | ⟨reference symbol⟩ | ⟨short key⟩
⟨additional parameter⟩ ::= ⟨constant⟩

Concerning their sequence, meaning, and number the parameters of the parameter list are also fixed by the definitions standing above.

4.3. Parameters

Numbers or logical values are assigned to the corresponding parameters of subroutines of the SIMTRA — run-time system by the entering of molecular genetic parameters and specification parameters into the parameter lists. The entry of the parameters can be carried out directly by entering of numbers into the corresponding parameter subfields or, indirectly, by references. One distinguishes three different types of parameters:

⟨parameter⟩ ::= ⟨constant⟩ | ⟨reference symbol⟩ | ⟨short key⟩

Note:

1. All entries in the parameter fields are made without dimensions.
2. The parameters are all unsigned.

4.3.1. Constants

One distinguishes three different types of constants in SIMTRA:

⟨constant⟩ ::= ⟨integer constant⟩ | ⟨real constant⟩ | ⟨special constant⟩

Integer constants

Integer constants have in SIMTRA-statements the following syntactic structure:

⟨integer constant⟩ ::= ⟨digit⟩ [⟨integer constant⟩]
⟨digit⟩ ::= 0|1|...|9

The set of the possible ⟨integer constants⟩ is restricted in a special implementation. Thus, it is a subset of natural numbers.

Note:
The set of all integer constants is characterized in the following by C_N.

Real constants

Real constants have in SIMTRA-statements the following syntactic structure:

⟨real constant⟩ ::= .⟨digit sequence⟩ | ⟨digit sequence⟩.|
⟨digit sequence⟩ . ⟨digit sequence⟩
⟨digit sequence⟩ ::= ⟨digit⟩[⟨digit sequence⟩]
⟨digit⟩ ::= 0|1|...|9

The set of the possible ⟨real constants⟩ is restricted in a special implementation. Thus, it is a subset of the positive real numbers.

Note:
The set of all real constants is characterized in the following by C_R.

Special constants

The special constants of the programming system SIMTRA have the following syntactic structure:

⟨special constant⟩ ::= ⟨integer constant⟩ ⟨specification symbol⟩
⟨specification symbol⟩ ::= + | −

Note:
The set of all special constants is characterized in the following by C_S.

4.3.2. Reference symbols

The functions, arithmetical variables, pools, lists, scalars, and random number generators of the programming system SIMTRA, realized by the definition statements, can be directly referenced in some definition, block and output statements by the so-called reference symbols (RFS). The latter have the following syntactic structure:

⟨reference symbol⟩ ::= ⟨standard symbol⟩ $ ⟨identifier⟩|
⟨special reference symbol⟩
⟨standard symbol⟩ ::= FN|LI|PL|RN|SC|VR
⟨identifier⟩ ::= ⟨identification number⟩ | ⟨symbolic name⟩
⟨special reference symbol⟩ ::= TM$CLOCK|TM$DELTA

The type of element referred to is fixed by the standard symbol. The identifier specifies the special element of the type of element concerned by means of the identification number or the symbolic name. During the processing, the reference symbols will be replaced by the values, which the corresponding elements posses at the simulation time point concerned.

Identification number

The identification number has the following syntactic structure:

⟨identification number⟩ ::= ⟨digit⟩ [⟨identification number⟩]
⟨digit⟩ ::= 0|1|...|9

The value j of the identification number depends on the type of element and ranges from 1 to j^* (see table 3).

Table 3. The reference symbols of the programming system SIMTRA.

type of element	statement	RFS	j^*
function	FUNCTION	FN$j	92
list	LIST	LI$j	100
protein pool	POOL	PL$j	100
random number generator	RNGENERATOR	RN$j	15
scalar	SCALAR	SC$j	50
arithmetic variable	VARIABLE	VR$j	120

Note:

With the help of the special reference symbol TM$CLOCK the user can directly call the system clock in order to obtain the actual simulation time and by means of the special reference symbol TM$DELTA he can refer to the simulation time interval ΔT.

Symbolic name

The symbolic name (sn) has the following syntactic structure:

⟨symbolic name⟩ ::= ⟨letter⟩ [⟨symbolic name⟩] | ⟨symbolic name⟩ ⟨digit⟩
⟨letter⟩ ::= A|B|...|Z
⟨digit⟩ ::= 0|1|...|9

Symbolic names denoting distinct element types must differ over their first characters.

Note:

The set of all possible symbolic names sn is characterized in the following with S_{name}.

4.3.3. Short keys

Short keys (SK) are short informations consisting of a letter sequence and having a certain meaning. They are used as parameters in definition, block and output statements. The short key, which is a parameter of a statement, depends strongly on the semantics of this statement.

The following letter sequences are defined as short keys in the programming system SIMTRA:

⟨short key⟩ ::= ADD|AE|AEF|ALL|AMP|ATF|BOTH|CAP|EF|EL|EQ|
EZ|GE|GL|GT|LE|LT|NE|NRPM|NRPT|NTI|NUC|PRP|
RP|RPM|RPT|SOR|SPR|SQ|STO|SUB|TF|TOA|TRP|TV

4.4. Statement sequence in SIMTRA-programs

The simulation of molecular genetic systems with the help of the programming system SIMTRA corresponds with the establishment of a program flow chart adjusted to SIMTRA, with the flow chart defining the time succession of all program steps for the input, storage, processing and output. Above all, a SIMTRA-program must contain those statements, which are necessary for the modelling of a molecular genetic system; these are definition and block statements. The order of the block statements determines the logic flow in the simulated system. In general, each SIMTRA-program has the following syntactic structure:

⟨SIMTRA-program⟩ ::= ⟨program head⟩
⟨definition statements⟩
⟨block statements⟩
⟨output statements⟩
END

The sequence of the different statement types is defined by the above syntactic rule. Additionally there are regime statements (see section 5.1). The position of regime statements corresponds in the SIMTRA-program with their function. They can stand before, between or after definition, block or output statements.

The other syntactic constructions are given by:

⟨program head⟩ ::= PROGRAM ⟨program identifier⟩
⟨program identifier⟩ ::= ⟨letter⟩[⟨program identifier⟩]|
⟨program identifier⟩ ⟨digit⟩
⟨letter⟩ ::= A|B|...|Z
⟨digit⟩ ::= 0|1|...|9

Note:

Program identifiers denoting distinct programs must differ over their first 6 characters.

Note the following definitions for the structure, the translation and the execution of SIMTRA-programs:

1. Regime statements are executed during the compilation. After having been carried out the regime statements and the operations connected with them have no longer any influence on the further execution of the SIMTRA-program.
2. Block statements are executable statements. The fixed order of the block statements does not change during the execution of the SIMTRA-program.
3. The output statements and the definition statements does not influence the execution of a SIMTRA-program.
4. At least four definition statements, three different block statements and one output statement are needed for a sensible simulation of a molecular genetic system.

Additional comments are possible in a SIMTRA-program. The comments may comprise any letters, digits, or symbols available on the machine, including blanks of course. It is not significant for the processing of a SIMTRA-program.

4.5. Some basic coding rules

Only one punched card is used for each SIMTRA-statement. Exceptions are definition statements; for their specifications so-called following (or sequence or continuation) cards are needed. The sequence of the information in tqe punched card is significant, e.g. the statements must be coded in punched cards corresponding to their syntactic structure (see section 4.2). On principle there are in SIMTRA two possibilities available for the user to code his SIMTRA-statements: the so-called standard coding format (see fig. 6) and the so-called free coding format. The standard coding format can be summarized as follows:

1	2 - 7	8	9 - 20	21	22 - 71	72	73 - 80
	Identifier field		Operation field		Parameter field		Card identification field

Fig. 6. Standard coding format in SIMTRA for eighty-column cards. The columns 1, 8, 21, and 72 contain blank characters.

The identifier field, if not blank, is used to specify the special element of the definition statement by the programmer.

The operation field contains the key word that identifies the statement type. This word is indicative of the operation or function of the card, and it must begin in column 9.

The paramerer field consists of a variable field. The variable field starts in column 22 and consists of as few as zero or as many as eight parameter subfields which are separated by commas. The number of parameter subfields having entries on a given card depends on the statement. The parameter subfields are denoted by the letters A, B, ..., H. Incidentally, the term 'parameter subfield' is often shortened to 'parameter field' for expedience (see also section 4.2). No blank characters are permitted among the entries in a card's variable field, because the SIMTRA-compiler interprets the end of the variable field to occur as soon as it encounters a blank character. Using the standard coding format the blanks are significant.

The so-called following cards for the statement specification of the definition statements have an asterisk in column 1 and a flash in column 2.

The free coding format can be summarized as follows:

In that way the SIMTRA-statements can be coded unformattedly on punched cards with help of the FORMAT-statement (see section 5.1.3). The blank characters do not have then a syntactic meaning, only the order of the information on punched cards is significant.

5. SIMTRA-statements

In the following we deal with the most important SIMTRA-statements (a full description is given by Kreischer, 1977). An unitary representation scheme is used of their definition. The starting point for this scheme is the standard coding format for SIMTRA introduced in section 4.5. The letters A, B, ..., H in the representation scheme are assigned to the parameters of the parameter lists. Additionally, the identifier of each definition statement is characterized by the letter I. Furthermore, the notations introduced in chapter 4 are applied. A vertical double line in the representation schemes divides the necessary from the additional parameters in the parameter lists.

5.1. Regime statements

The regime statements of a SIMTRA-program initiate global regime states, which influence in the following the mode of processing of the SIMTRA-compiler. At the beginning of the compilation of a program a so-called standard state is set up, which can be modifed by regime statements. Regime statements refer to the error analysis by the SIMTRA-compiler and to the mode and form of the object program listing of the SIMTRA-program.

5.1.1. LISTING- and NO LISTING-statement

With the unitary representation scheme the LISTING-statement can be described as follows:

9
LISTING

The LISTING-statement can set up the printing regime for the SIMTRA-compiler. In the LISTING-regime are printed:

- head line
  - compiler version
  - date
- source program with decimal line numbers
- end line
  - name of the program
  - number of input cards
  - compilation time
  - number of errors
- error information

If no LISTING-statement is programmed the printing-regime of the SIMTRA-compiler won't be set up. This state corresponds with the state generated by the NO LISTING-statement:

```
| 9
|-------------
| NO LISTING
```

By using the NO LISTING-statement the logging of the SIMTRA-program is totally supressed; the end line mentioned above is printed only. Errors being detected during the compilation are compulsorily logged and the corresponding error message is printed.

5.1.2. PERCENT LISTING- and NO PERCENT LISTING-statement

The NO PERCENT LISTING-statement

```
| 9
|--------------------
| NO PERCENT LISTING
```

gives the user the possibility to mask the comment cards coded in the SIMTRA-program, so that they don't appear in the object program listing. By using the PERCENT LISTING-statement

```
| 9
|-----------------
| PERCENT LISTING
```

the printing regime for the comment cards is set up.

5.1.3. FORMAT-statement

By using the FORMAT-statement

```
| 9
|--------
| FORMAT
```

the user can code his SIMTRA-statements unformattedly on punched cards. Only if the SIMTRA-compiler works in the so-called FORMAT-regime it is possible to decode unformattedly coded SIMTRA-statements. However, in this case the translation process is rather time-consuming.

5.1.4. CHECK-statement

The CHECK-statement

9
CHECK

gives the user the possibility to bypass restrictions of the error handling in the programming system SIMTRA. In the CHECK-regime

- the syntactic analysis of a SIMTRA-statement will be carried out till the end of the statement even when an error was detected in this statement before.
- an automatic limited error debugging is carried out,
- the error limit is increased to 100 listed errors.

5.2. Definition statements

Definition statements are useful for the definition of functions, scalars, variables, lists, random number generators and pools. With the help of these elements it is possible to describe functional connections, to make numerical computations, to store intermediate results, to fix initial values, to connect model parameters of different parameter classes with each other etc. The elements reflect molecular genetic facts, for instance biological active substances are defined with their help, computation rules fixed, initial values assigned to the pools of active proteins etc.

5.2.1. LIST-statement

In the programming system SIMTRA lists can be defined with the help of LIST-statements. A list L consists of an ordered sequence of groups G_i:

$$L = \{G_1, G_2, \ldots, G_m\} \quad \text{with} \quad m \in \{1, \ldots, 5\}$$

Each group consists of an ordered sequence of elements:

$$G_i = \{E_{i1}, E_{i2}, \ldots, E_{ik_i}\} \quad \text{with} \quad k_i \in \{1, \ldots, 250\}$$

The sequence and number of groups and elements within the groups of a list are defined by the notation rules for the parameters of the statements.

The LIST-statements are useful for the

- Enumeration of model parameters

  Model parameters, which belong to a common parameter class, are often summarized in lists in the programming system SIMTRA, so that they can be referred to in

SIMTRA-statements. If there is no direct connection between these parameters and the parameters of another parameter class within the statement, e.g. if it is only an enumeration of parameters, the entries are always carried out only within one group. In this case is $m = 1$. Note the structure of such a list L:

$$L = \{G_{i1}\} = \{E_{i1}, E_{i2}, \ldots, E_{ik_i}\} \quad \text{with} \quad k_i \in \{1, \ldots, 250\}$$

In the notation rules for the corresponding parameters of the statements at an exceptional enumeration of model parameters this fact is characterized by the term 'SQ-list'.

— Coupling of model parameters

In a SIMTRA-statement the correspondence between model parameters of different parameter classes, which are fixed in lists corresponding to their class, can be expressed by the coupling of these lists. Concerning a coupling of the lists one distinguishes between an element- and group-assigment:

- **Element-assignment**

  In SIMTRA only lists consisting of one and only one group ($m_j = 1$), e.g. which were inscribed into the model parameters of a common parameter class corresponding to the above definitions, can be element-like assigned with each other. As to this kind of assignment the k-th element of the list $L_i(L_i = \{G_{i1}\})$ is uniquely assigned to the k-th element of the list $L_j(L_j = \{G_{j1}\})$:

$$E_{j1k} \leftrightarrow E_{i1k} \Leftrightarrow E_{jk} \leftrightarrow E_{ik}$$

  The number of elements of the lists L_i and L_j must be necessarily equivalent, e.g. $k_i = k_j$.
  In the notation rules for the corresponding parameters of the statements the element-assignment between two lists is characterized by the term 'EL-list'.

- **Group-assignment**

  The presumtion for the group-assignment of two lists $L_i = \{G_{i1}, \ldots, G_{im_i}\}$ and $L_j = \{G_{j1}, \ldots, G_{jm_j}\}$ is, that the groups of list L_i contain only one element:

$$G_{ij} = \{E_{ij1}\} \quad \text{with} \quad j \in \{1, \ldots, m_i\}$$

  As to group-assignment the k-th group of the list L_j is biuniquely assigned to the k-th one-element group of list L_i:

$$G_{ik} \leftrightarrow G_{jk}$$

  The number of groups of the lists L_i and L_j must be necessarily equivalent, e.g. $m_i = m_j$.

  In the notation rules for the corresponding parameters of the statements the group-assignment between two lists is characterized by the term 'GL-list'.

  The syntactic structure of the LIST-statement is given by:

⟨LIST-statement⟩ ::= ⟨list declaration⟩
⟨list specification⟩

5.2.1.1. List declaration

The list declaration defines the structure of the list corresponding to the above notations and contains the identifier of the LIST-statement:

2	9	22
I	LIST	A
$j \vee$ sn		SK

I — Identifier of the list,
with $j \in \{1, \ldots, 100\}$ or sn $\in \mathcal{S}_{name}$

A — Parameter for the definition of the list structure, with
SK $\in$ {SQ, EL, GL}
SQ: Enumeration of model parameters (no coupling)
EL: Coupling of model parameters (element-like)
GL: Coupling of model parameters (group-like)

Note:

The reference to a list is carried out in the SIMTRA-statements by using the reference symbol LI$j or LI$sn.

5.2.1.2. List specification

The list specification contains the list elements, with the latter entering the list specification in a defined order.

The syntactic structure of the list specification is given by:

⟨list specification⟩ ::= ⟨group⟩ / [⟨list specification⟩]
⟨group⟩ ::= ⟨list element⟩ [⟨group⟩]
⟨list element⟩ ::= ⟨constant⟩ | ⟨reference symbol 1⟩
⟨reference symbol 1⟩ ::= ⟨standard symbol 1⟩ $ ⟨identifier⟩|
TM$CLOCK | TM$DELTA
⟨standard symbol 1⟩ ::= FN|PL|RN|SC|VR

Note:

The list specification is coded in the sequence cards. A maximum of 10 sequence cards are allowed for a list specification.

5.2.2. FUNCTION-statement

The description of functional relations between certain molecular genetic parameters takes place in the programming system SIMTRA with the help of so-called functions, which are realized by the FUNCTION-statements. The conception of the function used in SIMTRA corresponds with the mathematical one in such a way, that a function in SIMTRA is a unique mapping of two sets, i.e. a certain set of ordered pairs (x, y) with the attribute, that each x is exactly assigned one y. The user of the programming system has to carry out this assignment himself by describing the functional relations with the help of a sequence of points $P_j(x_j, y_j)$. Additionally, the functions defined in SIMTRA contain information, which are adapted by the biological specifics. The syntactic structure of the FUNCTION-statement reflects this state of affairs:

$$\langle \text{FUNCTION-statement} \rangle ::= \langle \text{function declaration} \rangle \\ \langle \text{function specification} \rangle$$

The function specification contains the necessary pairs (x, y) of the sequence of points needed for the description of the functional relations, and the function declaration characterizes the additional molecular genetic attributes.

5.2.2.1. Function declaration

The fixation

- of the type of the interaction probability of the elements to their respective binding partners in dependence of the concentration, and
- of the saturation concentration, i.e. the concentration from which on the probability of an interaction of the respective molecules with the corresponding partners becomes a maximum which is carried out for the elements of molecular groups of a molecular genetic system by means of the function declaration. The function declaration can be described with the help of the unitary representation scheme as follows:

2	9	22				
I	FUNCTION	A	B	C	D	E
$j \vee$ sn		$k \vee$ RFS	SK	k	k	$k \vee$ RFS

I — Identifier of the function, with $j \in \{1, \ldots, 92\}$ or sn $\in S_{\text{name}}$

A — Number k, $k \in \{1, \ldots, 5\}$, of the transcription system, in which the molecules of the defined function act **or** GL-list (LI$\$i_1 \vee$ LI$\$$sn$_1$) of numbers of the transcription systems, in which the molecules of the defined function simultaneously act

B — Parameter for the characterization of the function, with SK $\in$ {POL, SUB, CAP, AMP, EF, AEF, NUC, TF, ATF, TV}
POL: polymerase
SUB: polymerase subunit
CAP: catabolic activation protein
AMP: 3′, 5′-cyclic AMP
EF: effector (repressor, activator)
AEF: anti-effector (inducer, anti-activator)
NUC: nucleotide
TE: termination factor
ATF: antitermination factor
TV: transcription velocity of the polymerases along the transcription unit

C — Typ k, $k \in \{1, \ldots, 8\}$, of the concentration-dependent interaction probability (see fig. 7)

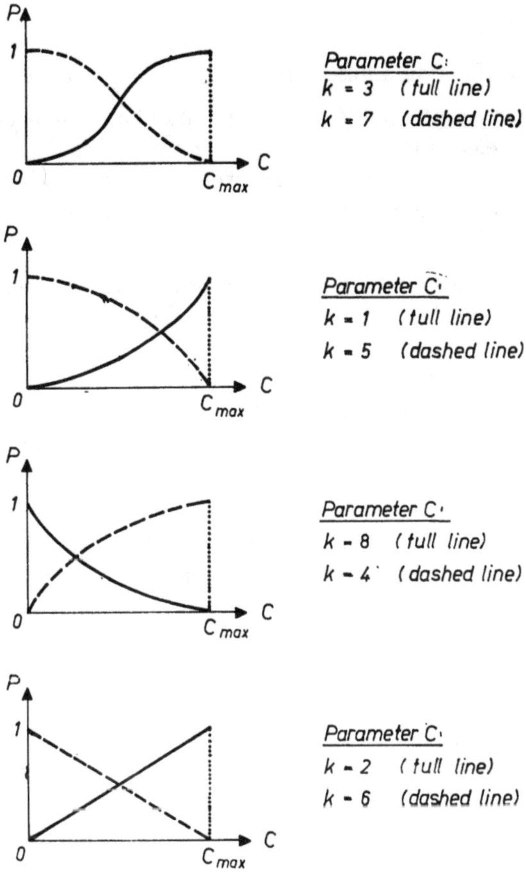

Fig. 7. Types of the concentration-dependent interaction probability P (parameter C of the function declaration).
The dashed lines are only used for repressor in case of a negative operator-localized control or antiactivator in case of a positive operator-localized control.

D — Saturation concentration k, given in number of molecules per volume with $k \in C_N$ or in percents with $k = 100.0$

E — Number k, $k \in \{1, \ldots, 20\}$, of the operator or terminator, if the molecules of the defined function interact with it **or** GL-List (LI\$$i_2$ ∨ LI\$$sn_2$) of numbers of the operators or terminators, if the molecules of the defined function interact with them

*** **Parameter *E* is omitted:**
* The molecules of the defined function do not interact with
* operators or terminators!

Note:

1. The reference to a function is carried out in the SIMTRA-statements by using the reference symbol FN\$j or FN\$sn.
2. If the short key TV is entered in the parameter field B, the parameters C, D and E will be omitted.
3. In the following the terms 'CAP-function', 'Polymerase-function', 'Terminator-function', etc. are used in the notation rules of the block statements, if for these molecule groups a corresponding element was defined by means of a FUNCTION-statement.

5.2.2.2. Function specification

By means of the function specification a description

- of the time course of the concentration $C = C(t)$, i.e. number of molecules per volume over the time for the elements of the molecules groups, and
- of the course of the transcription velocity V (nucleotides per second) along the appropriate transcription unit (with the length l) for the polymerases, i.e. $V = V(l)$,

is carried out.

From the functional dependences of these molecular genetic quantities follows, that in SIMTRA only considerations in the first quadrant of the Cartesian coordinate system are relevant.

For the representation of the functional relationship between the above molecular genetic quantities there are different possibilities, depending on the kind of connection of two points $P_i(x_i, y_i)$ and $P_{i+1}(x_{i+1}, y_{i+1})$ of the sequence of points $P_j(x_j, y_j)$:

D-function specification

At a D-function specification a step-function (piece-wise constant function) is produced by means of the points of the sequence $P_j(x_j, y_j)$ in such a way, as shown in fig. 8.

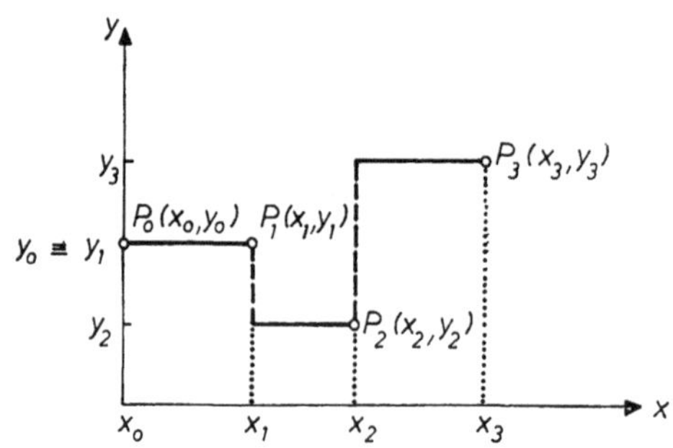

Fig. 8. Example for the D-function specification.

The syntactic structure of this specification is given by:

⟨D-function specification⟩ ::= ⟨origin values⟩ / ⟨sequence values⟩ / |
⟨D-function specification⟩
⟨sequence values⟩ /
⟨origin values⟩ ::= ⟨x_0 — value⟩, ⟨y_0 — value⟩
⟨x_0 — value⟩ ::= 0 | 0.0
⟨y_0 — value⟩ ::= ⟨integer constant⟩|⟨real constant⟩
⟨sequence values⟩ ::= ⟨x-value⟩, ⟨y-value⟩
⟨x-value⟩ ::= ⟨integer constant⟩|⟨real constant⟩
⟨y-value⟩ ::= ⟨integer constant⟩|⟨real constant⟩

Note:

The x,y-coordinates specified on so-called function following cards are successive points on the curve of the function. The values of the independent variable x must be monotonically increasing, i.e. $x_{i+1} > x_i$, but values of y may trend either way.

C-function specification

At a C-function specification the points $P_i(x_i, y_i)$ and $P_{i+1}(x_{i+1}, y_{i+1})$ of the sequence of points $P_j(x_j, y_j)$, $j = 0, 1, \ldots, N$ are connected by continous pieces of curve. Between the points mentioned above either oblique lines or curves with a determined form (see fig. 11) are generated in such a way, as displayed in fig. 10. The user fixes with a so-called course-value (see fig. 9) the shape of the curve between two points. With the D-function specification one can only produce continous functions.

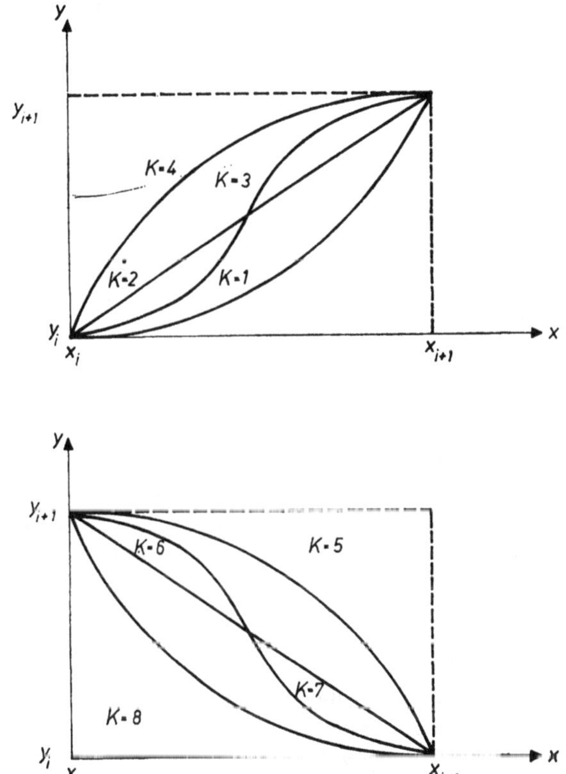

Fig. 9. Courses of the curves in correspondence to the course-value k, $k \in \{1, \ldots, 8\}$.

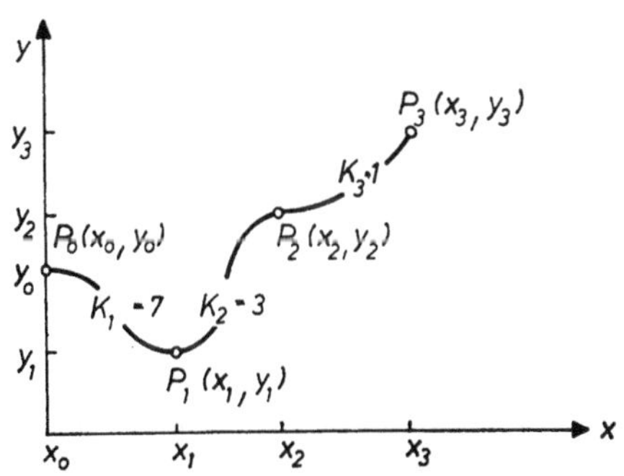

Fig. 10. Example for the C-function specification.

The syntactic structure of this specification is given by:

⟨C-function specification⟩ ::= ⟨origin values⟩ / ⟨sequence values⟩ / | ⟨C-function specification⟩ ⟨sequence values⟩ /
⟨sequence values⟩ ::= ⟨x-value⟩, ⟨y-value⟩, ⟨course-value⟩
⟨course-value⟩ ::= 1|2|...|8

The other syntactic constructions were already carried out above.

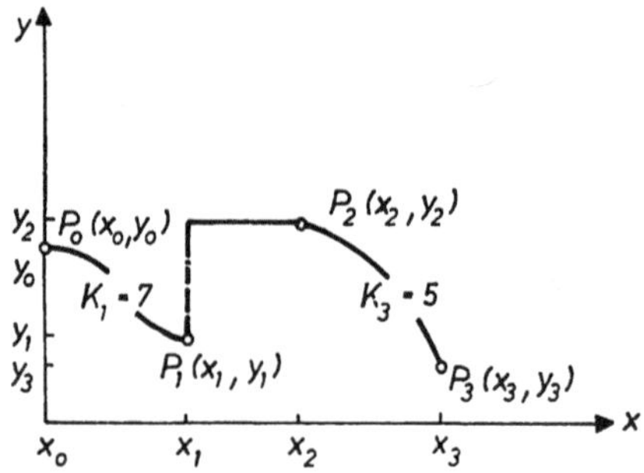

Fig. 11. Example for the DC-function specification.

DC-function specification:

The DC-function specification is a combination of both function specifications, described above. This means, that concerning a DC-function specification — at presumed course-value — the points are connected either by continous pieces of curve, which have a determined form, or the points are starting-points for generating constant pieces of curve in direction to the origin of the coordinate system.

The syntactic structure of this specification is given by:

⟨DC-function specification⟩ ::= ⟨origin values⟩ / ⟨sequence values⟩ / | ⟨DC-function specification⟩ ⟨sequence values⟩ /
⟨sequence values⟩ ::= ⟨x-value⟩, ⟨y-value⟩[, ⟨course-value⟩]

The other syntactic constructions were already carried out above.

Examples:

Several examples (see fig. 12) demonstrate, that the description of functional relations by the user with the help of the function specifications is relatively uncomplicated. The function specifications of the displayed functional connections are desribed here in correspondence to their coding on the function following cards.

5.2.3. VARIABLE-statement

In the programming system SIMTRA the so-called arithmetic variables can be defined with the help of the VARIABLE-statement. These are elements for the computation of the value of quantities corresponding to given rules of computation during the program execution. They are useful

- for the direct realization of the coupling strategies (see section 2), because with their help the relations between the output (mRNA) of a transcription system and the input (regulatory molecules and nucleotides) of the same transcription system and/or of one ore more other transcription systems can be mathematically easy formulated, and

- as calculation elements in the TEST-, COMPUTE- and/or PRINT-statements. Thereby the user is able to generate arithmetic expressions from the composition of biological model parameters.

The syntactic structure of the VARIABLE-statement is given by:

⟨VARIABLE-statement⟩ ::= ⟨variable declaration⟩
⟨variable specification⟩

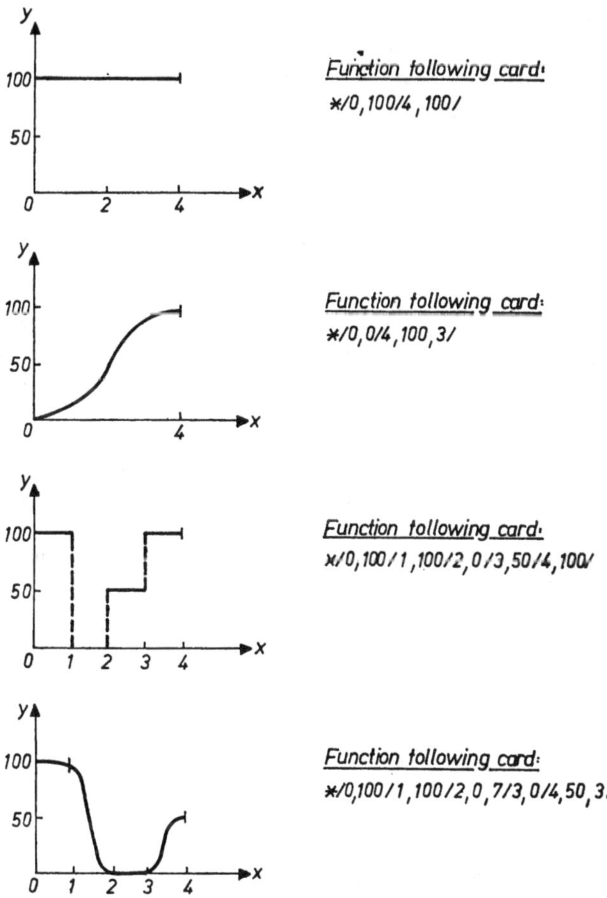

Fig. 12. Examples for the function specification.

5.2.3.1. Variable declaration

The variable declaration characterizes the protein pools in the arithmetic expression of the VARIABLE-statement and contains its identifier.

2	9	22
I	VARIABLE	A
$j \vee$ sn		SK

I — Identifier of the arithmetic variable, with $j \in \{1, \ldots, 120\}$ or sn $\in S_{\text{name}}$
A — Parameter for the characterization of the protein pools in the arithmetic expression with SK $\in$ {EZ, RP, AE}

EZ: The molecules of the protein pools (in form of their RFS) in the arithmetic expression are enzymes.
At constant concentration of the protein pools the concentration of the corresponding regulary molecules, fixed in the COUPLE-statement (parameter B), decreases or increases proportionally with the time.
(The turnover activity of the enzymes is fixed by the turn over number, defined in the corresponding POOL-statements.)

RP: The molecules of the protein pools (in form of their RFS) in the arithmetic expression are regulation proteins.
At constant concentration of the protein pools the concentration of the corresponding regulatory molecules, fixed in the COUPLE-statement (parameter B), doesn't change.

AE: The functional attributes of the proteins in the arithmetic expression are not specified, because they are of no interest for the computation of the value of the variable.

Note:

1. The short key EZ or RP must be entered in the parameter subfield A in the case of the direct realization of the coupling strategies. In such a way the arithmetic expression must contain reference symbols of protein pools.
2. The short key AE must be entered in the parameter subfield A, if the arithmetic expression is exclusively used as calculation element (without preprogrammed biological background).
3. The reference to an arithmetic variable is carried out in the SIMTRA-statements by using the reference symbol VR$j or VR$sn.

5.2.3.2. Variable specification

The variable specification contains the arithmetic expression and is essentially an algebraic combination of RFS's, which is defined by the user. The syntactic structure of the variable specification is defined as follows:

⟨variable specification⟩ ::= ⟨arithmetic expression⟩ /
⟨arithmetic expression⟩ ::= [⟨adding operator⟩] ⟨term 1⟩ |
⟨arithmetic expression⟩ ⟨adding operator⟩ ⟨term 2⟩
⟨adding operator⟩ ::= + | −
⟨term 1⟩ ::= ⟨reference symbol 1⟩ | ⟨minimum operator⟩ (⟨RFS-sequence⟩) | ⟨term 2⟩ ⟨multiplying operator⟩ ⟨term 1⟩

⟨term 2⟩ ::= ⟨integer constant⟩ | ⟨real constant⟩
((arithmetic expression⟩)

⟨reference symbol 1⟩ ::= ⟨standard symbol 1⟩ $ ⟨identifier⟩ |
TM$CLOCK | TM$DELTA

⟨standard symbol 1⟩ ::= FN|PL|RN|SC|VR

⟨minimum operator⟩ ::= MIN

⟨RFS-sequence⟩ ::= ⟨reference symbol 1⟩, ⟨reference symbol 1⟩ |
⟨RFS-sequence⟩, ⟨reference symbol 1⟩

⟨multiplying operator⟩ ::= *|/|DIV

Table 4 contains the operators, which are possible in the arithmetic expressions of the VARIABLE-statements.

Table 4. Operators of the arithmetic expressions in SIMTRA.

operator	operation	type of operands	type of result
*	multiplication	real, integer	real, integer
/	division	real, integer	real
DIV	division with truncation	integer	integer
+	addition	real, integer	real, integer
—	subtraction	real, integer	real, integer
MIN	minimum-determination	real, integer	real, integer

5.2.3.3. Examples

1. Half of the enzyme molecules, diminished by 20 molecules, degradate a pool of regulatory molecules corresponding to the turn over number of the POOL-statement (parameter D). For the VARIABLE-statement it follows then:

```
            :
1       POOL          2,3,0,4
            :
8       VARIABLE   EZ
*/ —(0.5 * PL$1 — 20)/
            :
```

2. In fig. 13 and 14 the effect of different short key notations (EZ and RP) of the parameter A of the variable declaration for the synthesis and inactivation of regulatory molecules is demonstrated. In general the following shall be valid for both cases considered:

   Given a transcription system TS1, which is feeded back over a function (FN$1) through a protein pool (PL$1). If the molecules of the pool are enzymes, the former ones shall turn over one substrate molecule per time interval, respectively. The initial concentration of the pool is zero and grows at the time point $t = t_1$ up to one molecule. The feedback of the transcription system is in the following described with the

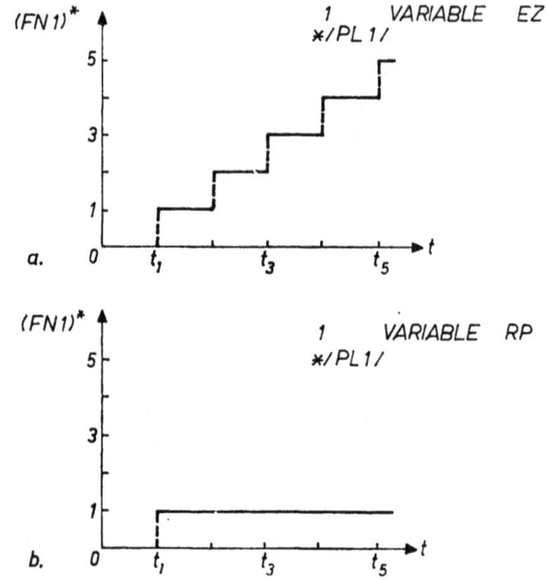

Fig. 13. Synthesis of regulatory molecules.

a. The synthesized proteins of pool PL$1 (enzymes) synthesize themselves the regulatory molecules, defined by FN$1.

b. The synthesized proteins of pool PL$1, which directly or indirectly act as regulatory proteins, are the regulatory molecules themselves, defined by FN$1.

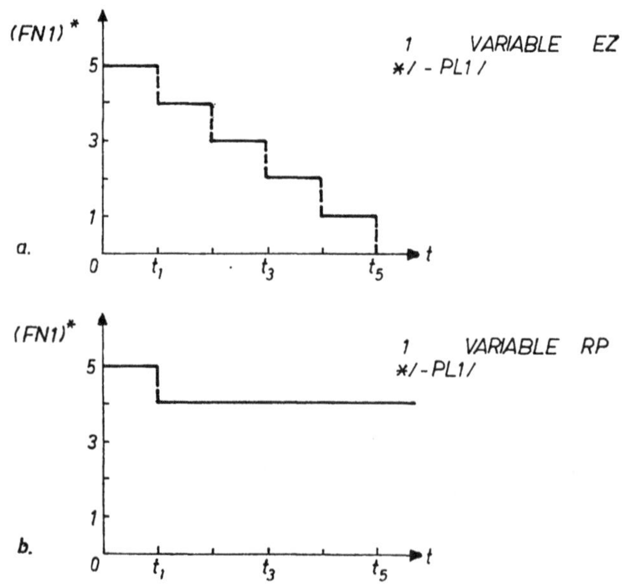

Fig. 14. Inactivation of regulatory molecules.

a. The synthesized proteins of pool PL$1 (enzymes) degrades the regulatory molecules, defined by FN$1.

b. The synthesized proteins of pool PL$1, which act directly as regulation proteins compensate the regulatory molecules, defined by FN$1.

help of the arithmetic variable VR$1. The concentration function FN$1 modified by the proteins of the pool is denoted by (FN$1)*. For the synthesis (fig. 13) the function FN$1, which describes the group of regulatory molecules, is zero during the observation time. But it contains five molecules during the observation time for the inactivation (fig. 14). The facts represented in fig. 14 can be demonstrated for instance by the following SIMTRA-statements:

```
        :
1       POOL        1,3,0,1
        :
1       VARIABLE    EZ
*/ — PL$1 /
        :
1       FUNCTION    1,AEF,3,100,2
*/0,5/360,5/
        SIMULATE    360
        :
        COUPLE      1,FN$1, VR$1
        :
```

5.2.4. RGENERATOR-statement

For the realization of random events SIMTRA incorporates fifteen random number generators. Each generator provides, independent of each other, a sequence of uniformly distributed pseudorandom numbers between 0 and 1 (see Hartley, 1975 and Karplus, 1976). Using different initiation values for the random number generators the user has the possibility to generate different sequences of random numbers. The user can simulate the stochastic behaviour of a molecular genetic model by generating a random number and then testing its value or converting it (with the help of an arithmetic variable) to a form that is appropriate to the situation.

22	9	22
I	RGENERATOR	A
$j \vee sn$		k

I — Identifier of the random number generator, with $j \in \{1, \ldots, 15\}$ or $sn \in S_{name}$

A — initiation values k, $0.0 < k < 1.0$, of the random number generator

Note:

The reference to a random number generator is carried out in the SIMTRA-statements by using the reference symbol RN$j or RN$sn.

5.2.5. SCALAR-statement

The scalars defined in the programming system SIMTRA are elements, which are suitable for the storage of integer or real intermediate results during the execution of a SIMTRA-program. Morevover, the scalars can be used as calculation elements.

With the unitary representation scheme one can describe the SCALAR-statement as follows:

2	9	22
I	SCALAR	A
$j \vee$ sn		k

I — Identifier of the scalar, with $j \in \{1, \ldots, 50\}$ or sn $\in S_{\text{name}}$
A — Value k, $k \in C_N \cup C_R$, of the scalar at the beginning of the simulation

Note:

The reference to a scalar is carried out in the SIMTRA-statements by using the reference symbol SR$j or SR$sn.

5.2.6. POOL-statement

The pools of the active proteins or protein subunits, which are generated by the analysed transcription systems and by the appropriate posttranscriptional processes, have to be defined in the dependence on the coded cistrons and the corresponding transcription systems with the help of POOL-statements. Furthermore, these are useful for:

— the assignment of initial values for the protein pools,
— the fixation of the turn over number.

With the unitary representation scheme the POOL-statement can be described as follows:

2	9	22			
I	POOL	A	B	C	D
$j \vee$ sn		k	k	k	k

I — Identifier of the protein pool, with $j \in \{1, \ldots, 100\}$ or sn $\in S_{\text{name}}$
A — Number k, $k \in \{1, \ldots, 5\}$, of the transcription system, containing the cistron, which codes the proteins of the pool
B — Number k, $k \in \{1, \ldots, 20\}$, of the coding cistron
C — Number k, $k \in C_N$, of active molecules of the defined protein pool at time point $T_0 = 0$
D — Number k, $k \in C_N \cup C_R$, of turned over substrate molecules per enzyme molecule in one minute

*** **Parameter *D* is omitted:**

* Protein molecules are no enzymes of the defined pool!

Note:

The reference to a pool is carried out in the SIMTRA-statements by using the reference symbol PL$j or PL$sn.

5.3. Block statements

The block statements are the operative part of a SIMTRA-program. Describing the molecular genetic structures and processes they serve for the representation of the logic and time sequence of the molecular genetic problem to be simulated. One can classify the block statements as follows:

- statements for the fixation of the simulation time,
- statements for the definition of the running molecular genetic processes, including the corresponding control mechanisms,
- statements for the control of the SIMTRA-program flow.

The order of the block statements is significant in a SIMTRA-program.

5.3.1. Block statements for the fixation of the simulation time

5.3.1.1. TIME-statement

By using the parameters, which are entered in the parameter fields of the TIME-statement, the calculation of the simulation time interval ΔT is carried out (see section 3.2).

9	22		
TIME	A	B	C
	k	k	k

A — Number k, $k \in \{1, \ldots, 5\}$, of the transcription system

B — Maximal transcription velocity k, $k \in C_N \cup C_R$, of the polymerases along the transcription unit (k shall be given in nucleotides per second!)

C — Polymerase diameter k, with $k \in C_N \cup C_R$ (k shall be given in Ångstrøm!)

Note:

1. Going out from the demand, that the polymerase diameter shall not be less than a nucleotide, a sensible calculation of the time interval ΔT by the programming system SIMTRA can only be carried out, if for each transcription system the following relation

$$T_{\text{COND}} \geqq 3{,}4$$

is fulfilled. The quantity T_{COND} is given by:

$$T_{\text{COND}} = \frac{PD_i}{TV_i/\min(\forall\, TV_i)}$$

where

PD_i — Polymerase diameter of the i-th transcription system

TV_i — Maximal transcription velocity of the polymerase of the i-th transcription system

2. In a SIMTRA-program the TIME-statements have to take place prior to the block statements characterizing the molecular genetic processes of the transcription systems concerned.

5.3.1.2. SIMULATE-statement

With the help of the SIMULATE-statement the simulation duration of the molecular genetic model programmed in the SIMTRA-language is fixed.

9	22
SIMULATE	A
	k

A — Duration k, $k \in C_N \cup C_R$, of the simulation run (model time).
k shall be given in seconds!

Note:

If there is no SIMULATE-statement, only a translation will take place, but no execution of the SIMTRA-program. With the message

*** SIMULATE-CARD IS ABSENT — NO SIMULATION

the translation of the SIMTRA-program will be closed.

5.3.2. Statements of the description of the elementary mechanisms on the transcription level

Starting from the determined sequence of the stages of the transcription process itself,

- the primary binding of the polymerase to a specific DNA sequence (promoter),
- the initiation closed by formation of the first phosphodiester bond after a series of processes,
- the elongation (actual phase of the RNA synthesis),
- the termination, including the release of the RNA product and of the polymerase, after transcription of a defined DNA sequence,

and the control mechanisms to the influence of the transcription frequence, four mechanisms are to be distinguished. Four appropriate SIMTRA-statements are defined for these elementary mechanisms.

5.3.2.1. INDUCE-statement

The INDUCE-statement describes among the structure of the operator region operator-localized control mechanisms being of

- negative kind, accomplished by operator-repressor-inducer interactions, and/or
- positive kind, accomplished by operator-activator-antiactivator interactions.

9	22					
INDUCE	A	B	C	D	E	F
	k	RFS	RFS	$k \vee$ RFS	$k \vee$ RFS	k

A — Number k, $k \in \{1, \ldots, 5\}$, of the transcription system
B — Effector-function (FN\$$j_1$ $\vee$ FN\$$sn_1$)
or
EL-list (LI\$$i_1$ $\vee$ LI\$$sn_2$) of effector-functions

C — Antieffector-function (FN\$$j_2$ ∨ FN\$$sn_3$)

or

EL-List (LI\$$i_2$ ∨ LI\$$sn_4$) of antieffector-functions

D — Binding stability k, $0.0 \leq k \leq 1.0$, of the operator-effector-complex

or

EL-list (LI\$$i_3$ ∨ LI\$$sn_5$) of binding stabilities of the effectors to the concerning operators

E — Length k, $k \in C_N$, of the operator followed by a specification symbol

or

EL-list (LI\$$i_4$ ∨ LI\$$sn_6$) of lengths of the operators with specification symbols

(The length k of the operator shall be given in nucleotide pairs!)

Specification symbol:

+: Operator cucceeds to the promoter region in transcription direction

—: Operator is enclosed by the promoter region (in the case of the "preinitiator" model)

*** **Parameter *E* is omitted:**

* There are only "overlapping" operators!

F — Number k, $k \in \{1, \ldots, 20\}$, of the operators having "overlapping" character (structural or functional) regarding to promoter region

*** **Parameter *F* is omitted:**

* There are no "overlapping" operators!

Note:

The INDUCE-statement can only be defined if the transcription system is subject to an operator-localized control.

Examples:

1. For a transcription system (see fig. 15), being assigned the number 2, it is assumed:

The length of the operator shall amount to 20 nucleotide pairs. The binding of the repressor to the operator is to be absolutely stable in the absence of the inducer. During the whole period of the experiment (60 seconds) the concentration of the repressor is to be above the saturation level. The inducer concentration is to amount to one

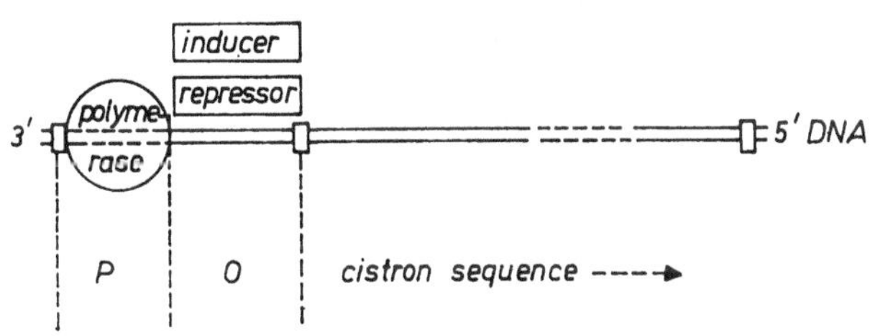

Fig. 15. Abstract model of a transcription system in a formal representation, the transcription unit of which contains an operator region O and a promoter region P.

half of the saturation level. The concentration-dependent probability of interaction of the repressor molecules to the operator is described by a curve, which is characterized in fig. 7 by the value $k = 7$.
For the INDUCE-statement follows then:

```
            :
43    FUNCTION      2,EF,7,100,1
*/0,100/60,100/
 1    FUNCTION      2,AEF,3,100,1
*/0,50/60,50/
            :
      INDUCE        2,FN$43,FN$1,1.0,20+
            :
```

2. An "overlapping" between promoter and operator is to be assumed for the transcription system given in example 1. With the above defined functions for the INDUCE-statement it follows then:

```
      INDUCE        2,FN$43,FN$1,1.0,1
```

The parameter E is omitted here!

5.3.2.2. INITIATE-statement

The INITIATE-statement reflects the initiation variants and promoter-localized control mechanisms. The latter are

- promoter-polymerase interactions,
- promoter-CAP/cAMP-(polymerase) interactions.
- polymerase-polymerase subunit (σ) interactions.

9	22							
INITIATE	A	B	C	D	E	F	G	H
	k	k	k	k	RFS	RFS	RFS	RFS

A — Number k, $k \in \{1, \ldots, 5\}$, of the transcription system
B — Number k, $k \in \{1, \ldots, 5\}$, of the initiation variant[1]
C — Number k, $k \in \{1, \ldots, 15\}$, of the promoters followed by a specification symbol

Specification symbol:

+: At the beginning of the simulation experiment the promoters are occupied by polymerases

−: At the beginning of the simulation experiment the promoters are *not* occupied by polymerases

D — Time period k, $k \in C_N \cup C_R$, for the initiation of a polymerase molecule at the promoter region
(k shall be given in seconds!)

---

[1] An exactly definition of the initiation variants is given in this issue by NEUMANN and KREISCHER (1982).

E — Polymerase-function (FN\$$j_1$ ∨ FN\$$sn_1$)
F — Polymerase subunit-function (FN\$$j_2$ ∨ FN\$$sn_2$)

***** Parameter *F* is omitted:**

*** Polymerase subunit does not exist in the defined system!**

G — CAP-function (FN\$$j_3$ ∨ FN\$$sn_3$)

***** Parameter *G* is omitted:**

*** CAP does not exist in the defined system!**

H — cAMP-function (FN\$$j_4$ ∨ FN\$$sn_4$)

***** Parameter *H* is omitted:**

*** cAMP does not exist in the defined system!**

Note:
The time course of the nucleotide concentration must be defined in the ELONGATE-statement.

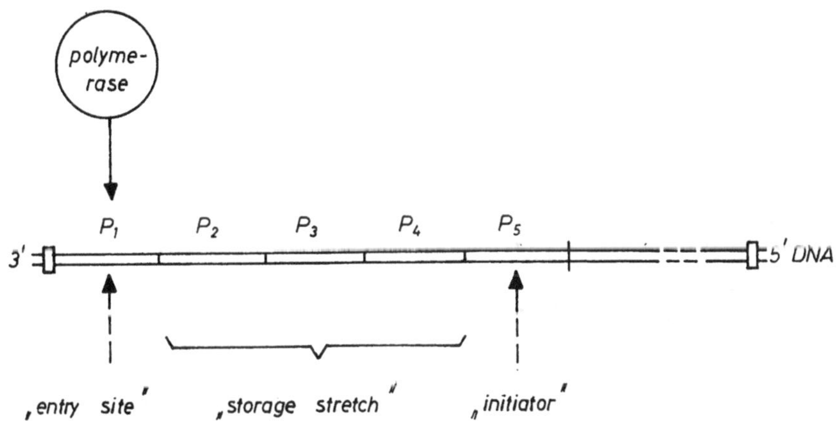

Fig. 16. Abstract model of a transcription system with a complicated structure of the promoter region ("preinitiator" model).

Examples:

1. For a transcription system (see also fig. 16), being assigned the number 2, is assumed:

   During the whole period of the experiment (120 seconds) the polymerase (holoenzyme)-concentration is above the saturation level. The time period for the initiation of a polymerase molecule is to amount to 14.8 seconds. The initiation shall be the variant 4, i.e. at every promoter $p_1, \ldots, p_i$ $(i > 1)$ one polymerase is able to stay until the labile complex is formed, but can perform the next steps towards the initiation only at p_i. The promoter region consists of five promoters ("preinitiator" model). The system is not subject to the mechanism of catabolic repression. The probability of the interaction of a polymerase molecule to the promoter is dependent on the polymerase concentration and is described by a curve, which is characterized in fig. 7

by the value $k = 3$. At the beginning of the experiment the promoters are not occupied by polymerases. For the INITIATE-statement it follows then:

```
              ⋮
11    FUNCTION     2,POL,3,100
*/0,100/120,100/
              ⋮
      INITIATE     2,4,5—,14.8,FN$11
              ⋮
```

2. The transcription system, given in example 1 of this section, is subject to the control of catabolic repression mechanism. Whereas the polymerase (coreenzyme)- and the CAP-concentration is above the saturation level for the time interval of the experiment, the concentration of the polymerase subunit (σ) is to amount to one half of the saturation level. The cAMP-concentration is to decrease exponentially to zero. The concentration-dependent probability of interaction of all named regulatory molecules to their binding partners is given by curves, which are characterized in fig. 7 by the value $k = 3$. At the beginning of the experiment the promoter region is occupied by preinitiated polymerases.
   For the INITIATE-statement it follows then:

```
              ⋮
13    FUNCTION     2,SUB,3,100
*/0,50/120,50/
15    FUNCTION     2,CAP,3,100
*/0,100/120,100/
16    FUNCTION     2,AMP,3,100
*/0,100/120,0,8/
11    FUNCTION     2,POL,3,100
*/0,100/120,100/
              ⋮
      INITIATE     2,4,5+,14.8,FN$11,FN$13,FN$15,FN$16
              ⋮
```

5.3.2.3. ELONGATE-statement

In the ELONGATE-statement there is described the actual RNA synthesis. The velocity of the synthesis is dependent on

- the substrate level (nucleoside triphosphate concentration), and
- the transcription velocity of the initiated polymerase molecules along the transcription unit.

9	22			
ELONGATE	A	B	C	D
	k	RFS	RFS	$k \vee$ RFS

A — Number k, $k \in \{1, \ldots, 5\}$, of the transcription system
B — Nucleotide-function (FN\$$j_1 \vee$ FN\$$sn_1$)

C — Function ($FN\$j_2 \vee FN\$sn_2$) for the describing of the transcription velocity of the polymerases along the transcription unit

*** **Parameter *C* is omitted:**

* The polymerases transcribe along the transcription
* unit constant with the velocity, which is entered
* in parameter field B of the TIME-statement

D — Position k, $k \in C_N$, of a transcribing polymerase
or
SQ-list ($LI\$i_1 \vee LI\$sn_3$) of positions of transcribing polymerases on the transcription unit of the beginning of the simulation
(k shall be given in nucleotide pairs!)

*** **Parameter *D* is omitted:**

* It exists no transcribing polymerases at the
* beginning of the simulation!

Examples:

1. For a transcription system, being assigned the number 2, it is assumed:
During the whole period of the experiment (120 seconds) the nucleotide concentration is to be above the saturation level. The concentration-dependent probability of the nucleotide incorporation to the RNA chain elongation is given by a curve, which is characterized in fig. 7 by the value $k = 3$. The length of the transcription unit being worked off by a polymerase molecule is to amount to 1000 nucleotide pairs. The transcription velocity of the polymerases is to decrease linearly to the half of the start velocity (40 nucleotides/sec.). At the beginning of the experiment the transcription unit is not occupied by transcribing polymerases.
For the ELONGATE-statement it follows then:

```
                 :
 17    FUNCTION     2,NUC,3,100
*/0,100/120,100/
 19    FUNCTION     2,TV
*/0,40/1000,20,6/
                 :
       ELONGATE     2,FN$17,FN$19
                 :
```

2. At the beginning of the experiment the transcription system of the example 1 is characterized by the position pattern of transcribing polymerases on the transcription unit, which is displayed in fig. 17.

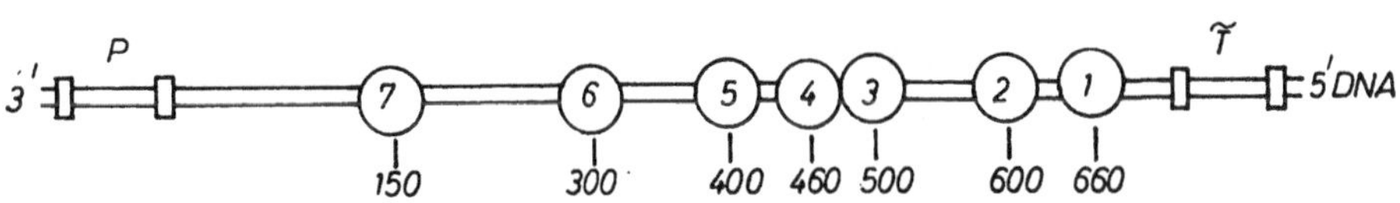

Fig. 17. Abstract model of a transcription system, the transcription unit of which has the displayed position pattern of transcribing polymerases.

With the above defined functions for the ELONGATE-statement it follows then:

```
              ⋮
     51    LIST           SQ
    */660,600,500,460,400,300,150/
              ⋮
           ELONGATE    2,FN$17,FN$19,LI$51
              ⋮
```

5.3.2.4. TERMINATE-statement

The TERMINATE-statement describes

- terminator-localized control mechanisms of the transcription accomplished by terminator-termination factor-antitermination factor interactions, and
- interruptions of the RNA chain elongation produced by the DNA mutations, which effects a release of transcribing polymerases.

9	22						
TERMINATE	A	B	C	D	E	F	G
	k	RFS	RFS	$k \vee$ RFS	k	$k \vee$ RFS	$k \vee$ RFS

A — Number k, $k \in \{1, \ldots, 5\}$, of the transcription system

B — Termination factor-function (FN$$j_1 \vee$ FN$$sn_1$)
or
EL-list (LI$$i_1 \vee$ LI$$sn_2$) of termination factor-functions

C — Antitermination factor-function (FN$$j_2 \vee$ FN$$sn_3$)
or
EL-list (LI$$i_2 \vee$ LI$$sn_4$) of antitermination factor-functions

D — Termination site k, $k \in C_N$,
or
EL-list (LI$$i_3 \vee$ LI$$sn_5$) of termination sites
(k shall be given in nucleotide pairs!)

E — Parameter for the characterization of the restart possibility of the polymerases with $k \in \{1, 2, 3\}$:

$k = 1$: Restart of released polymerases is not possible
$k = 2$: Restart of released polymerases is possible
$k = 3$: Released polymerases are no longer considered

F — Mutation site k, $k \in C_N$,
or
EL-list (LI$$i_4 \vee$ LI$$sn_6$) of mutation sites

*** **Parameter *F* is omitted:**

* No mutation!

G — Probability k, $0.0 \leqq k \leqq 1.0$, of release of the polymerase at the mutation site, given in parameter F,
or

EL-list ($LI\$i_5 \vee LI\$sn_7$) of probabilities of release of the polymerases at the mutation sites, given in parameter F

*** **Parameter *G* is omitted:**

* No mutation sites wouldn't be defined in the parameter field F!

Examples:

1. The transcribing polymerases reaching the end terminator of a transcription system being assigned the number 2 are to be released in the presence of a termination factor, the concentration of which shall be above the saturation level for the period of experiment. The concentration-dependent probability of interaction is to be given by a curve, which is characterized in fig. 7 by the value $k = 3$. The antitermination factor concentration over the period of the experiment is zero. The end terminator is to be located around the nucleotide pair 800. Released polymerases are no longer considered.
For the TERMINATE-statement it follows then:

```
            :
24      FUNCTION      2,TF,3,100,1
*/0,100/120,100/
34      FUNCTION      2,ATF,3,100,1
*/0,0/120,0/
            :
        TERMINATE     2,FN$24,FN$34,800,3
            :
```

2. The transcription system of the example 1 is characterized by the mutation sites as shown by fig. 18.

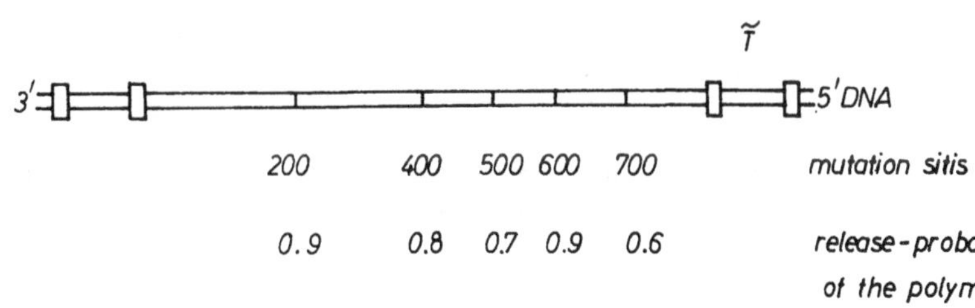

Fig. 18. Abstract model of a transcription unit with five mutation sites and the corresponding release-probabilities of the polymerases at the latter.

For the TERMINATE-statement it follows then:

```
            :
70      LIST          EL
*/200,400,500,600,700/
71      LIST          EL
*/0.9,0.8,0.7,0.9,0.6/
            :
        TERMINATE     2,FN$24,FN$34,800,3,LI$70,LI$71
            :
```

5.3.3. Statements for the description of posttranscriptional processes

For a sufficient representation of the transcription system it is necessary to consider the posttranscriptional processes. These processes are formulated only generally, that is, without giving a detailed representation of structural elements and regulation mechanisms (see section 2).

5.3.3.1. TRANSLATE-statement

The TRANSLATE-statement describes the transformation of the genetic information of a nucleotide sequence (mRNA) into an aminoacid sequence (polypeptide chain) and accordingly the actual protein synthesis. The translation parameters can be specified for each mRNA pool produced by a determined transcription system and for each kind of cistron copies of this system.

9	22							
TRANSLATE	A	B	C	D	E	F	G	H
	k	k	$k \vee$ RFS	$k \vee$ RFS	$k \vee$ RFS	$k \vee$ RFS	k	k

A — Number k, $k \in \{1, \ldots, 5\}$, of the transcription system of which mRNA chains (produced by this system itself) are to be translated

B — Translation velocity k, $k \in C_R$, of the ribosomes (*k* shall be given in aminoacid per second!)

C — Translation frequency k, $k \in C_N$, of the cistron copy of a mRNA chain (monocistronic messenger)
or
EL-list (LI\$$i_1$ ∨ LI\$$sn_1$) of translation frequencies of all cistron copies of a mRNA chain (polycistronic messenger)

Note:

In case of polycistronic translation all cistron copies of a mRNA chain are translated with the same frequency.

D — Translation distance k, $k \in C_N$, between two neighbouring ribosomes for the cistron copy of a mRNA chain (monocistronic messenger)
or
EL-list (LI\$$i_2$ ∨ LI\$$sn_2$) of translation distances between two neighbouring ribosomes for *all* cistron copies of a mRNA chain (polycistronic messenger)
(*k* shall be given in nucleotides!)

Note:

In case of polycistronic translation all cistron copies of a mRNA chain are translated with the same distance of ribosomes.

E — End k, $k \in C_N$, of the cistron copy of a mRNA chain (monocistronic messenger)
or

EL-list (LI\$$i_3$ ∨ LI\$$sn_3$) of the ends of all cistron copies of a mRNA chain (polycistronic messenger)
(k shall be given in nucleotides!)

F — Beginning k, $k \in C_N$, of the cistron copy of a mRNA chain (monocistronic messenger)
or
EL-list (LI\$$i_4$ ∨ LI\$$sn_4$) of the beginnings of *all* cistron copies of a mRNA chain (polycistronic messenger)
(k shall be given in nucleotides!)

Note:

1. In case of the polycistronic translation:
   The beginning of all cistron copies is adequate the beginning of the RNA copy of the first cistron!
2. In case of the monocistronic translation:
   The beginnings of all cistron copies must be inscribed!

G — Minimum length k, $k \in C_N$, of a mRNA chain for the binding of the first ribosome in case of the translation nascent mRNA chains
(k shall be given in nucleotides!)

*** **Parameter *G* is omitted:**
* Translation of complete mRNA chains!

Examples:

1. The entries in the parameter fields E and F of the TRANSLATE-statement are to be demonstrated by means of the transcription system shown in fig. 19. For the two parameters E and F of the TRANSLATE-statement it follows from this:

Parameter E:

```
8       LIST        EL
*/790,1870,2730/
```

Parameter F:

a. polycistronic translation

```
19      LIST        EL
*/60,60,60/
```

b. monocistronic translation

```
19      LIST        EL
*/60,790,1870/
```

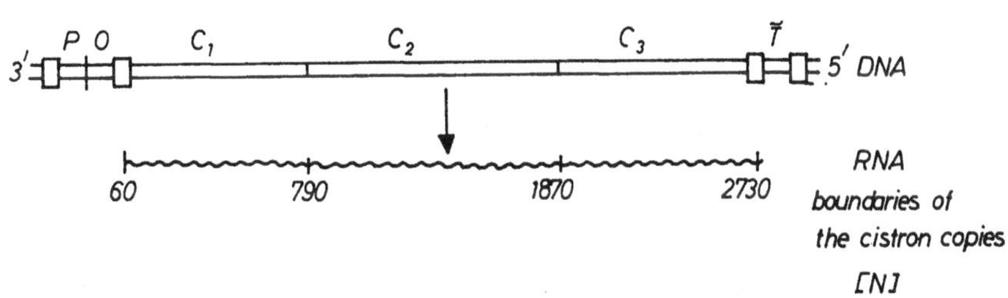

Fig. 19. Abstract model of a transcription unit with the corresponding cistron copies.

2. The transcription system of the preceding example, being assigned the number 2, is expanded as follows:

   The translation velocity is to amount to 15 aminoacids per second. Nascent mRNA chains are to the translated polycistronicly. Each cistron copy is to be translated tenfold. The translation distance is to amount to 110 nucleotides and the minimum length of a nascent mRNA chain for the binding of the first ribosome 200 nucleotides. For the TRANSLATE-statement it follows then:

```
          :
 8        LIST             EL
*/790,1870,2730/
 7        LIST             EL
*/110,110,110/
 2        LIST             EL
*/10,10,10/
 19       LIST             EL
*/60,60,60/
          :
        TRANSLATE     2,15,LI$2,LI$7,LI$8,LI$19,200
          :
```

5.3.3.2. AGGREGATE-statement

The AGGREGATE-statement describes the aggregation of several of the same polypeptide chains coded by a determined cistron of the simulated transcription system. These polymeric molecules can be either active proteins themselves or subunits for the formation of the latter.

9	22	
AGGREGATE	A	B
	k	$k \vee$ RFS

,

A — Number k, $k \in \{1, \ldots, 5\}$, of the transcription system cistrons of which codes for polypeptide chains, composed to the active protein molecules

B — Number k, $k \in C_N$, of the polypeptide chains for the formation of an active protein or of a protein subunit, if this chain is coded monocistronicly
or
EL-list (LI$i_1 \vee$ LIsn_1) of numbers of the polypeptide chains for the formation of active proteins or of protein subunits, if these chains are coded polycistronicly

Note:

If all active proteins or protein subunits consist of only one polypeptide chain, coded by the modelled transcription system, the definition of the AGGREGATE-statement is unnecessary.

Examples:

1. For a transcription system, structured in such a way as shown in fig. 20, the AGGREGATE-stratement must be omitted!

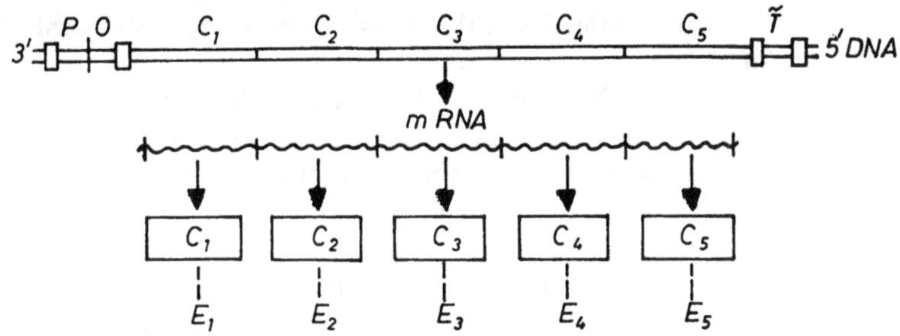

Fig. 20. The displayed abstract model demonstrates the case that all active proteins consist of only one polypeptide chain.

2. For a transcription system, being assigned the number 2 and being structured in such a way as shown in fig. 21, the AGGREGATE-statement must be defined. For the AGGREGATE-statement it follows from this:

```
          :
20      LIST              EL
*/1,2,1,2,2/
          :
        AGGREGATE  2,LI$20
          :
```

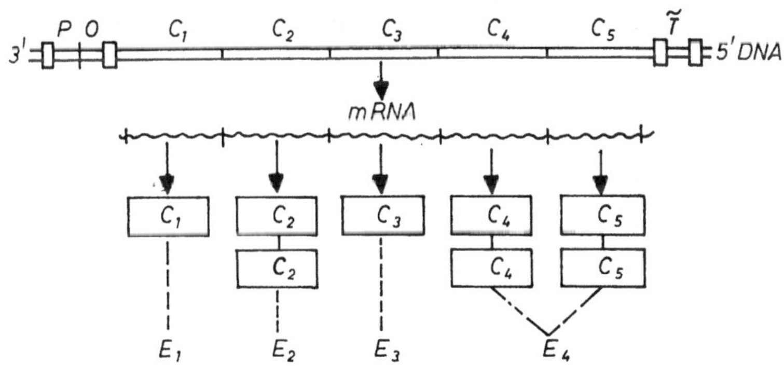

Fig. 21. The displayed abstract model demonstrates the case that some active proteins consist of different polypeptide chains and more than one of the same, respectively.

5.3.3.3. INACTIVATE-statement

The INACTIVATE-statement describes the time limited efficacy of the active proteins or protein subunits. A decrease of the concerned pools can be simulated by this statement.

9	22	
INACTIVATE	A	B
	k	$k \vee$ RFS

A — Number k, $k \in \{1, \ldots, 5\}$, of the transcription system coding those proteins or protein subunits, which are subject to an inactivation

B — Mean period k, $k \in C_N \cup C_R$, between the synthesis and inactivation of a molecule, which belongs to the concerned pool of active proteins or protein subunits, if the transcription system codes only this protein
or
EL-list (LI\$$i_1$ ∨ LI\$$sn_1$) of mean periods between the synthesis and inactivation of the molecules from *all* pools of active proteins or protein subunits coded by the transcription system

Example:

The mean periods for the inactivation of the molecules from five pools of proteins coded by five adequate cistrons of a transcription system, being assigned the number 1, are 107, 120, 138, 390, and 87 seconds.

For the INACTIVATE-statement it follows then:

```
       ⋮
 8     LIST           EL
*/107,120,138,390,87/
       ⋮
       INACTIVATE  1,LI$8
       ⋮
```

5.3.3.4. COUPLE-statement

The COUPLE-statement realizes the coupling (including feedback) of the transcription systems corresponding with certain coupling strategies (see section 2) realized by arithmetic variables.

9	22		
COUPLE	A	B	C
	k	RFS	RFS

A — Number k, $k \in \{1, \ldots, 5\}$, of the (passive) transcription system, which is to be coupled

B — Function (FN\$$j_1$ ∨ FN\$$sn_1$), by means of which the transcription system is coupled
or
EL-list (LI\$$i_1$ ∨ LI\$$sn_2$) of functions, by means of which the transcription system is coupled

C — Arithmetic variable (VR\$$l_1$ ∨ VR\$$sn_3$), which realizes the coupling strategy corresponding to the function mentioned above
or
EL-list (LI\$$i_2$ ∨ LI\$$sn_4$) of arithmetic variables which realize the coupling strategies corresponding to the functions given in the above EL-list.

Example:

A transcription system, being assigned the number 2, represents the passive transcription system, which is subject to a feedback mechanism. This system is to be coupled with the (active) transcription systems TS1, TS3, and TS4 in such a way as shown in fig. 22.

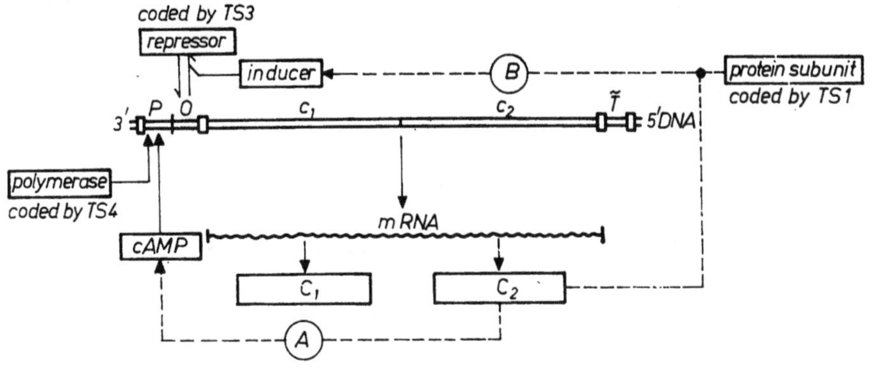

(A) C_2 lowers the cAMP-concentration

(B) The protein consisting of C_2 and the protein subunit coded by TS1 reduces the inducer

Fig. 22. Coupling strategy of the transcription system TS2.

Excerpts from a SIMTRA-program that uses the COUPLE-statement might look as follows:

```
          PROGRAM     COUPLING
             :
48        POOL          1,4,0
63        POOL          2,2,0
31        POOL          3,1,10
57        POOL          4,3,1200
17        VARIABLE      EZ
*/ — MIN(PL$63,PL$48) /
 9        VARIABLE      RP
*/ — 0.5 * PL$63/
48        VARIABLE      RP
*/PL$57 /
22        VARIABLE      RP
*/PL$31 — 20/
             :
11        FUNCTION      2,POL,2,100.0
*/0,100.0/3600,35.8,6/
16        FUNCTION      2,AMP,2,100.0
*/0,100.0/1200,100.0/3600,80.0/
 3        FUNCTION      2,AEF,3,100.0,1
*/0,100.0/3600,100.0/
 8        FUNCTION      2,EF,7,100.0,1
*/0,50.0/3600,50.0/
```

```
          :
  1   LIST          EL
*/FN$11,FN$16,FN$3,FN$8/
  2   LIST          EL
*/VR$48,VR$9,VR$17,VR$22/
          :
      COUPLE        2,LI$1,LI$2
          :
      END
```

5.3.4. Statements for the control of the SIMTRA-program flow

The execution of a SIMTRA-program is usually carried out in the static sequence of its record, i.e. from one block statement to the next one. In order to change the sequence of the execution or to repeat a determined program part, block statements are defined, which are used for the control of the SIMTRA-program flow.

5.3.4.1. TEST-statement

With the help of the TEST-statement the test of an arithmetic condition is carried out. A transfer operation is connected with it. If the arithmetic condition is fulfilled, the program will be continued at the branch destination; if not, the next block statement will be executed.

9	22			
TEST	A	B	C	D
	RFS	SK	$k \vee$ RFS	k

A — Left operand
Arithmetic variable **or** function **or** pool **or** scalar **or** random number generator **or** system clock (in form of their RFS)

B — Relational operator,
with SK $\in$ {LT, LE, EQ, NE, GT, GE}
LT: Less then
LE: Less than or equal to
EQ: Equal to
NE: Not equal to
GT: Greater than
GE: Greater than or equal to

C — Right operand
Constant k, $k \in C_N \cup C_R$
or
Arithmetic variable **or** function **or** pool **or** scalar **or** random number generator (in form of their RFS)

D — Transfer address k, $k \in \{1, \ldots, 999\}$

Note:

The left operand is compared with the right operand in such a way, as determined by the relational operator. The six relational operators offer the user the possibility to represent all quantity-relations between the model parameters.

Examples:

```
        TEST            PL$3,GT,PL$4,63
        TEST            TM$CLOCK,LT,600,587
```

5.3.4.2. CONTINUE-statement

As the block statements are unlabelled, one needs in connection with the TEST-statement a statement, which is the branch destination. For this purpose the CONTINUE-statement was introduced. Its syntactic structure differs from those possessing usually the block statements.

2	9
L	CONTINUE
k	

L — Label (or in other words transfer address) k, $k \in \{1, \ldots, 999\}$

Note:

The leading zeros of a label are ignored by the SIMTRA-compiler.

Example:

```
          ⋮
        TEST            FN$45,LE,80.0,1
          ⋮
  1       CONTINUE
          ⋮
```

5.3.4.3. COMPUTE-statement

The COMPUTE-statement enables the change and the storage of the actual values of functions, pools, arithmetic variables, scalars and random numbers.

9	22		
COMPUTE	A	B	C
	RFS	SK	$k \vee$ RFS

A — First operand
Arithmetic variable or function or pool or scalar or random number generator (in form of their RFS)

B — Linkage operator,
with SK $\in$ {ADD, SUB, STO}
ADD: Addition
SUB : Subtraction
STO : Storage

C — Second operand
Constant k, $k \in C_N \cup C_R$
or
Arithmetic variable **or** function **or** pool **or** scalar **or** random number generator **or** system clock **or** simulation time interval (in form of their RFS)

Examples:

```
COMPUTE    FN$50,ADD,FN$POLYM1
COMPUTE    SC$17,STO,TM$DELTA
COMPUTE    PL$2,SUB,PL$PROT
```

5.3.4.4. PRINT-statement

The PRINT-statement is a statement giving the user the possibility to print the actual values of the functions, pools, scalars, arithmetic variables, random number generators, and of the system clock and simulation time interval, respectively.

9	22
PRINT	A
	RFS

A — Output quantity
Arithmetic variable **or** function **or** pool **or** scalar **or** random number generator **or** system clock **or** simulation time interval (in form of their RFS)
or
SQ-list (LI$$i_i$ $\vee$ LI$$sn_1$) of the above output quantities
(in form of their RFS)

Note:

1. The PRINT-statement is, considered from the syntactic point of view, a block statement and an output statement from the standpoint of its semantics.
2. The system clock time and the simulation time interval ΔT are put out in seconds!
3. The output, which is realized by using the PRINT-statement, has the following form:

$$*** \langle \text{output time} \rangle *** \langle oq_1 \rangle = \langle v_1 \rangle, \langle oq_2 \rangle = \langle v_2 \rangle, \ldots \langle oq_n \rangle = \langle v_n \rangle$$

Hereby mean:

output time — actual output time point (simulation time)
oq_i — the i-th output quantity
v_i — the value, which is assigned to the i-th output quantity

Example:

The time-behaviour of several protein pools will be considered: the corresponding program part is given as follows:

```
               :
14     LIST              SQ
*/PL$6,PL$PROT1,PL$PROT7/
               :
       PRINT             LI$14
               :
```

The print might look as follows:

*** 37.5 *** PL$6 = 270, PL$PROT1 = 0, PL$PROT7 = 150

5.4. Output statements

The output statements enable the output of characteristic result data of the simulated molecular genetic problem. The output can take place in standardized form but also in a form chosen by the user. By means of SYSTEMOUTPUT- or SELECTOUTPUT-statement the output of certain characteristic data in standardized form is carried out automatically by the programming system SIMTRA. In this case the format and the order of the data are fixed. But the user is not restricted to this fixed form, because he can establish his own output with the help of USEROUTPUT-statement.

5.4.1. SYSTEMOUTPUT-statement

By means of the SYSTEMOUTPUT-statement certain data are printed, which describe the state of the transcription system and give information concerning the efficiency of the system and the posttranscriptional processes.

9	22		
SYSTEMOUTPUT	A	B	C
	k	k	RFS

A — Number k, $k \in \{1, \ldots, 5\}$, of the transcription system, for which the data shall be printed

B — Time interval k, $k \in C_N \cup C_R$, between the cyclic consecutive output time points at which the print of the result data shall take place (output cycle-time)

Note:

In the case of omitted parameter C, k is the output cycle-time defined in seconds; if the user refers to the simulation time interval ΔT in the parameter C, he fixes the output cycle-time in ΔT-time intervals, i.e. as product $k * \Delta T$ with k being an integer factor ($k \in C_N$)

C — Simulation time interval ΔT (TM$DELTA) as reference quantity for the output cycle-time

*** **Parameter *C* is omitted:**

* The output cycle-time is not defined in ΔT-time intervals!

Note:

The standard output is suppressed by using the USEROUTPUT-statements.

Examples:

1. SYSTEM OUTPUT 1,1,TM$DELTA

For the transcription system with the number 1 the standard output is carried out for each simulation time interval ΔT.

2. SYSTEMOUTPUT 1,30

For the transcription system with the number 1 the standard output is carried out in cycles with a thirty seconds time distance.

Content and form of the output

With the help of the SYSTEMOUTPUT-statement the following result data are printed side by side in a fixed order:

- output time point (simulation time)
- state of the operator region
- state of the promoter region
- nucleotide total incorporation
- turnover amount
- position pattern of transcribing polymerases
- protein pool pattern

A complete output of these data is carried out only for the respective transcription system, if in the program both the INDUCE-statement and the block statements for the description of the posttranscriptional processes are included. If the latter statements are not programmed, the output of the protein pool pattern and the turnover amount is omitted.

The above molecular genetic output data have the following meaning:

— **State of the operator region**

An operator o_m of the operator region can take on two states $s(o_m)$:

$$s(o_m) = \begin{cases} 1, & \text{operator } o_m \text{ is closed} \\ 0, & \text{operator } o_m \text{ is opened} \end{cases}$$

— **State of promoter region**

A promoter p_l of the promoter region can take on eleven states $s(p_l) \in \{1, \ldots, 11\}$. Each state describes the promoter in the initiation process of a polymerase molecule in a different way. A detailed information about these facts is given in this issue by NEUMANN and KREISCHER (1982).

– **Nucleotide total incorporation**

The total number of those nucleotides are put out, which are incorporated up to the respective output time points in the synthesized RNA chains. This number follows from the sum of pairs of nucleotides, which were worked off by the already released and by the transcribing polymerases.

– **Turnover amount**

The turnover amount means here the number of those nucleotides knotted to RNA chains and not split off from the latter. The split-process is the result of the translation of the RNA chains by ribosomes.

– **Position pattern of transcribing polymerases**

The positions of the polymerases on the transcription unit are printed at the respective output time points. In addition to the latter, numbers are put out, which refer to the order of the polymerases concerning their initiation. The positions are given in nucleotide pairs. Polymerases, which are released, will not be printed.

– **Protein pool pattern**

For each cistron of a considered transcription system the number of those active protein molecules that or the polypeptide chains of which are coded by the respective cistron, will be printed.

The form of the standard output depends on the number of simulated transcription systems. If only one transcription system is considered, the output of the result data is uncomplicated because the output can be carried out without data buffering in this case. The data of the transcription system are the available for the user in a closed form. At a simultaneous modelling of several transcription systems by SIMTRA the result data for each output time point at unbuffered output (on the premises of equal output time points) should be printed alternately. But this would destroy the lucidity of the log and would complicate its interpretation by the user. Because of this fact there is a buffered output in SIMTRA. This makes it possible to print the result data for the corresponding transcription systems in so-called data complexes. The volume of the latter depends on the number of the output time points fixed by the user and is automatically calculated by the programming system. The output of the result data is initiated with the message

*** OUTPUT DATA OF THE PROGRAM ⟨program identifier⟩ ***

Then an identification of the transcription system, by which in the following the data are put out, takes place by the information

*** TRANSCRIPTION SYSTEM ⟨number⟩ ***

The print of the result data is carried out for each output time point in a compressed form, which depends on the degree of intricacy of each transcription system, i.e. on the number of the system elements. If the operator region of a transcription unit consists of not more than 10 operators and if the promoter region contains not more than 5 promoters the print of the data for each fixed output time point takes place in *one* print line (127 print positions). If these boundaries mentioned just now are exceeded *two* print lines will be used. Because of these facts, in several cases only a reduced number of positions of transcribing polymerases can be printed, since maximal 250 polymerases simultaneously can occupy the transcription unit. However, the programming system

Table 5. Example for the print of the result data for a transcription system arbitrarily chosen in standardized form.

*** OUTPUT DATA OF THE PROGRAM MODEL 1 ***
*** TRANSCRIPTION SYSTEM 1 ***

0.25	0	3	3	0				
0.50	0	5	5	0				
0.75	0	5	5	0				
1.00	0	7	7	0				
1.25	0	7	7	0				
1.50	0	9	9	0				
1.75	0	11	10	0	0 1			
2.00	0	11	10	10	10 1			
2.25	0	11	10	20	20 1			
2.50	0	11	10	30	30 1			
2.75	0	11	10	40	0 2	40 1		
3.00	0	3	10	60	10 2	50 1		
3.25	1	5	10	80	20 2	60 1		
3.50	1	5	10	100	30 2	70 1		
3.75	1	7	10	120	40 2	80 1		
4.00	1	7	3	140	50 2	90 1		
4.25	1	9	4	160	60 2	100 1		
4.50	1	11	4	180	70 2	110 1		
4.75	1	11	4	200	80 2	120 1		
5.00	1	11	4	220	90 2	130 1		
5.25	1	11	4	240	100 2	140 1		
5.50	1	11	5	250	150 1			
5.75	1	11	5	260	160 1			
6.00	1	11	7	270	170 1			
6.25	1	11	7	280	180 1			
6.50	1	11	9	290	190 1			
6.75	1	11	10	300	0 3	200 1		
7.00	1	11	10	310	0 3	210 1		
7.25	1	11	10	320	0 3	220 1		
7.50	1	11	10	330	0 3	230 1		
7.75	1	11	10	340	0 3	240 1		
8.00	0	11	10	360	10 3	250 1		
8.25	0	11	10	380	20 3	260 1		
8.50	0	11	10	390	30 3			
8.75	0	11	10	400	0 4	40 3		
9.00	0	3	10	420	10 4	50 3		
9.25	0	5	10	440	20 4	60 3		
9.50	0	5	10	460	30 4	70 3		
9.75	0	7	10	480	40 4	80 3		
10.00	0	7	3	500	50 4	90 3		
10.25	0	9	5	520	60 4	100 3		
10.50	0	11	5	540	70 4	110 3		
10.75	0	11	7	560	80 4	120 3		
11.00	0	11	7	580	90 4	130 3		
11.25	0	11	9	600	100 4	140 3		
11.50	0	11	10	620	0 5	110 4	150 3	
11.75	0	11	10	650	10 5	120 4	160 3	
12.00	0	11	10	680	20 5	130 4	170 3	
12.25	0	11	10	710	30 5	140 4	180 3	
12.50	0	11	10	740	0 6	40 5	150 4	190 3

provides the user with the actual information about the position pattern by printing the positions and numbers of the five first initiated and of the five last initiated transcribing polymerases. Certain restrictions also exist for the protein pool pattern.

Example:

In Tab. 5 the result data for a chosen transcription system with the number 5, for which no posttranscriptional processes are modelled, are shown in the form of the standard output by using the following SYSTEMOUTPUT-statement:

SYSTEMOUTPUT 5,1,TM$DELTA

The operator region of the observed transcription system shall consist only of one operator and the promoter region of two promoters. The print of the following data is carried out:

- output time point
- state of the operator region, that is $s(o_1)$
- state of the promoter region, that is $s(p_1)$ and $s(p_2)$
- nucleotide total incorporation
- position pattern of transcribing polymerases

In fig. 23a visual interpretation of these printed data is carried out at two selected time points, which describe the position pattern of transcribing polymerases.

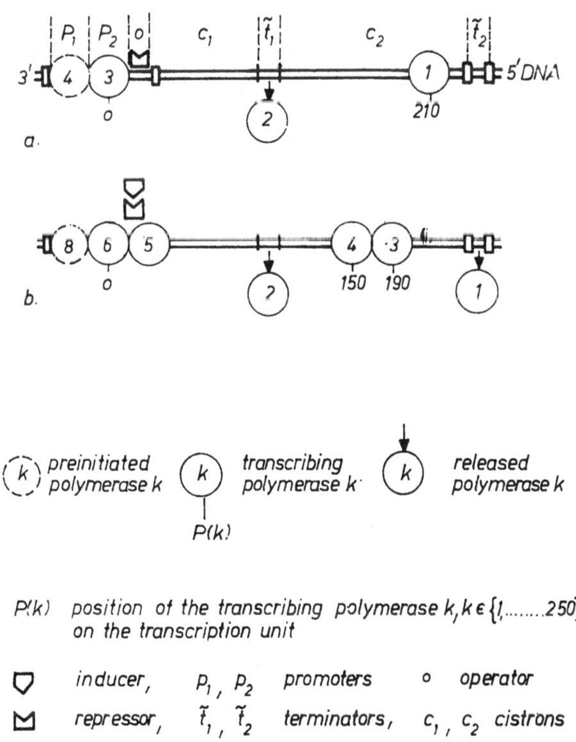

Fig. 23. Display of two different system states of a transcription system, the data of which are given in table 5, after a simulation time of 7.0 seconds and 12.5 seconds, respectively.

5.4.2. SELECTOUTPUT-statement

The SELECTOUTPUT-statement gives the user the possibility to by-pass the restrictions mentioned above.

9	22			
SELECTOUTPUT	A	B	C	D
	k	SK	k	RFS

A — Number k, $k \in \{1, \ldots, 5\}$, of the transcription system, for which the data shall be printed

B — Parameter for the characterization of the output data, with SK $\in$ {TRP, PRP, BOTH}

TRP: Print of the position pattern of *all* transcribing polymerases for the transcription system fixed by parameter A

PRP: Print of the complete protein pool pattern

BOTH: Print as the position pattern as the complete protein pool pattern

C — Time interval k, $k \in C_N \cup C_R$, between the cyclic consecutive output time points at which the print of the result data shall take place (output cycle-time)

Note:

See also the notation rule for the parameter B of the SYSTEMOUTPUT-statement!

D — Simulation time interval ΔT (TM$DELTA) as reference quantity for the output cycle-time

*** **Parameter *D* is omitted:**

* The output cycle-time is not defined in ΔT-time intervals!

Note:

The standard output is suppressed by using the USEROUTPUT-statements.

Content and form of the output

Analogous to the output mentioned above the print of the result data is here also initiated with the message

*** OUTPUT DATA OF THE PROGRAM ⟨program identifier⟩ ***

Then an identification of the transcription system and a characterization of the output quantities takes place by the information

*** TRANSCRIPTION SYSTEM ⟨number⟩ ⟨output data characterization⟩ ***

The data put out in the following are not subjected to restrictions, however it is necessary to take into consideration the volume of the data. At a simultaneous modelling of several transcription systems the respective patterns are printed in connected data complexes (compare section 5.4.1.).

Example:

In Tab. 6 the complete protein pool pattern for two chosen transcription systems (number 2 and 5), for which posttranscriptional processes are modelled, is shown for four consecutive output time points. The corresponding output statements should be given as follows:

```
SELECTOUTPUT 2,PRP,1,TM$DELTA
SELECTOUTPUT 5,PRP,1,TM$DELTA
```

In this example the simulation time interval ΔT runs to 0.25 seconds.

Table 6. Example for the print of the complete protein pool pattern of two transcription systems arbitrarily chosen at four consecutive time points.

```
*** OUTPUT DATA OF THE PROGRAM TRIGGER ***
                      :
*** TRANSCRIPTION SYSTEM 2 — PROTEIN POOL PATTERN ***
                      :
109.75  573  112  609  433  297
110.00  573  112  610  434  297
110.25  574  112  611  434  298
110.50  574  112  611  435  298
                      :
*** TRANSCRIPTION SYSTEM 5 — PROTEIN POOL PATTERN ***
                      :
109.75  703  750  760  743  768  789  738  740  711
110.00  704  750  760  743  768  790  739  740  712
110.25  704  751  761  743  769  791  740  741  712
110.50  704  751  761  744  770  792  741  741  713
                      :
```

5.4.3. USEROUTPUT-statement

In addition to the standard output a so-called user output exists in SIMTRA. One can refer to this output-possibility by using the USEROUTPUT-statement. The user output enables

— the print of additional result data, and
— the selection of output quantities and their formation to an own output.

9	22				
USEROUTPUT	A	B	C	D	E
	k	SK	k	RFS	$k \vee$ RFS

A — Number k, $k \in \{1, \ldots, 5\}$, of the transcription system, for which the data shall be printed

B — Parameter for the characterization of the output data, with SK $\in$ {NRPM, NRPT, SOR, SPR, PRP, TRP, NTI, TOA}

NRPM: Number of released polymerases at the mutation sites
NRPT: Number of released polymerases at the terminators
SOR: State of certain operators of the operator region
SPR: State of certain promoters of the promoter region
PRP: Protein pool pattern of the respective coding cistrons
TRP: Position pattern of certain transcribing polymerases
NTI: Nucleotide total incorporation
TOA: Turnover amount

Note:

It is obvious, that for these output quantities corresponding specifications (see parameter E) are necessary, with the exception of the turnover amount and of the nucleotide total incorporation.

C — Time interval k, $k \in C_N \cup C_R$, between the cyclic consecutive output time points at which the print of the result data shall take place (output cycle-time)

Note:

See also the notation rule for the parameter B of the SYSTEMOUTPUT-statement!

D — Simulation time interval ΔT (TM$DELTA) as reference quantity for the output cycle-time

*** **Parameter *D* is omitted:**

* The output cycle-time is not defined in ΔT-time intervals!

E — Number k, $k \in \{1, \ldots, s\}$, of the mutation site **or** of the terminator **or** of the operator **or** of the promoter **or** of the coding cistron **or** of the first initiated or last initiated transcribing polymerases, respectively **or**
SQ-list (LI$$i_1$ $\vee$ LI$$sn_1$) of numbers of the mentioned above molecular genetic parameters

Note:

$s \in C_N$ is the possible number, defined in SIMTRA, for the molecular genetic parameters mentioned above

*** **Parameter *E* is omitted:**

* A specification for the output data is not necessary!

Examples:

1. USEROUTPUT 5,TRP,20,10

For the transcription system with the number 5 the position of the ten first initiated and of the ten last initiated transcribing polymerases shall be put out in cycles with a twenty seconds time distance.

2.

```
        :
7       LIST            SQ
*/1,3,5,6/
        :
        USEROUTPUT      5,NRPT,10,TM$DELTA,LI$7
        :
```

For the transcription system with the number 5 the print of the released polymerases at the terminators 1, 3, 5, and 6 of the transcription unit is carried out for each tenth simulation time interval.

3.

```
        USEROUTPUT      5,NTI,50
```

For the transcription system with the number 5 the nucleotide total incorporation shall be put out cyclicly with a fifteen seconds time distance.

Content and form of the output

Analogous to the standard output the print of the result data is initiated in case of the user output also with the message

*** OUTPUT DATA OF THE PROGRAM ⟨program identifier⟩ ***

Then an identification of the transcription system and a characterization of the output quantities takes place by the information

*** TRANSCRIPTION SYSTEM ⟨number⟩ ⟨output data characterization⟩ ***

Especially for the latter a detailed description is carried out in correspondence to the parameter E of the USEROUTPUT-statements. The order of the data, which are put out in the following, is independent of the order of the programmed USEROUTPUT-statements. The SIMTRA-compiler ascertains this succession under the aspect of an optimal line limitation, in which the lucidity of the issue list is the essential criterion. Because the user has the possibility to print molecular genetic data at some different output time points by using the user output, the lucidity of the log can be lost at the modelling of already one transcription system. In such a case the parameters of equal classes are collected and are printed in connected data complexes. For this purpose the volume of the data, which are put out, is calculated automatically from the simulation duration, the output cycle-time and the simulation time interval ΔT by the programming system prior to the beginning of the print.

At a simultaneous modelling of several transcription systems the output data are printed in such a way as described in both sections mentioned above.

6. Application examples

In this chapter two examples shall demonstrate, that it is possible to describe more complicated molecular genetic systems with simple programming means by using the programming system SIMTRA. Moreover, these two examples show the syntax of complete SIMTRA-programs. The examples are hypothetical ones.

6.1. Model of a single transcription system

For a single transcription system represented in fig. 24 is assumed:

1. The transcription velocity TV of the polymerases along the transcription unit is given by the function displayed in fig. 24.

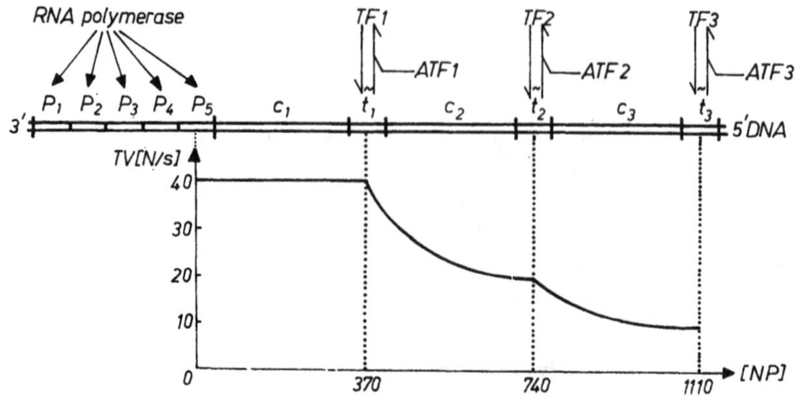

Fig. 24. Abstract model of a single transcription system.

$p_1, \ldots, p_5$	promoters	TF1. TF2. TF3	termination factors
$c_1, \ldots, c_3$	cistrons	ATF1 ATF2, ATF3	antitermination factors
$\tilde{t}_1, \tilde{t}_2$	intermediate terminators	l[NP]	length of the cistron sequence
$\tilde{t}_3$	end terminator		
TV [N/s]	transcription velocity of the polymerase molecules along the transcription unit in nucleotides per second		

2. The diameter of the polymerase is to be 144 Å. The time period for the initiation of a polymerase molecule at the promoter region is to amount to 8.0 seconds. The initiation shall be carried out with initiation variant 4, i.e. at every promoter $p_1, \ldots, p_i$ ($i > 1$) one polymerase is able to stay until the labile complex is formed, but can perform the next steps towards the initiation only at p_i.
3. During the whole period of the experiment (300 seconds) the concentration of the polymerases, the nucleotides and the termination factors is above the saturation level.
4. The concentration of the antitermination factors, which act at the intermediate terminators, are simulation variables, i.e. different simulation experiments with different however during the experiment constant ATF1- and ATF2-concentrations are carried out. The concentration of the antitermination factor ATF3, which acts at the end terminator, shall be zero.
5. The concentration-dependent probability of interaction of all molecule groups mentioned above to their binding partners is given by curves, the shape of which is characterized in fig. 7 by the value $k = 3$.
6. The output of the simulation results shall take place every ten seconds.

The SIMTRA-program for this transcription system, being assigned the number 5, looks as follows:

```
        PROGRAM     SINGLE
        LISTING
        PERCENT LISTING
%
% SIMULATION OF A SINGLE TRANSCRIPTION SYSTEM
%
% TIME COURSE OF THE POLYMERASE-CONCENTRATION
  1     FUNCTION     5,POL,3,100.0
*/0,100.0/300,100.0/
%
```

```
% COURSE OF THE TRANSCRIPTION VELOCITY
  3      FUNCTION      5,TV
*/0,40/370,40/740,20,8/1110,10,8/
%
% TIME COURSE OF THE TF1 – CONCENTRATION
  4      FUNCTION      5,TF,3,100.0,1
*/0,100.0/300,100.0/
% TIME COURSE OF THE TF2 – CONCENTRATION
  5      FUNCTION      5,TF,3,100.0,2
*/0,100.0/300,100.0/
% TIME COURSE OF THE TF3 – CONCENTRATION
  6      FUNCTION      5,TF,3,100.0,3
*/0,100.0/300,100.0/
% TIME COURSE OF THE ATF1 – CONCENTRATION
  7      FUNCTION      5,ATF,3,100.0,1
*/0,50.0/300,50.0/
% TIME COURSE OF THE ATF2 – CONCENTRATION
  8      FUNCTION      5.ATF,3,100.0,2
*/0,50.0/300,50.0/
% TIME COURSE OF THE ATF3 – CONCENTRATION
  9      FUNCTION      5,ATF,3,100.0,3
*/0,0.0/300,0.0/
% TIME COURSE OF THE NUCLEOTIDE-CONCENTRATION
  2      FUNCTION      5,NUC,3,100.0
*/0,100.0/300,100.0/
%
% LIST OF THE TF – FUNCTIONS
  1      LIST          EL
*/FN$4,FN$5,FN$6/
% LI$T OF THE ATF – FUNCTIONS
  2      LIST          EL
*/FN$7,FN$8,FN$9/
% LIST OF THE TERMINATOR SITES
  3      LIST          EL
*/370,740,1110/
%
% BLOCK STATEMENTS FOR THE DESCRIPTION OF THE TRANSCRIPTION
%
         SIMULATE      300
         TIME          5,40,144
         INITIATE      5,1,5-,8.0,FN$1
         ELONGATE      5,FN$2,FN$3
         TERMINATE     5,LI$1,LI$2,LI$3
%
% OUTPUT IN STANDARDIZED FORM
%
         SYSTEMOUTPUT 5,10
%
         END
```

The simulation results are shown in fig. 25. They underline, that a strong decrease of the transcription velocity by 75 percent between the terminators — at an increase of the ATF1 — and ATF2-concentrations — causes a "jam" of polymerases. If the ATF-concentrations are zero, this means that all polymerases are already released at

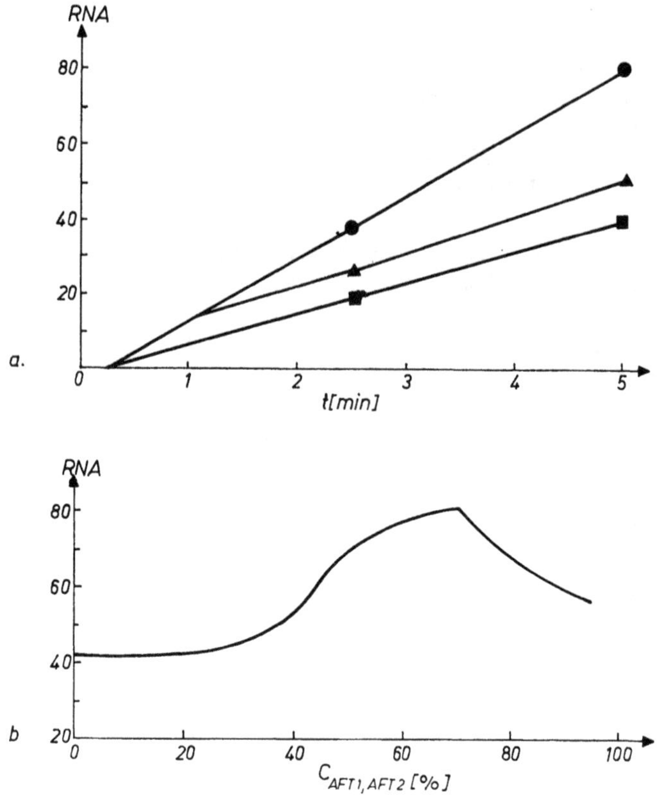

Fig. 25. Behaviour of a single transcription system given in fig. 24 at different concentrations of the antitermination factors ATF1 and ATF2.

a. Time course of the RNA concentration [number of knotted nucleotides $\times 10^3$] at zero (■), 70 (●) and 100 percent (▲) of the ATF1- and ATF2-saturation concentrations, respectively.

b. Mutual dependence of the RNA synthesis [number of knotted nucleotides $\times 10^3$] upon the ATF1- and ATF2-concentration after a simulation time of five minutes.

the intermediate terminator $\bar{t}_1$. An increase of the antitermination factor concentration leads to a decrease of the release-probability of the polymerases at the terminators, which enables an increase of the RNA synthesis. At ATF-concentrations of 70 percent one reaches the maximal synthesis efficiency of the transcription system. At concentrations above 70 percent the synthesis efficiency should — at least formally — increase because the release-probability of the polymerases at the intermediate terminators converges to zero. But at these concentrations the "jam" of polymerases is so momentous, that the RNA synthesis efficiency decreases. This effect is caused by the strong decrease of the transcription velocity and by the high ATF-concentrations produced by the increased seizure-density of the transcription unit with polymerases.

6.2. Model of two coupled transcription systems

For the two coupled transcription systems TS1 and TS2 represented in fig. 26 is assumed:

1. The polymerases of the transcription system TS1 (polymerase 1) transcribe along the transcription unit with the constant velocity of 40 nucleotides per second and the polymerases of the transcription system TS2 (polymerase 2) do the same with the constant velocity of 200 nucleotides per second.

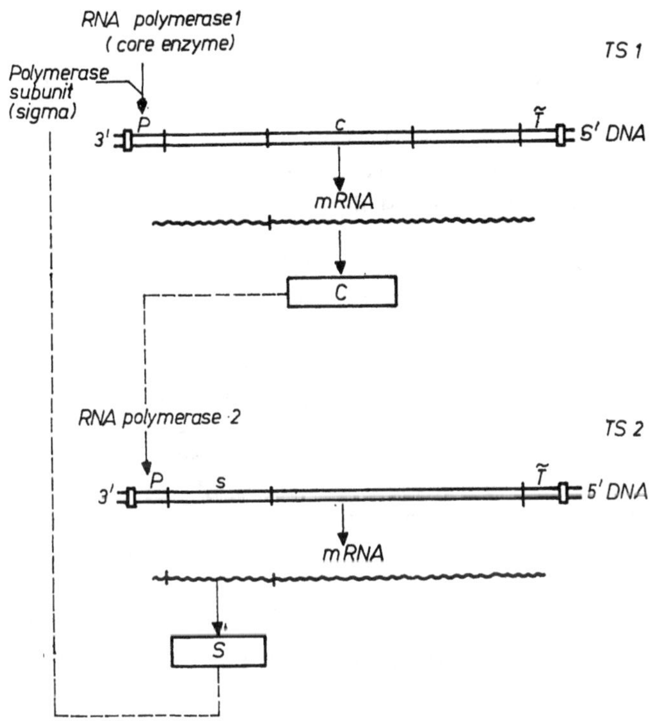

Fig. 26. Abstract model of two coupled transcription systems (TS).

P "classical" promoter
T end terminator
c, *s* cistrons

2. The diameter of a "polymerase 1 molecule" is to be 163 Å and the diameter of a "polymerase 2 molecule" shall be 102 Å. The length of each promoter corresponds to the respective polymerase diameters.
3. The time period for the initiation of a "polymerase 1 molecule" at the promoter region is to amount to 13.5 seconds and the corresponding period for a "polymerase 2 molecule" is to amount to 2,7 seconds. The initiation in both cases shall be carried out with initiation variant 1, i.e. at every promoter $p_1, \ldots, p_i$ one polymerase is able to initiate.
4. At the beginning of the simulation experiment the promoter of both systems is not occupied by polymerases.

5. The cistron *c* of the TS1 is localized between the nucleotide pair (NP) 2316 and 4998 of the transcription unit, the length of which shall be 7200 nucleotide pairs; the cistron *s* of the transcription system TS2, the length of which shall be 30000 NP, is localized directly after the promoter of the transcription unit. The length of this cistron is to amount to 1020 nucleotide pairs.
6. Released polymerase at the end terminator of the both systems are no longer considered.
7. The translation takes place polycistronicly at nascent RNA chains, the length of which shall be 150 nucleotides, before they can be translated by the first ribosome. The translation velocity of the ribosomes is to amount to 13.3 aminoacids per second.
8. Each RNA chain is to be translated tenfold and the translation distance between two neighbouring ribosomes is to amount to 108 nucleotides. Hereby the RNA chains shall not be degraded.
9. The synthesized proteins consist in their active form of one subunit, respectively. They shall not be subjected to a degradation process.
10. Each protein molecule of the pool *S* shall turn over 8000 polymerase subunit molecules per minute.
11. During the whole period of the experiment (660 seconds) the concentration of the polymerase 1, the nucleotides, the termination factors and the antitermination factors is above the saturation level.
12. The initial concentration of the polymerase 2 is zero. The saturation concentration, i.e. the concentration at which an immediate binding of the polymerases at the promoter can take place, is already reached with one molecule per volume unit.
13. The initial concentration and also the saturation concentration of the polymerase subunit is to amount to 100 molecules per volume unit.
14. The concentration-dependent probability of interaction of all molecule groups mentioned above to their binding partners is given by curves, the shape of which is characterized in fig. 7 by the value $k = 3$.
15. The output of the simulation results shall take place every ten seconds.

The SIMTRA-program for these transcription systems, being assigned the number 1 and the number 2, looks as follows:

```
          PROGRAM      COUPLE
          LISTING
          PERCENT LISTING
%
% SIMULATION OF TWO COUPLED TRANSCRIPTION SYSTEMS
%
% TIME COURSE OF THE POLYMERASE 1 - CONCENTRATION
  1     FUNCTION      1,POL,3,100.0
*/0,100.0/660,100.0/
% TIME COURSE OF THE POLYMERASE 2 - CONCENTRATION
  2     FUNCTION      2,POL,3,1
*/0,0/660,0/
% TIME COURSE OF THE SUBUNIT - CONCENTRATION
  3     FUNCTION      1,SUB,3,100
*/0,100/660,100/
```

```
% TIME COURSE OF THE NUCLEOTIDE - CONCENTRATION
  4     FUNCTION      LI$1,NUC,3,100.0
*/0,100.0/660.100.0/
% LIST OF THE POLYMERASE - FUNCTIONS
  1     LIST          GL
*/1/2/
% TIME COURSE OF THE TERMINATION FACTOR - CONCENTRATION (TS1)
  6     FUNCTION      1,TF,3,100.0
*/0,100.0/660,100.0/
% TIME COURSE OF THE TERMINATION FACTOR - CONCENTRATION (TS2)
  7     FUNCTION      2,TF,3,100.0
*/0,100.0/660,100.0/
% TIME COURSE OF THE ANTITERMINATION FACTOR - CONCENTRATION
% (TS1)
  8     FUNCTION      1,ATF,3,100.0
*/0,100.0/660,100.0/
% TIME COURSE OF THE ANTITERMINATION FACTOR - CONCENTRATION
% (TS2)
  9     FUNCTION      2,ATF,3,100,0
*/0,100.0/660,100.0/
% ARITHMETIC VARIABLES FOR THE REALIZATION OF THE COUPLING
% STRATEGIES
  1     VARIABLE      EZ
*/ - PL$2/
%
  2 VARIABLE          RP
*/PL$1/
% DEFINITION OF THE POOL C
  1     POOL          1,1,0
% DEFINITION OF THE POOL S
  2     POOL          2,1,0,8000
%
% FIXATION OF THE SIMULATION DURATION
        SIMULATE      660
%
% BLOCK STATEMENTS FOR THE TS1
%
        TIME          1,40,163
        INITIATE      1,1,1-,13.5,FN$1,FN$3
        ELONGATE      1,FN$4
        TERMINATE     1,FN$6,FN$8,7200,3
        TRANSLATE     1,13.3,10,108,4998,2316,150
        COUPLE        1,FN$3,VR$1
%
% BLOCK STATEMENTS FOR THE TS2
%
        TIME          2,200,102
        INITIATE      2,1,1-,2.7,FN$2
        ELONGATE      2,FN$4
```

```
        TERMINATE   2,FN$7,FN$9,30000,3
        TRANSLATE   2,13.3,10,108,1020,1,150
        COUPLE      2,FN$2,VR$2
%
% OUTPUT IN STANDARDIZED FORM
        SYSTEMOUTPUT 5,10
%
        END
```

Fig. 27 illustrates the time course of the mRNA concentrations and of the concentration of the C-protein and the S-protein, which are produced by the defined overall system during the experiment. After the beginning of the mRNA synthesis for the

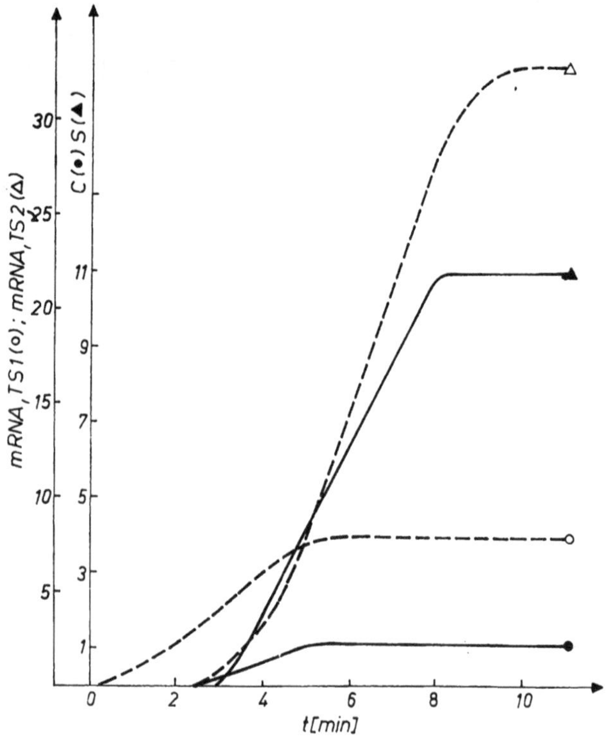

Fig. 27. Behaviour of two coupled transcription systems (TS) given in fig. 26.
mRNA, TS1 (number of knotted nucleotides $\times 10^4$)
mRNA, TS2 (number of knotted nucleotides $\times 10^5$)
C, S (number of molecules $\times 10^2$)

TS1 the synthesis of the protein C, the polymerase 2, starts with a certain delay. These polymerases transcribe the transcription unit 2, which contains the cistron *s*, with the latter coding the protein S. The protein S inactivates in an increasing measure the molecules of the polymerase subunit, which are essential for the transcription of the TS1. Thus, the RNA synthesis for the TS1 is repressed. This effect results in a decreased synthesis of C-protein molecules, so that even the transcription process of the TS2 comes to an end.

Thus, the regulation circuit

polymerase 1 → mRNA (TS1) → protein C → mRNA (TS2) → protein S → polymerase 1

leads to a "switch off" of the overall system.

References

BELLMANN, K. (1976): Algorithmische Simulationsmodelle zur Analyse komplexer, nichtlinearer biologischer Systeme in: Mathematische Modellbildung in Naturwissenschaft und Technik Akademie-Verlag Berlin, pp. 197—208.

FRANK, M. (1977): Einige Forderungen an Simulationssysteme für Fertigungsprozesse. Die Technik 32, 7, pp. 378—382.

HARTLEY, M. G. (1975): Digital simulation methods. IEE Monograph Series Vol. 15. Peter Peregrinus Ltd. Stevenage.

KARPLUS, W. J. (1976): The spectrum of mathematical modelling and simulation in: Simulation of systems North-Holland Publishing Company Delft, pp. 5—13.

KREISCHER, U. (1977): Internal ZfR-report about the programming system SIMTRA, AdW der DDR, Zentrum für Rechentechnik Berlin.

KREISCHER, U. (1978): SIMTRA — ein Programmiersystem zur Systemanalyse in der Molekulargenetik, Humboldt-Universität Berlin, Vortrag auf dem III. Wissenschaftlichen Kolloquium zur Organisation der Informationsverarbeitung 1/1978. Kurzfassung in ZfR-Information 77.14.

KREISCHER, U., and A. KNIJNENBURG (1976): Zur Methodologie der Simulation eines molekulargenetischen Systems. Biometr. Zeitschrift 18, 1, pp. 69—84.

MAURER, H. (1969): Theoretische Grundlagen der Programmiersprachen. Theorie der Syntax. Bibliographisches Institut Mannheim · Wien · Zürich.

MIHRAM, G. A. (1972): Simulation: statistical foundations and methodology. Math. in Science and Engineering Vol. 92. Academic Press New York · San Francksco · London.

NEUMANN, H., and U. KREISCHER: Mathematico-cybernetical model of procaryotic transcription systems, these proceedings.

SCHWARZE, G. (1977): Studie Simulation mittels Digitalrechnern ZfR-Information 77.05.

7. Modelling and simulation in analysis of eukaryotic transcription control

K. Bellmann, J. Born, R. Lindigkeit, R. Schulz

Contents

Introduction

The first step of gene expression is the transcription process. Therefore, regulation of transcription is of crucial importance in the chain of events leading to the controlled expression of distinct genes. It is generally believed and it seems to be reasonable that the control mechanisms which decides on the species of messenger RNA to be synthesized are effective at the level of initiation. We can assume that in this control both the activity of the RNA synthesizing enzyme, the RNA polymerase, and the conformation of the chromatin template in the nearer and broader region of the initiation site are involved. Because of the complicated chromatin organization in eukaryotes, however, one should expect that also along the chromatin fibre further control mechanisms are effective. They might involve conformational changes of the chromatin fibre necessary to transform the inactive template into an active state. In which way controls leading to the initiation of new RNA chains, on the one hand, and allowing the RNA polymerase to travel along the template, on the other hand, cooperate is an open question as yet. The understanding of the rather complex transcription process in eukaryotic cells and of its control processes require profound knowledge of chromatin organization.

Two types of chromatin are present in the nuclei. These are the transcriptionally inactive condensed chromatin and the transcriptionally active decondensed chromatin (for a review see PEDERSON, 1978). The transcriptionally active chromatin exists probably as an extended 10 mm fibre whereas the inactive chromatin exhibits compact 20 mm fibres which are further coiled or folded to give a very compact structure.

In the last few years much progress in understanding of the organization of the chromatin structure has been reached. So it has been shown that the basic 10 mn chromatin fibre is organized in regular repeating structural units consisting of about 200 base pairs (bp) of DNA wounded around a core of 8 histones, formed by each two molecules of H2A, H2B, H3 and H4. Independently of the species each 145 bp of DNA is stronger associated to the histone core so that after digestion experiments with micrococcal nuclease the so-called core particles are obtained. To the repeated unit of 200 bp of DNA belongs beside the core histones also each one molecule of histone H1 anyhow bound to the linker DNA connecting each two core particles. This complete unit has been termed nucleosome (for reviews see: KORNBERG, 1977; FELSENFELD, 1978; BRADBURY et al., 1978; OUDET et al., 1978).

By the package of the DNA in form of nucleosomes its transcription is widely repressed. So it has been shown that nucleosomes inhibit initiation and elongation of RNA chains by RNA polymerase B under low salt conditions in vitro (WASYLYK and CHAMBON, 1979). In vitro the inhibition of elongation can be prevented at high ionic strength (WASYLYK et al., 1978; CEDAR and FELSENFELD, 1973; GISSINGER et al., 1974; WILLIAMSON and FELSENFELD, 1978). At the present time little is known about the mechanism allowing RNA polymerase to pass through the nucleosomes in vivo. It is, however, now well established that in transcriptionally active chromatin regions the nucleosomes exhibit a changed conformation. So they are more sensitive against deoxyribonuclease I than those of inactive chromatin (WEINTRAUB and GROUDINE, 1976). On the basis of this deoxyribonuclease I sensitivity it has been shown that transcriptionally active nucleosomes contain modified histones; especially the arginin rich histones H3 and H4 occur in acetylated form thereby weakening their binding to the DNA. In last time it has been shown that also defined low molecular weight nonhistones such as the HMG 14 and 17 (for rev. see GOODWIN et al., 1977) are associated especially with the nucleosomes of the transcriptionally active chromatin segments (WEISBROD et al., 1980). Very recently it has been shown by REEVES and CANDIDO, 1980, that the HMG 14 and 17 prevent deacetylation of the histone so elevating the level of acetylation in the transcription active nucleosomes. The exact conformational changes of the active nucleosomes provoked by acetylation or other modifications of the histones is not yet known. Two main types of DNA depended RNA polymerase activity have been distinguished in the cell nucleus. RNA polymerase B located in the nucleoplasm synthesizes the heterogeneous nuclear RNA including also the messenger RNA. Low concentrations of α-amanitin inhibit highly specific this type of RNA polymerase. RNA polymerase A occurs in the nucleolus and has been shown to synthesize the ribosomal RNA.

In isolated nuclei the RNA synthesis by the B-enzyme is small at low ionic strength and comes to an end already after a few minutes. Increasing ammonium sulfate concentrations (ASC) up to 0.4 M in the pres nse of Mn-ions stimulate the RNA synthesis by the endogenous RNA polymerase B (WIDNELL and TATA, 1966; BLACKBURN and KLEMPERE, 1967; RO and BUSCH, 1967; CHAMBON et al., 1968; POGO, 1969; MANDEL and CHAMBON, 1969). It is assumed that this stimulation concerns the elongation of RNA chains which were already involved in the RNA synthesis process of the cell. This suggestion is also supported by the fact that at the high salt concentration of 0.4 M ammonium sulfate frequently used in the stimulation experiments with isolated nuclei initiation of new RNA chains can be excluded (GISSINGER et al., 1974).

On the other hand, in small kidney epithelial cells which are transcriptionally inactive in vivo RNA synthesis can be induced in vitro by addition of 0.4 M ammonium sulfate (AUER et al., 1973; MOORE et al., 1974). The authors of this finding interpret this fact as indication that in the small cells RNA polymerase was already bound to DNA but was present there in a repressed state. In the presence of the high ASC of 0.4 M they regain the ability to incorporate ^{3}H-UTP into the RNA since by the salt extensive alterations in the structure of the chromatin occur which allow the in vivo bound RNA polymerase to travel in vitro as a consequence of a changed chromatin structure. According to the authors the existence of the high salt activity in small cells indicates a way in which genetic activity could be initiated, i.e., by the controlled stimulation of RNA chain elongation by RNA polymerase which was already bound to DNA but was present in a repressed state. In vivo the onset of genetic reactivation in such small inactive cells is probably regulated by proteins which enter the nucleus (MOORE et al., 1974).

Studies on cells, however, do not allow the determination of RNA synthesis at defined chromosomal sites. Experiments performed by SASS (1978) on isolated native giant chromosomes of *Chironomus tentans* on the salt stimulation of the RNA synthesis over distinct chromosomal regions (bands) as localized by autoradiography in the different bands, support the suggestions of MOORE et al., 1974. Earlier autoradiographic investigations of PELLING, 1969, have shown that the chromosom I of that species exhibits in the mean 45 transcriptionally active decondensed chromosomal bands, the so-called puffs. In contrast to that in vitro incubated chromosomes showed at high salt concentrations 60–80 bands (SASS, 1978). A more detailed consideration showed that the mass of the silver grains of the incorporated ^{3}H-UTP was located at other sites than is typical for in situ labelled chromosomes. Moreover, in contrast to the typical RNA labelling picture at which silver grains are preferrentially localized over puffs (PELLING, 1964, 1965), in the case of the cell-free incubated giant chromosomes the silver grains were found over condensed bands, whereas the less condensed bands showed rather poor labelling. However, also under in vitro conditions at high salt only a limited number of bands showed transcription activity. In an RNA synthesizing system with isolated polytene nuclei from *Chironomus tentans*, HAMEISTER, 1977, has obtained similar results and has discussed them in terms of activated and preactivated genes.

The results of SASS (1978) and HAMEISTER (1977) seem to indicate that there are three different types of bands differing in their state of transcription activity. The in vivo transcriptionally active decondensed bands or puffs appear to be inactivated under in vitro conditions at high salt concentration. Two types of condensed bands being transcriptionally inactive in vivo can be distinguished in so far as only one type of them can be activated by high salt concentrations in vitro. Since under these experimental conditions RNA polymerase does not bind to the template one should suggest that the in vitro stimulated bands contained already in vivo bound RNA polymerase molecules being, however, repressed there. If these RNA polymerase molecules are present in reactivated genes as suggested by HAMEISTER (1977) one should expect that they are present nearby the initiation point.

In order to analyse the salt stimulation effect on the RNA synthesis in vitro, the pattern of the RNA synthesized in isolated rat liver nuclei by the endogeneous RNA polymerase B has been investigated by centrifugation in an isokinetic sucrose gradient. Always a typical pattern of distinct RNA fractions has been found. The different RNA fractions were identified by their sedimentation coefficients (LINDIGKEIT et al., 1974, 1975). Action of nucleases have been prevented by performing nuclei preparation in the presence of the cytoplasmic ribonuclease inhibitor isolated by affinity chromatography. As checked at salt concentrations above 0.1 M ammonium sulfate ribonuclease activity in the in vitro RNA synthesis systems were completely inhibited. Moreover, chase experiments performed between 0.2 and 0.4 M ammonium sulfate, excluded the possibility that the distinct RNA fractions arose from a subsequent cleavage of originally longer RNA chains. On the contrary, the chase experiments indicated that the size of the labelled RNA increased with time in the inactive chase medium. This means that the RNA synthesized in vitro by the B-enzyme consisted of a collection of distinct RNA fractions with different chain lengths growing with time. The RNA patterns published by several authors of in vitro in isolated nuclei synthesized RNA indicate also discontinuity (ZYLBER and PENMAN, 1971). The in vivo in Balbiani rings synthesized RNA, however, shows continuous lengths (DANEHOLT et al., 1978). The continuous pattern of in vivo synthesized RNA might derive from the fact that in vivo all the inhibiting or blocking structures have changed their conformations by the process which makes the chromatin segments fully transcribable. Because on the other hand, as mentioned above, it seems that in vitro at least a part of the stimulation of the RNA synthesis by salt bases on conformational changes of in vitro condensed chromatin, containing, however, already RNA polymerase bound to the template one might suggest that the changes by salt along the template are incomplete so that some of the "blockers" are not removed. As a consequence, the RNA synthesis would stop at distinct sites at the template and would continue only after the "blocking" principle has been overcome.

Performing the in vitro RNA synthesis with purified nuclei between 0.2 and 0.4 M ammonium sulfate it has been shown that the relative amounts of the distinct RNA fractions synthesized in vitro by the B-enzyme depended on the salt concentration in the way that the more rapidly sedimenting RNA fractions increased with the ammonium sulfate concentration (LINDIGKEIT et al., 1974, 1975). WIDNELL and TATA (1966) were the first ones demonstrating that the in vitro synthesis of RNA in the presence of 0.4 M ammonium sulfate yielded much longer RNA chains than the synthesis in the absence of the salt. They, however, compared the average chain length of the ribosomal-like RNA which is mainly synthesized in the presence of Mg-ions at low salt concentrations with the product of the B-enzyme synthesized at ASC of 0.4 M in the presence of Mn-ions.

The results of time experiments at high ionic strength analysed in this report are of interest. On the one hand, the amount of long sized labelled RNA increase with time. On the other hand, in the same experiments besides an increase of rather short chains appeared with time which was not the result of subsequent degradation of longer RNA chains as proved by chase experiments. The impression arose that during the first 20 min of incubation, especially the more slowly sedimenting RNA fractions between 9 and 19 S increase. When prolonging the incubation time they are shifted to the fractions with higher sedimentation coefficients. This result might indicate that at the early time of in vitro RNA synthesis additional RNA chains are elongated deriving probably from RNA polymerase already bound to the chromatin template in the living cell but being present near the initiation site in an inactive state.

From the results of the analysis of the in vitro synthesized RNA it has been suggested that at least on parts of the transcription units "transcription blockers" are localized. These hypothetical blockers have been suggested by the result according to which in a high resolving sucrose gradient the in vitro synthesized RNA occurred always in form of a collection of distinct RNA fractions of different chain length. The discontinuity was obviously the result of the RNA synthesized by the RNA polymerase B as shown by the fact that after inhibition of the B-enzyme by α-amanitin the typical RNA fractions were not obtained. The RNA synthesized in vitro in the isolated nuclei is, of course, the result of the transcription of a big number of different transcription units. Therefore, one has to suggest that the hypothetical blockers should reflect a general organization principle of chromatin. The question arises on what type of structure the blocking effect might be based? If the so-called blockers are nucleosomes the distances between them should be 200 base pairs or multiples of them. We have determined, however, approximately 540 base pairs as the mean blocker distance. Based on the nucleosome arrangement along the chromatin fibre this would amount to the distance between each two nucleosomes and one core particle. Because of the relatively rough determination, the distances could also mean three nucleosomes. Whether this makes a real sense is an open question.

Although the nature of the postulated transcription blockers remains unclear, in the present approach we have modelled the in vitro transcription process under the assumption that at distinct points along the chromatin fibre the transcription process is blocked. These hypothetical transcription blockers may be partly removed or inactivated by increasing salt concentrations as well as with time at a distinct ASC of 0.1 M. Between two blockers the RNA polymerase may travel more or less unchecked (LINDIGKEIT et al., 1974). An other assumption induced in order to analyse the distribution of the RNA of different chain length is that the counts as measured in the sucrose gradient fractions indicate RNA chains labelled over the whole chain. This holds true of course only if the RNA polymerase starts in vitro with new RNA chains. However, at the high salt concentrations used in the experiments initiation of new RNA chains must be excluded. Yet under the assumption that a relatively big portion of the in vitro synthesized RNA derives from RNA polymerase which have already started RNA chains in vivo but was prevented from travelling because of the inactive chromatin template, the assumption of complete labeling over the chain seems to be a useful, although simplified, approximation.

Experimental results and the blocker hypothesis

The main results of this work were:

a) As indicated by the obvious peaks of the cpm (S) profile (as a result of isokinetic gradient centrifugation analysis of RNA synthetized by B-enzyme) distinct RNA fractions exist (see fig. 1).

b) With increasing ASC an increased activation of the transcription process can be observed [compare a) and b) of fig. 1].

c) Summarizing the results of all AS stimulated RNA-synthesis experiments (ASC: 0.2, ... 0.4 M) one can show that the RNA fractions with chains of the following

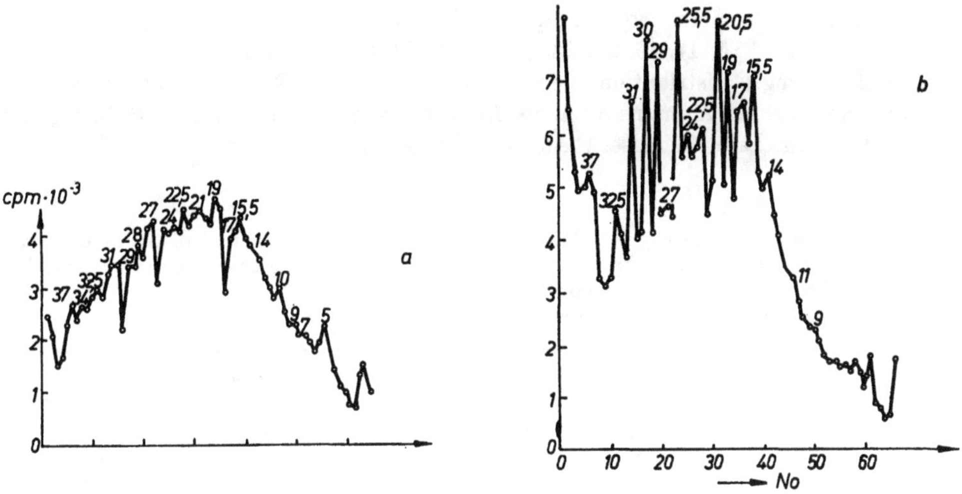

a) b)

Fig. 1. Isokinetic gradient centrifugation analysis of RNA synthesized in vitro in rat liver nuclei in the presence of (a) 0.05 M ammonium sulfate; (b) 0.3 M ammonium sulfate. From LINDIGKEIT et al., 1974.

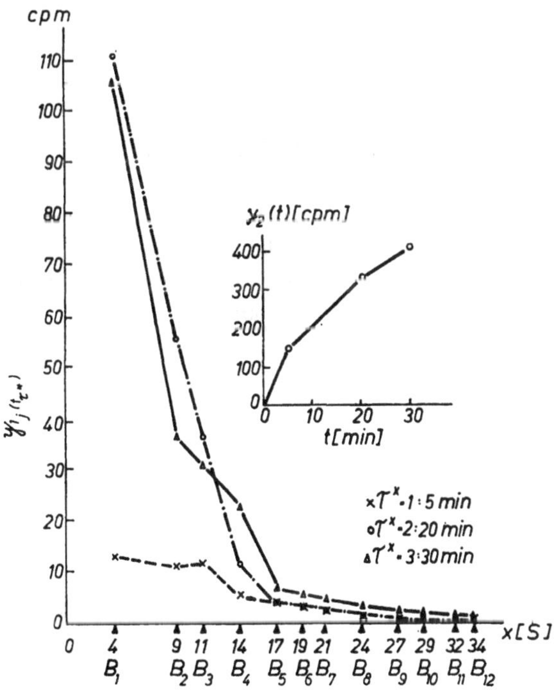

Fig. 2. Distribution $y_{1_j}(t_{\tau^*})$, in terms of cpm, of RNA chain lengths in a real experiment (isokinetic gradiant centrifugation analysis of RNA synthesized in vitro in rat liver nuclei in the presence of 0.4 M ammonium sulfate, 30 min, 30°C), and total amount $y_2(t_{\tau^*})$ RNA de novo produced: $\tau^* = 1$ (5 min), $\tau^* = 2$ (20 min), $\tau^* = 3$ (30 min). From BELLMANN et al., 1975.

lengths are much more frequent (approx. 6-fold) than the other chain lengths: 5, 7, 9, 11, 12.5, 14, 15.5, 17, 19, 21, 22.5, 24, 25.5, 27, 29 S (see tab. 1).

d) The chain length distributions depends on time (see fig. 2). Generally, at each instant the short RNA chains are more frequent than the long ones. With increasing time the long chains increase, the short ones decrease.

Table 1. Frequency (n) of the appearance of RNA fractions with distinct sedimentation coefficients (S) related to 40 runs of different RNA preparations synthesized between 0.2 and 0.4 M ammonium sulfate.
From LINDIGKEIT et al., 1974.

Approx. S	n	Approx. S	n	Approx. S	n
5	**15**	14	**28**	21.5	3
7	**27**	15.5	**27**	22.5	**27**
8	3	16	3	24	**34**
9	**35**	17	**30**	25.5	**26**
10	8	18	8	26.5	2
11	**30**	19	**33**	27	**27**
12.5	**24**	19.5	3	28	3
13	2	20	6	29	**30**
13.5	3	21	**37**		

Fractions with higher sedimentation coefficients are not listed. The fat values represent RNA fractions with distinct chain lengths that are characteristic of the in vitro RNA synthesis product of RNA polymerase B.

2. Logical analysis of the real system and its formalization

The model of the real transcription system and process was to be constructed in such a way that the decondensation of the chromatin structures (blockers) is a time consuming process that leads to such a polymerase travel dynamics which is compatible with the time depending chain lengths distribution observed experimentally.

2.1. Transcription apparatus

2.1.1. Template set

The transcription system is defined by a set of templates $T_\varrho \in \tau$. At each T_ϱ RNA synthesis is carried out by polymerase molecules ($\varrho = 1, \ldots, r$; r: nbr. of templates, r is finite but unknown, r is assumed to be of order 10^x, $3 \leq x \leq 5$). Each T_ϱ is assumed to be divided in u units

$$U_{\varrho 0}, U_{\varrho 1}, U_{\varrho 2}, \ldots, U_{\varrho i}, \ldots, U_{\varrho u}; \quad i \in I = (0, 1, \ldots, u),$$

which are of the same length of 200 nucleotides (1 S). In our experiments it proved suitable to set $u = 34$ (see fig. 4).

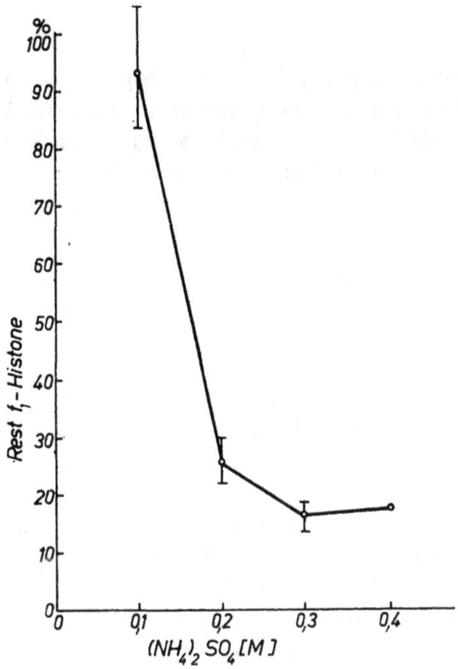

Fig. 3. Extractability of histone f_1 from salt treated chromatin. From LINDIGKEIT et al., 1974.

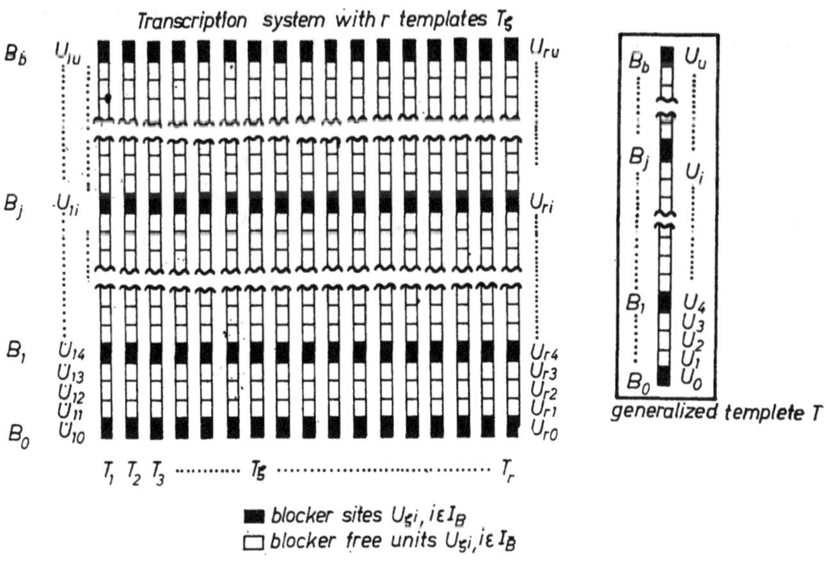

Fig. 4. Transcription system with templates T_ϱ, $\varrho = 1, \ldots, r$, consisting on the segments $U_{\varrho i}$, $i = 1, \ldots, u$, some of them being blockers $U_{\varrho i}$, $i \in I_B$.

2.1.2. Blocker sites

Each template $T\varrho \in \tau$ shows at special units $U_{\varrho i}$, $i \in I_B$, the property to stop travelling polymerase molecules temporably or finally. These units are called **blocker sites** $B_{\varrho j}$ ($j = 0, 1, 2, \ldots, b$; in our experiments $b = 12$), arranged at the sites $A(B_{\varrho j}) = A_j$ on the templates $T\varrho$, I_B is the index set of units that are blockers. The chromatin of each template $T\varrho$ was assumed to be organized in the same way, i.e. the A_j are independend on DNA sequence of $T\varrho$:

$$A_j = A_{1j} = A_{2j} = \ldots, A_{\varrho j}, \ldots, A_{rj}; \quad j = 0, 1, \ldots, b.$$

It is

$$\boldsymbol{A} = (A_0, A_1, A_2, \ldots, A_j, \ldots, A_b)$$

the **blocker position vector**, definable by experiments. The blocker sites U_i, $i \in I_B$, were assumed to be arranged below the peaks and "shoulders" of the cpm (nbr.) profiles of the chain lengths distribution (as shown in fig. 1: result of isokinetic gradient centrifugation analysis of RNA).

In order to get information about the uncertainty of the obviously rough procedure of identifying the blocker sites B_j the question arose: what is the highest possible experimental error in isokinetic gradient centrifugation analyses (IGCA) allowing a correct identification of all blocker sites with a certainty of $Q \geqq 0.95$?

Therefore, a small simulation model for the calculation of cpm (nbr.) profiles consisting of b superposed (chain length-) distributions (each described by mean and variance) was developed. The b distributions correspond to distinct RNA-chain sets with the chains $K_{i\varrho} \in K_i, i \in I_B$, each $K_{i\varrho}$ of length $X_i = X_{i\varrho} = X_{i\varrho}$; $\varrho, \varrho' = 1, \ldots, N_i$ (N_i: nbr. of chains $K_{i\varrho}$). Since X_i is not estimable by IGCA without an error ε_ϱ, X_i becomes a random variable with realizations $x_{i\varrho} = X_i + \varepsilon_\varrho$; ε: $N(0, \sigma)$; $i \in I_B$. By simulation of superposed normal distributions ϱ_i: $N(X_i, \sigma)$, $i \in I_B$, it was found that the A_j can be determined with a certainty $Q \geqq 0.95$ if $\sigma \leqq 0.7$ S (Bellmann 1975, unpublished); see fig. 5: Q is the probability that all blockers were identified correctly, i.e. at their real template site A_j, $j = 0, 1, 2, \ldots, b$.

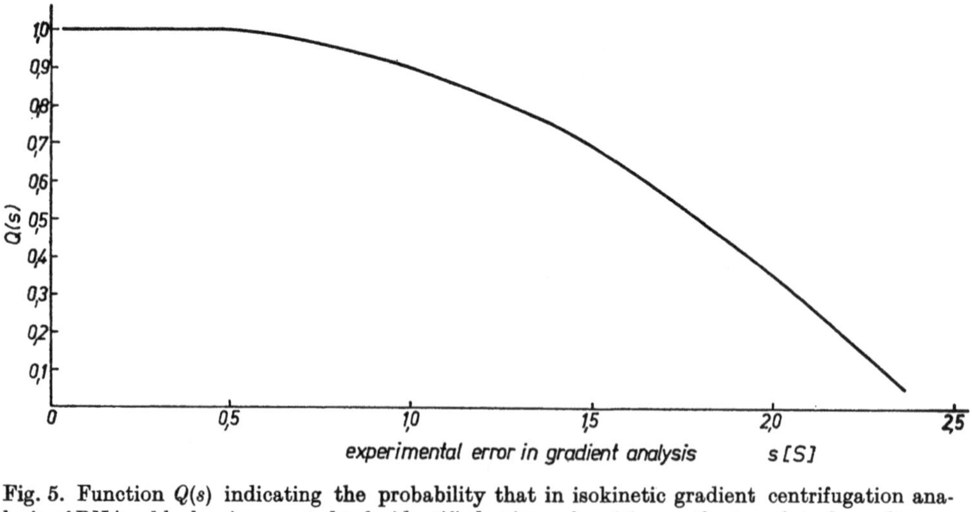

Fig. 5. Function $Q(s)$ indicating the probability that in isokinetic gradient centrifugation analysis of RNA a blocker is expected to be identified at its real position on the template depending on the experiment error σ in gradient analysis (σ in terms of S-units).

We used for most simulation experiments the estimated blocker position vector

$$\hat{A} = (A_0 = 0\text{S},\ A_1 = 4\text{S},\ A_2 = 9\text{S},\ A_3 = 11\text{S},\ A_4 = 14\text{S},\ A_5 = 17\text{S},\ A_6 = 19\text{S},\ A_7 = 21\text{S},\ A_8 = 24\text{S},\ A_9 = 27\text{S},\ A_{10} = 29\text{S},\ A_{11} = 31\text{S},\ A_{12} = 34\text{S}),$$

i.e. it is

$I_B = (0, 4, 9, 11, 14, \ldots, 31, 34)$: index set for blocker sites

$I_{\bar{B}} = (1, 2, 3, 5, 6, 7, 8, 10, 12, 13, \ldots, 32, 33)$: index set of blocker free sites.

2.1.3. State changes of blocker sites

The blocker sites U_i, $i \in I_B$, may occur in 2 states $\{0, 1\}$:

$$D_i = D(U_i, i \in I_B) = \begin{cases} 1, & \text{if } U_i \text{ is deblocked}, \\ 0, & \text{if } U_i \text{ is blocked}, \end{cases} \qquad i \in I_B$$

$D(U_{\varrho i})_{t_\tau} = 1$ (deblocked) means that a polymerase travelling along T_ϱ may pass the blocker site $U_{\varrho i}$, $i \in I_B$, unchecked at instant t_τ; $D(U_{\varrho i}) = 0$ means, that this polymerase is stopped at instant t_τ. For a non blocker site U_i, $i \in I_{\bar{B}}$, it is $D(U_{\varrho i})_{t_\tau} = 1$ for all ϱ and τ $\tau = 0, 1, \ldots, T$, i.e. a polymerase molecule may pass $U_{\varrho i}$, $i \in I_B$, unchecked at each instant (no restriction in polymerase movement between two subsequent blocker sites).

We assume that (i) at t_0 (start of experiment) all blocker sites $U_{\varrho i}$, $i \in I_B$, are not permeable for a polymerase molecule synthetizing RNA, i.e.

$$D_0(t_0) = D_1(t_0) = \cdots = D_i(t_0) = \cdots D_u(t_0) = 0, \quad i \in I_B$$

which is equivalent to

$$D_0(t_0) = D_1(t_0) = \cdots = D_j(t_0) = \cdots D_b(t_0) = 0$$

(ii) the blocker sites B_j become deblocked independend on ϱ via loosing of bindings due to effects of AS ions, that means the probability $Pr(D_j(t_\tau) = 1)$ that at instant t_τ the site B_j is deblocked is an (generally) increasing function $F1_j(t_\tau)$ with

$$Pr(D_j(t_0) = 1) = 0 \quad \text{and}$$

$$Pr(D_j(t_\tau) = 1) = F1_j(t_\tau, \text{ASC} = e) \leqq 1.$$

For the present, in the model no assumption is made about the shape of $F1_j(t_\tau)$; $0 \leqq F1_j(t_\tau) \leqq 1$. $F1_j(t_\tau)$ is called **blocker removal function** indicating the relative amount (probability) of blocker sites $B_{\varrho i}$ at instant t_τ being in the state $D_j = 1$, i.e. in a state allowing a polymerase molecule to pass the site $B_{\varrho i}$. The function $F1_j(t_\tau)$ yields information about the permeability state of site B_j at instant t_τ. By means of $F1_j(t_\tau)$, the intensity of the deblocking process is described, thereby reblocking is not excluded (see fig. 6).

In order to overcome the obstacle $B_{\varrho j}$ by a polymerase it is necessary that $B_{\varrho j}$ becomes dissolved, i.e. more exactly $D_{\varrho j} = 0$ is to change to $D_{\varrho j} = 1$. The relative amount of $B_{\varrho j}$ being closed (blocked) and become dissolved (decondensed) is described by the function

$$F2_j(t_\tau) = \frac{F1_j(t_{\tau+1}) - F1_j(t_\tau)}{1 - F1_j(t_\tau)}, \quad j = 1, \ldots, b.$$

This function is utilized only if $F1_j(t_\tau) \geqq F1_j(t_\tau)$. $F2_j(t_\tau)$ is the probability that a blocker site $B_{\varrho j}$ being closed up to t_τ becomes open (deblocked) at that moment t_τ and remains being deblocked furtheron for all t_τ, $\tau < \bar{\tau} \leqq T$, providing that $F1_j(t_\tau)$ is not

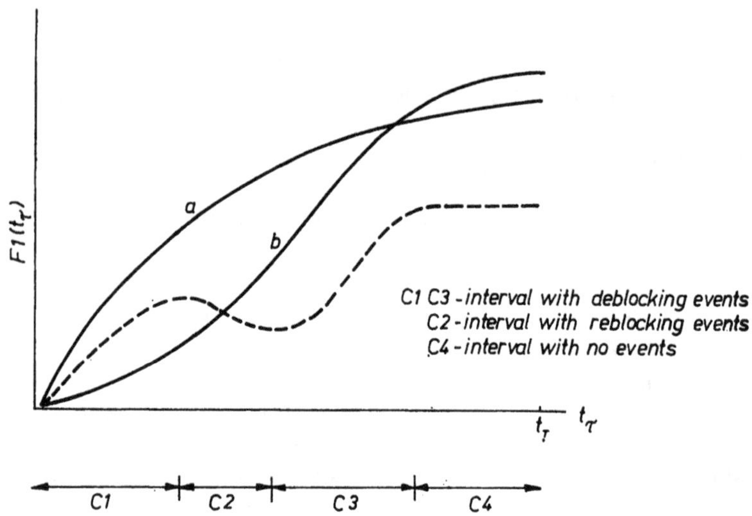

Fig. 6. Possible $F1_j(t_\tau)$ shapes, e.g. (a), (b) without reblocking events, (c) shape indicating different intervals with deblocking, reblocking, and no events (C1, C3; C2; C4 resp.).

decreasing at any instant t_τ. $F2_j(t_\tau)$ is called **blocker removal rate.** Consequently, $F2_j(t_\tau)$ is the probability that a polymerase waiting at $B_{\varrho j}$ up to t_τ is now allowed to move on towards $B_{\varrho j+1}$. Therefore $F2_j(t_\tau)$ is also called **moving-on-function.**

Only if $F1_j(t_\tau)$ decreases, $F1_j(t_{\tau+1}) < F1(t_\tau)$, a function

$$F3_j(t_\tau) = \frac{(F1_j(t_{\tau+1}) - F1(t_\tau)}{F1_j(t_\tau)}$$

is used. $F3_j(t_\tau)$ is called **reblocking rate,** indicating the relative amount (probability) of $B_{\varrho j}$ changing from state $D_j(t_\tau) = 1$ to state $D_j(t_{\tau+1}) = 0$ during the time interval $(t_\tau, t_{\tau+1})$, i.e. an deblocked (open) site becomes impermeable for a polymerase molecule by recondensation of the template at site $B_{\varrho j}$.

2.2. Polymerase onset

The site $U_0 = B_0$ possesses a special importance for the dynamics of the transcription process. U_0 is considered to be a template region where the polymerases are already bounded and starting there to synthesize RNA, i.e. U_0 be the onset site for inflow of polymerases into the transcription system if $U_0 = B_0$ is "open" (deblocked). At instant t_0 sites $B_{\varrho 0}$ be in the state $D_{\varrho 0} = 1$. At the begin of the experiment all $B_{\varrho 0}$ are assumed to be inactive ($D_{\varrho 0} = 0$) and change from $D_{\varrho 0} = 0$ to $D_{\varrho 0} = 1$ with a certain probability at t_τ.

The **inflow function** (initiation function, source function) $F1_0(t_\tau)$ indicates the probability that a site $B_{\varrho 0}$ is in state $D_0 = 1$ and consequently able to allow the onset of polymerases.

2.3. Transcription process

2.3.1. Polymerase movement (elongation)

The movement of the polymerase molecule along the templates $T\varrho$ leads to RNA growth, which was assumed to be unrestricted between two subsequent blocker sites B_j, B_{j+1}. In the model the RNA synthesis proceeds from the (left) onset site $B_{0\varrho}$ of each template $T\varrho$ to $B_{0\varrho}$ under the restricting conditions of temporary (or total) stop of polymerase movement. A RNA chain is growing during the total time interval (t_0, t_τ). The elongation process depends on (i) the polymerase **velocity** v, assumed to be $v = 34$ nucleotides/sec. (i.e. 7.5 sec. are needed for synthesis of a S unit U_i), (ii) on the **waiting time** at a closed blocker site $B_{\varrho j}$, and (iii) a the polymerase **survival probalitity**[1]. A polymerase $PM_{\varrho 1}$ moving along a template $T\varrho$ is stopped at $B_{\varrho j}$ if $D_{\varrho j}(t_\tau) = 0$ and is to wait until $B_{\varrho j}$ becomes dissolved $\left(D_{\varrho j}(t_{\tau'}) = 1,\ \tau' > \tau\right)$. Than, the waiting time of $PM_{\varrho j1}$ is equal to $\tau' - \tau$. The polymerases $PM_{\varrho 2j}, PM_{\varrho 3j}, \ldots$ moving along the same interval between $B_{\varrho j-1}$ and $B_{\varrho j}$ of the template T_ϱ behind $PM_{\varrho 1j}$ are stopped at $B_{\varrho j}$ too. ($PM_{\varrho 1j}$, $PM_{\varrho 2j}, \ldots$) form a polymerase platoon (ensemble) waiting at instant t_τ at $B_{\varrho j}$.[2]

The set of the platoons of all templates $T\varrho$ is called "waiting platoon at B_j".[3] The nbr. $y_j(t_\tau)$ of this platoon is a state variable of the system (see **2.3.2.**).

Furthermore, a **mortality probability function** $M(t_{\tau'})$ was introduced, describing polymerase decay. $M(t_\tau)$ is the probability $Pr(M = 1)$, that a randomly chosen polymerase molecule, independent on its momentary position, is unable to move on further on. It is

$$M = \begin{cases} 1 \text{ if polymerase is died (inactive)} \\ 0 \text{ if polymerase is vital} \end{cases}$$

$M(t_\tau)$ is an increasing function with

$$M(t_0) = 0 \quad \text{and} \quad M(t_\tau) > 0,\ \tau = 1, \ldots, T.$$

2.3.2. Polymerase distribution on the template (state of the system, state transition)

The state of the transcription system is defined by the positions of the polymerases on the templates $T\varrho \in \tau$ and described by means of a discrete probability function

$PrPM(t_\tau)$ on the set $T\varrho \in T$.

---

[1] A situation like that one observed in urban traffic flow.
[2] Like a platoon of cars at a traffic light of a lane.
[3] Like the set of cars waiting at a traffic light in the queues of a multilane road way.

$PrPM(t_\tau)$ is the probability, that at instant t_τ at template site U_i a polymerase molecule is localized, i.e. $PrPM_i(t_\tau)$ is the relative amount of polymerase at site U_i at instant t_τ.

It is

$$z(t_\tau) = PrPM(t_\tau) = (PrPM_0(t_\tau), PrPM_1(t_\tau), \ldots, PrPM_i(t_\tau), \ldots, PrPM_u(t_\tau);$$

$$z_i(t_\tau) = PrPM_i(t_\tau).$$

Starting with the initial state $PrPM(t_0)$ the transcription system realizes a state sequence

$$PrPM(\cdot) = \big(PrPM(t_0), PrPM(t_1), \ldots PrPM(t_\tau), \ldots, PrPM(t_\tau)\big)$$

It was assumed $PrPM_i(t_0) = 0, \quad i \in I_B$.

The state transision of the transcription system from $z(t_\tau)$ to $z(t_{\tau+1})$ is described recursively:

$$z(t_{\tau+1}) = H\big(z(t_\tau); \quad A, v, L(t_\tau), \{F1_j(t_\tau, e, \theta 1_j \mid j = 0, 1, \ldots, b)\}\big)$$

where by H is the state transition algorithm (it is the core of the model), yielding the connection between the old and new state; the shapes of blocker removal functions $F1_j$ (and consequently the shapes of the control functions $F2_j$ and $F3_j$) determine the state transitions in a high degree. The functions $F1_j$, $j = 0, 1, \ldots, b$ may be described by a table or by a suitable function with parameters $\theta 1_j$.

2.3.3. Effect of cell milieu on blockers (input)

The qualitative shapes of the functions $F1_j(t_\tau)$, $j = 0, 1, \ldots, b$, are fixed according to the hypotheses about the type of the decondensation process of the chromatin, e.g. assuming irreversible deblocking of the B_j a monotonuously increasing function is used, if not one can choose a suitable non monotonuous function. The functions $F1_j$, fixed in this manner are modified (controlled) by ionic strength, i.e. the ASC of the cell milieu. The ASC is considered as the input e, $0 \leqq e \leqq 1$, into the transcription system controlling $F1_j(t_\tau)$:

$$F1_j(t_\tau, e) = eF1_j(t_\tau \mid e = 1)$$

$e = 1$ means optimal ASC for decondensation,

$e = 0$ means no application of AS.

We used a linear function with

$e = 1$ if ASC $= 0.4$ M and

$e = 0$ if no AS is applied.

In this way the velocity of the AS depending decondensation process of the template is described. With decreasing e in the model the blocker removal process is inhibited more and more.

2.3.4. **Result of the transcription process** (output, output function)

Two output functions calculated on the basis of the system state were used:

(i) Considering the discrete random variable X with its realizations $x_i(i = 0, 1, \ldots, u, u = 34)$ the first output variable is the **chain length distribution** at instant t_τ

$$\boldsymbol{y}_1(t_\tau) = PrX(t_\tau) = \big(Prx_0(t_\tau), Prx_1(t_\tau), \ldots, Prx_i(t_\tau), \ldots, Prx_u(t_\tau)\big).$$

The $Prx_i(t_\tau)$ are calculated by the output function

$$Prx_i(t_\tau) = \big(PrPM_i(t_\tau), PrPM(t_0)\big).$$

Comparing the results of many simulation experiments with the behaviour of the real system we realized that we have to consider only the movement of polymerase molecules already bounded in front of template site B_0, i.e. we assumed $PrPM_0(t_0) = 1$; all polymerases are waiting in front of B_0.

With this it is $Prx_i(t_\tau) = PrPM_i(t_\tau)$.

(ii) The second output variable is the total amount of de novo synthesized RNA

$$y_2(t_\tau) = Y(t_\tau) = \sum_{i=1}^{u} Prx_i(t_\tau)\, x_i.$$

Consequently, comparisons between simulation model outputs $\boldsymbol{y}_1(t_\tau)$, $y_2(t_\tau)$ and the corresponding real system outputs $\hat{\boldsymbol{y}}_1(t_\tau)$ and $\hat{\boldsymbol{y}}_2(t_\tau)$ resp. we used the time series

$$\boldsymbol{y}_1(\cdot) = \{\hat{\boldsymbol{y}}_1(t_{\tau*})\}$$

$$\hat{\boldsymbol{y}}_1(\cdot) = \{\hat{\boldsymbol{y}}_1(t_{\tau*})\}$$

and

$$y_2(\cdot) = \{y_2(t_{\tau*})\}$$

$$\hat{y}_2(\cdot) = \{\hat{y}_2(t_{\tau*})\}$$

for a set of $\tau^* \in I_{\tau*}$.

3. The simulation model EUTRANSCRIPT 1

3.1. Formal notations

In scheme 1 the parameters, variables, and functions introduced so far are listed. In a very compact form the model is noted as

$$M = \big(\underbrace{R_1, R_2, R_3; DT; F1(t_\tau, e), L(t_\tau);}_{\text{parameters and functions describing the logical structure of the system and the time regime}} \quad \underbrace{z, \boldsymbol{y}_1(\cdot), y_2(\cdot); H, G;}_{\text{state, output state transition and output function}} \quad \underbrace{z_0}_{\text{initial state}}\big)$$

Scheme 1. Parameters and functions of EUTRANSCRIPT 1

(1) Parameters describing the architecture of the transcription system.

* $R1 = u$ — length of the templates T_Q, length of the longest chain, expressed in sedimentation units.
* $R2 = \boldsymbol{A} = A(B_j)$ — blocker position vector, $j = 1, \ldots, b$; $1 < b \leqq 34$.

(2) Parameters describing the template conformation (deblocking, reblocking) and the polymerase onset.

* $F1_j(t_\tau, e, \theta_j)$ — blocker removal functions of the blockers B_j, $j = 1, \ldots, b$, depending on the rel, ionic strength e (ASC), $0 \leqq e \leqq 1$, and the parameters θ_j.
* $F2_j(F1_j(t_\tau), F1_j(t_{\tau+1}))$ — blocker removal rate (moving-on-function) describing the deblocking process.
* $F3_j(F1_j(t_\tau), F1_j(t_{\tau+1})$ — reblocking rate, describing the reblocking process.
* $F1_0(t_\tau, e, \theta_0)$ — onset function describing the rate of polymerase introduced into the transcription system.

(3) Parameters describing polymerase features.

* $R3 = v$ — velocity of polymerase in RNA de novo synthesis, expressed in nucleotides per sec. $M(t_\tau)$ — polymerase mortality function.

(4) State parameters

$\boldsymbol{z}(t_\tau) = PrPM(t_\tau)$ — vector of probabilities that a polymerase is localized at template site $U_i (i = 0, 1, \ldots, u)$ or blocker site $B_j (j = 1, \ldots, b)$ rsp.

(5) Input parameter.

* e(ASC) — rel. ionic strength, linearly depending on ammonium sulfate conc.

(6) Output parameters.

* $\boldsymbol{y}_1(t_\tau) = PrX(t_\tau)$ — distribution of the RNA chain lengths x_i ($i = 0, 1, \ldots, 34$) at instant t_τ.
* $y_2(t_\tau)$ — total ammount of RNA de novo synthetized until t_τ.

(7) State transition function.

* H — algorithmic block for next state computation:

$$\boldsymbol{z}(t_{\tau+1}) = H(\boldsymbol{z}(t_\tau)); F1_j(t_\tau, e, \theta_j); F1_0(t_\tau, e, \theta_0); R1, R2, R3; M(t_\tau).$$

(8) Output function

* G — algorithmic block for output computation

$$\boldsymbol{y}_1(t_\tau) = G(\boldsymbol{z}(t_\tau), \boldsymbol{z}(t_0))$$

and

$$y_2(t_\tau) = \sum_{i=1}^{u} Pr_i(t_\tau)\, x_i.$$

(9) Time regime parameter.

* T — elementary time interval (time needed for synthesis of a single S-unit in sec.).

Starting with the initial state $\boldsymbol{z}(t_0) = z_0$, $F1_j(t_0) = 0$, $j = 0, 1, \ldots, b$, and $M(t_0) = 0$, the state transitions of the system are calculated recursively on the discrete time parameter t_τ, $\tau = 0, 1, \ldots, T$ (sec) by means of the algorithm H. Different scenarios are calculated with different hypotheses about $F1(t_\tau)$, $M(t_\tau)$ and z_0.

In all simulation experiments R_1, R_2, R_3, and DT were not varied. The smallest time interval, called tact $DT = t_{\tau+1} - t_\tau$ ($\tau = 0, 1, \ldots, T$, T = max nbr. of tacts of a single run) is defined by the real time needed by a polymerase molecule to move along a single template unit U_i, i.e. DT is the time needed to synthesize about 270 nucleotides of a RNA chain corresponding to a single S-unit.

3.2. State transition algorithm H

3.2.1. Blocker as abstract automaton

As mentioned above a blocker site U_1, $i \in I_B$, may appear in the states s_0 and $s_1 (s_0 \Leftrightarrow D = 0, s_1 \Leftrightarrow D = 1)$. The following features of a blocker site are assumed:

(1) $Pr(D_i(t_0) = 0) = 1$ for all $i \in I_B$, i.e. at instant t_0 all blocker sites U_i, $i \in I_B$, are blocked; that means for $B_0 = U_0$ that no onset of a polymerase molecule happend so far.

(2) In the interval $(t_\tau, t_{\tau+1})$ at a blocker site $U_{\varrho i}$ $i \in I_B$ of a template $T\varrho$ one of the 4 cases may happen locally

$$E_{00_i}(t_\tau) = (D_{\varrho i}(t_{\tau+1}) = 0 \mid D_{\varrho i}(t_\tau) = 0): U_{\varrho i} \quad \text{remains blocked}$$
$$E_{01_i}(t_\tau) = (D_{\varrho i}(t_{\tau+1}) = 1 \mid D_{\varrho i}(t_\tau) = 0): U_{\varrho i} \quad \text{becomes deblocked}$$
$$E_{10_i}(t_\tau) = (D_{\varrho i}(t_{\tau+1}) = 0 \mid D_{\varrho i}(t_\tau) = 1): U_{\varrho i} \quad \text{becomes reblocked}$$
$$E_{11_i}(t_\tau) = (D_{\varrho i}(t_{\tau+1}) = 1 \mid D_{\varrho i}(t_\tau) = 1): U_{\varrho i} \quad \text{remains deblocked}$$

with the probabilities of these four events

$$Pr(E_{00_i}(t_\tau)) = P_{00_i}(t_\tau), \quad Pr(E_{01_i}(t_\tau)) = P_{01_i}(t_\tau)$$
$$Pr(E_{10_i}(t_\tau)) = P_{10_i}(t_\tau), \quad Pr(E_{11_i}(t_\tau)) = P_{11_i}(t_\tau)$$

which depend on the 3 possible tendencies of the control function $F1_j$ in the interval $(t_\tau, t_{\tau+1})$, i.e. $F1_j$(i) being constant, (ii) increasing or (iii) decreasing. In the event orientated (discrete) algorithm the three corresponding cases are:

(i) $F1_j(t_{\tau+1}) - F1_j(t_\tau) = \Delta F1_j(t_\tau) = 0 \Leftrightarrow h_2(t_\tau)$

$$\Leftrightarrow P_{00_j}(t_\tau, h_{2_j}(t_\tau)) = 1$$
$$\Leftrightarrow P_{11_j}(t_\tau, h_{2_j}(t_\tau)) = 1$$

i.e. a closed blocker B_j remains closed and an open blocker remains open resp. iff $h_{2_j}(t_\tau)$ is true (nothing happens)

(ii) $\Delta F1_j(t_\tau) > 0 \Leftrightarrow h_1(t_\tau)$

$$\Leftrightarrow P_{01_j}(t_\tau, h_{1_j}(t_\tau)) = F2_j(t_\tau) > 0$$
$$\Leftrightarrow P_{11_j}(t_\tau, h_{1_j}(t_\tau)) = 1$$

i.e. a closed blocker B_j becomes open and an open blocker remains open resp. iff $h_{1_j}(t_\tau)$ is true (deblocking happens). $F2$ is the blocker removal rate (see above).

(iii) $\Delta F1_j(t_\tau) < 0 \Leftrightarrow h_2(t_\tau) \Leftrightarrow P_{01_j}(t_\tau, h_{2_j}(t_\tau) = F3_j > 0$

i.e. an open blocker becomes closed iff $h_{2_j}(t_\tau)$ is true (reblocking happens). $F3$ is the reblocking rate (see above).

(i) and (iii) are compressed to together:

$$h_1 \Leftrightarrow \Delta F1 > 0$$
$$h_2 \Leftrightarrow \Delta F1 \leq 0.$$

Consequently, a blocker may be described by an probability automaton with the states s_0 and s_1 and the inputs h_1 and h_2, as shown in fig. 7 and table 2. The $P_{\mu\nu}(\mu, \nu \in \{0, 1\})$ are interpreted as rel. amounts, e.g.:

if $F1_j(t_\tau)$ is increasing $P_{01_j}(t_\tau, h_1(t_\tau))$ is the rel. amount of $B_{\varrho j}$'s being in state s_0 and changing their state in s_1. Polymerases being stopped at instant $t_{\tau-1}$ (since $D_j(t_{\tau-1}) = 0$) may now continue moving over the site B_j. In the state transition algorithm H the polymerase movement along the T_ϱ are described by the actual values of $P_{\mu\nu}(h_\eta)$; $\mu, \nu \in \{0, 1\}$; $\eta \in \{1, 2\}$.

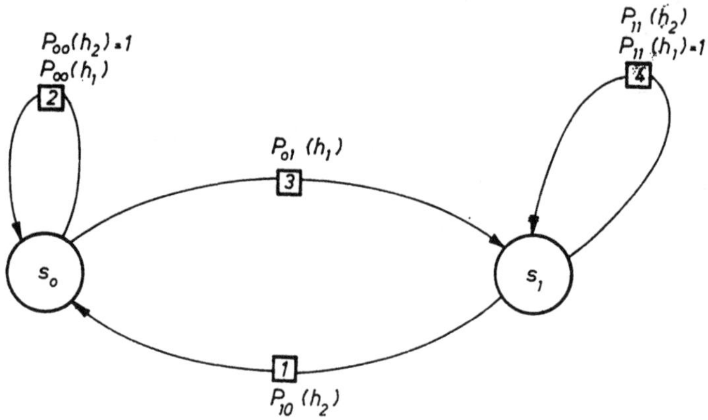

Fig. 7. Probability automaton of a blocker site.

Table 2. State transition probabilitlies of a blocker site.

input	h_1		h_2	
state t_τ \ $t_{\tau+1}$	s_0	s_1	s_0	s_1
s_0	$P_{00}(h_1) = 1 - P_{01}(h_1)$	$P_{01}(h_1)$	$P_{00}(h_2) = 1$	$P_{01}(h_2) = 0$
s_1	$P_{10}(h_1) = 0$	$P_{11}(h_1) = 1$	$P_{10}(h_2)$	$P_{11}(h_2) = 1 - P_{10}(h_2)$

$h_1 \Leftrightarrow \Delta F1 > 0$ $\quad P_{01}(h_1)_{t\tau} = F2(t_\tau)$, see text
$h_2 \Leftrightarrow \Delta F1 \leq 0$ $\quad P_{10}(h_2)_{t\tau} = F3(t_\tau)$, see text

3.2.2. Control of the polymerase platoons

Without blocker sites the movement of the polymerase molecules would be unrestricted and their destribution on the template would not exhibit specific pattern. Conversely, the presence of blockers leads to daming up of polymerases at a blocker site, where they are released as platoons (ensembles) for further movement. These moving platoons arrive at the next obstacle, a part of the arriving molecules may pass corresponding to $F1(t_\tau)$, the others are stopped again (just as cars moving along a street controlled by a badly organized green wave) and are allowed to continue corresponding to $F2(t_{\tau+1})$. By extensive simulation experiments we realized that the whole transcription process

Scheme 2. **Managing the polymerase platoons**

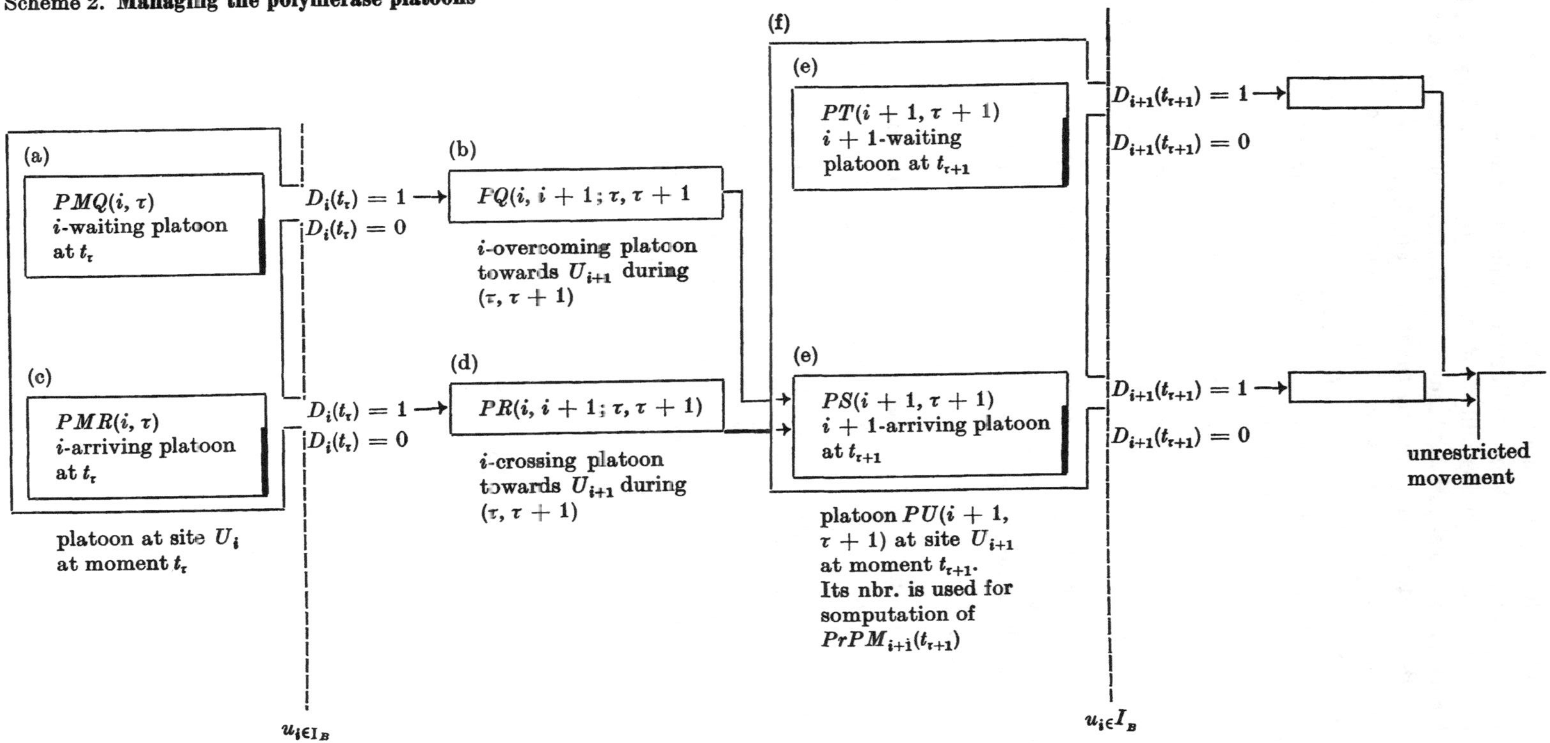

was mainly controlled by the functions $F2_j$ because mainly they govern essentially the release of the platoons out of the waiting ensemble at a blocker site.

The time dependent "permeability" of the site B_0 with $A(B_0) = U_0$ (onset site) is to be interpreted specifically: $F2_0(t_\tau)$ indicates the real. amount of polymerases "waiting" in front of B_0 in order to start during the interval $(t_\tau, t_{\tau+1})$.

Within the algorithm H some type of platoons are considered (see scheme 2):

a) i-waiting platoon PMQ, waiting at $U_i (i \in I_B)$ during the interval $(t_{\tau-1}, t_\tau)$.

b) The ratio of PMQ, that is allowed to pass anew U_i towards U_{i+1} calculated by means of $F2_{ij}$ be the i-overcoming platoon PQ:

$$PQ(i, i+1; t_\tau, t_{\tau+1}) = F2_i(t_\tau) \cdot PMQ(t_\tau);$$

is the rel. amount of polymerases of the waiting platoon that arrive at U_{i+1} at instant $t_{\tau+1}$; the i-overcoming platoon moves from U_i to U_{i+1} during the interval $(t_\tau, t_{\tau+1})$.

c) PMR be the i-arriving platoon, arriving at U_i at instant t_τ.

d) The ratio of PMQ, that is allowed to proceed without stop across U_i towards U_{i+1} is calculated by means of $F1_{ij}$ be the i-crossing platoon PR:

$$PR(i, i+1; t_\tau, t_{\tau+1}) = F1_i(t_\tau) \cdot PMR_i(t_\tau)$$

it is the rel. amount of polymerases of the i-arriving platoon that arrive at U_{i+1} at instant $t_{\tau+1}$; this platoon moves from U_i to U_{i+1} during the interval $(t_\tau, t_{\tau+1})$.

e) PQ and PR are united to PS:
PS is the $(i+1)$ — arriving, arriving at U_{i+1} at instant $t_{\tau+1}$, and PT is the $(i+1)$ — waiting platoon waiting at U_{i+1} during the interval $(t_\tau, t_{\tau+1})$.

f) $PU = PS \cup PT$ is the union of polymerases present at U_{i+1} at instant $t_{\tau+1}$. PU relative to total amount of polymerases yields $PrPM_{i+1}(t_{\tau+1})$. The $P_{\mu\nu}$, shown in tab. 2 are used to manipulate the arriving and waiting platoons, at which the moving on process of waiting platoons proved to be the most important process in increasing the RNA synthesis.

4. Simulation experiments

First of all, the aim was to investigate the effect of some variable model parameters on the polymerase movement effect of changing z_0, $F1(t_\tau)$, $M(t_\tau)$, and AS shortage on $y_1(\cdot)$, $y_2(\cdot)$), and secondly to identify the open parameters and functions so that one can observe minimal differences between $y_1(\cdot)$, $y_2(\cdot)$ and $\hat{y}_1(\cdot)$, $\hat{y}_2(\cdot)$ resp.

The following questions were of interest:

1. Is the blocker concept compatible with reality, more exactly: does the existence of blocker sites ($I_B \neq \emptyset$) lead consequently to distinct chain length distribution ($Prx_i(t_\tau) \gg \varepsilon$ for all $i \in I_B \wedge Prx_i(t_\tau) \approx 0$ for all $i \in I_{\bar{B}}$?

2a. Which shape and maximum of the $F1_j$ must be chosen to get correspondence between the y and $\hat{y}$? Does B_0 play a special role?

2b. Is it necessary to assume qualitatively different shapes of $F1_j$ due to AS shortage?

3. Have we to assume mortality of polymerase molecules essentially to find correspondence of y and $\hat{y}$?

4. Which start distribution of polymerases on the template excludes an accordance of $y_1(\cdot)$, $y_2(\cdot)$ and $\hat{y}_1(\cdot)$, $\hat{y}_2(\cdot)$ resp.
5. Which conclusions about the nature of the blocker sites may be drawn by means of the $F1_j$, $F2_j$, $F3_j$?

By variations of the system parameters in scenario simulations our understanding of possible mechanisms of the system behaviour in connection with the process of template decondensation was increased bit by bit.

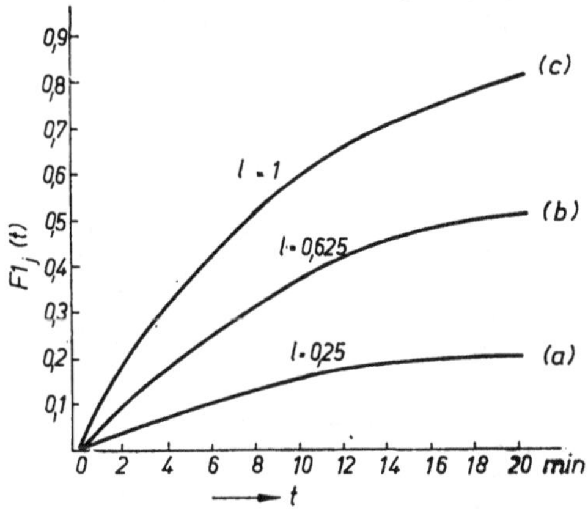

Fig. 8. Blocker removal functions $F1_j(t_\tau)$ fixed for the case of (a) low, (b) intermediate, and (c) high level of ammonium sulfate application.

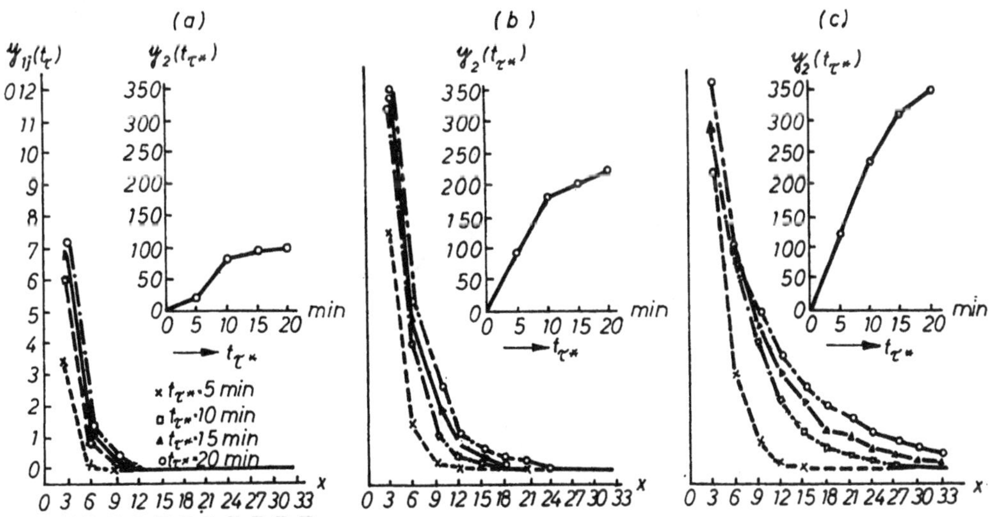

Fig. 9. Distribution $y_{1_j}(t_{\tau^*})$, in terms of rel. amounts of polymerases at B_j, of RNA chain lengths in a simulation experiment with EUTRANSCRIPT 1, depending on the ammonium sulfate concentration of (a) low, (b) intermediate, and (c) high level ($e = 0{,}25$; $e = 0{,}625$; $e = 1$ rsp.), corresponding to the functions $F1_j(t_\tau)$ of fig. 8, and total amount $y_2(t_{\tau^*})$ of RNA produced de novo during 20 min ($\tau^* = 5$, 10, 15, 20 min). From Bellmann et al., 1975.

Con. question 1:

A very simple simulation experiment was performed:

- $\boldsymbol{A} = (0, 3, 6, 9, 12, 15, 18, 21, 24, 27, 30, 33)$ i.e. equidistant blockers
- zero polymerase mortality: $M(t_\tau) = 0$ for $\tau = 0, 1, \ldots, T$.
- 3 *ASC* inputs: $e = 1$, $e = 0.8$, $e = 0.5$
- all $F1_j$ equal, $j = 0, 1, 2, \ldots, b$ (see fig. 8)
- $z_0 : PrPM_0(t_0) = 1$, i.e. all polymerases at B_0
- $T = 20$ min.

The result of chain growth is shown in fig. 9 indicating qualitative correspondence: many short chains, only some long chains, low *ASC* lead to a small amount of RNA; i.e. the blocker concept is compatible.

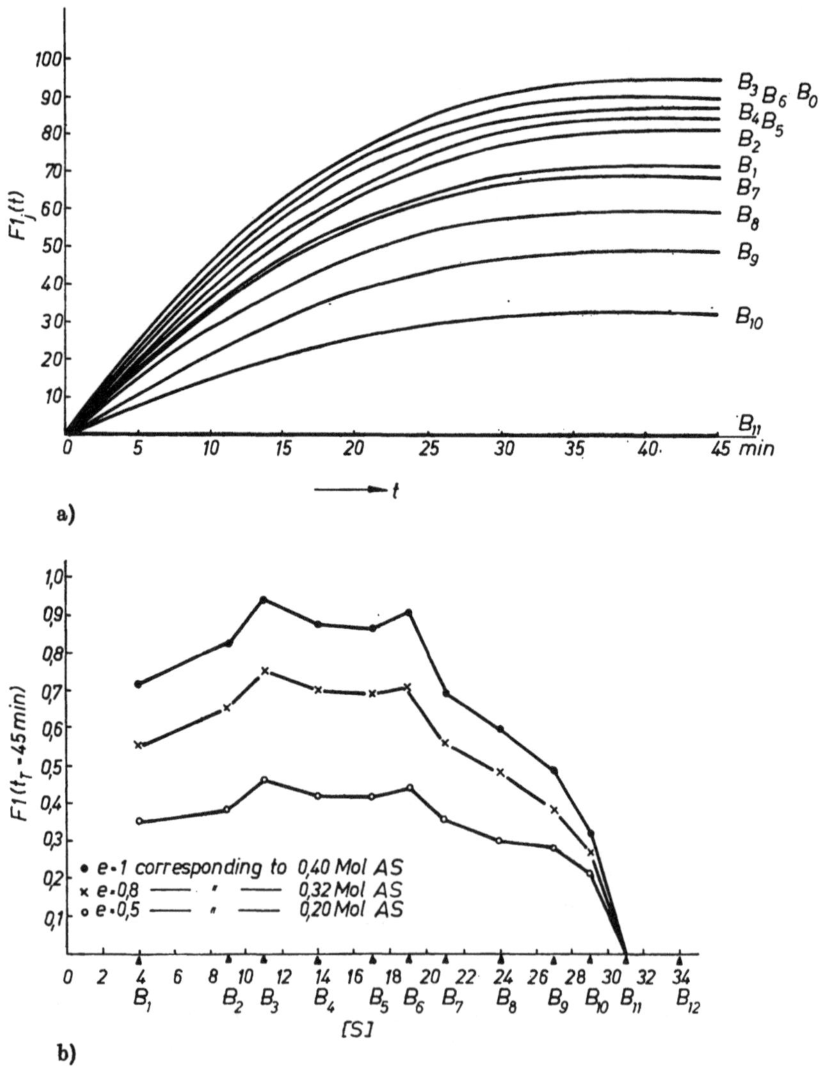

Fig. 10. (a) $F1_j(t_\tau)$ shapes utilized in the simulation experiments with results shown in figs. 11, 13, 15; (b) maximal values of $F1_j$, $j = 1, \ldots, 11$, depending linearly on e.

The results led to some insights I.

* (I1): if blocker sites with the desribed features are existing, distinct chain lengths occur, which are distributed qualitatively like such ones observed in the real experiments, as shown in fig. 10. —

In order to improve the coincidence between the simulated dynamics and the real one shown in fig. 2 further simulation runs were performed utilizing the blocker sites

$$\boldsymbol{A} = (0, 4, 9, 11, 14, 17, 19, 21, 24, 27, 29, 32, 34),$$

which were experimentally estimated. In these simulation runs (i) shapes and maxima of the blocker removal functions $F1_j(t_\tau)$ and (ii) of the mortality function $M(t_\tau)$, and (iii) the initial distribution $PrPM_j(t_0)$, $j = 0, 1, \ldots, b$ of the polymerase molecules on the templates $PrPM(t_0)$ were modified.

Con. question 2a:

Question 2a is involved with the functions $F1_i(t_\tau)$, $j = 0, \ldots, b$, and $F2_j(t_\tau)$, $j = 1, \ldots, b$.

From different runs of the simulation model we found that an improved coincidence between $\boldsymbol{y}_1(\cdot)$ and $y_2(\cdot)$ and $\hat{\boldsymbol{y}}_1(\cdot)$, $\hat{y}_2(\cdot)$ if the maxima of the blocker removal functions $F1_j(t_\tau)$, $j = 1, \ldots, b$, decrease with increasing j (fig. 10):

* (I2): It is

$$F1_1^{\max}, \ldots, F1_7^{\max} > F1_8^{\max} > \cdots > F1_b^{\max}, b = 11$$

assuming vital polymerases ($M(t_\tau) = 0$, $\tau = 1, \ldots, T$). —

By further simulations we found that the shapes of functions $F1_j(t_\tau)$ must be much steeper in the early phase indicating a very rapid deblocking process during the first 8 to 10 min, as shown in fig. 11:

* (I3): $t^{\max}\left(F1^{\max} = F1(t^{\max})\right)$ is reached not later than 15—20 min after starting the experiment. The shape of the $F1_j(t_\tau)$, $j = 1, \ldots, b$, is to be fixed such that $F2_j(t_\tau)$ is of a Poisson-type with the maximum after 4 min:

$$F2_j^{\max} = F2_j(4 \text{ min}), \text{ providing sufficient } ASC \ (\varepsilon = 1). \text{ —}$$

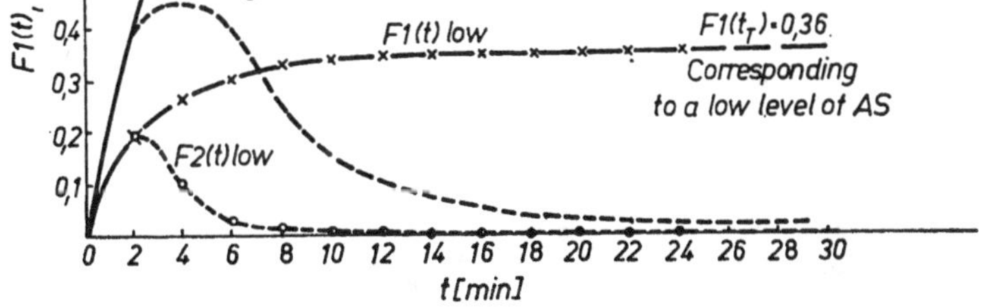

Fig. 11. Shapes of blocker removal functions $F1$ and blocker removal rates $F2$, corresponding to low and high ammonium sulfate concentrations (0.1 and 0.4 M rsp.), which were realized by simulation experiments.

For further improvement of the coincidence of the simulated and experimental results we focused our attention to the shape of the onset function $F1_0(t_\tau)$. Hitherto, $F1_0(t_\tau)$ was assumed to be monotonuously. By a lot of simulation experiments we got an optimum shape (fig. 12):

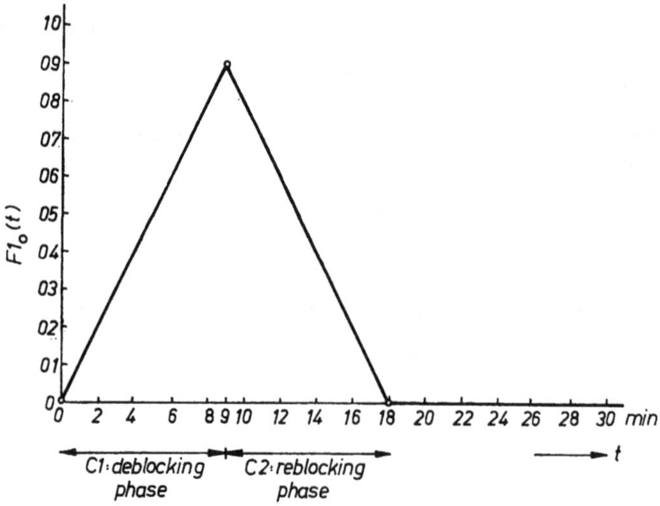

Fig. 12. Highly probable shape of the onset function $F1_0(t_\tau)$ guaranteeing chain length distribution similar that one observed in the real experiment (shown in fig. 2).

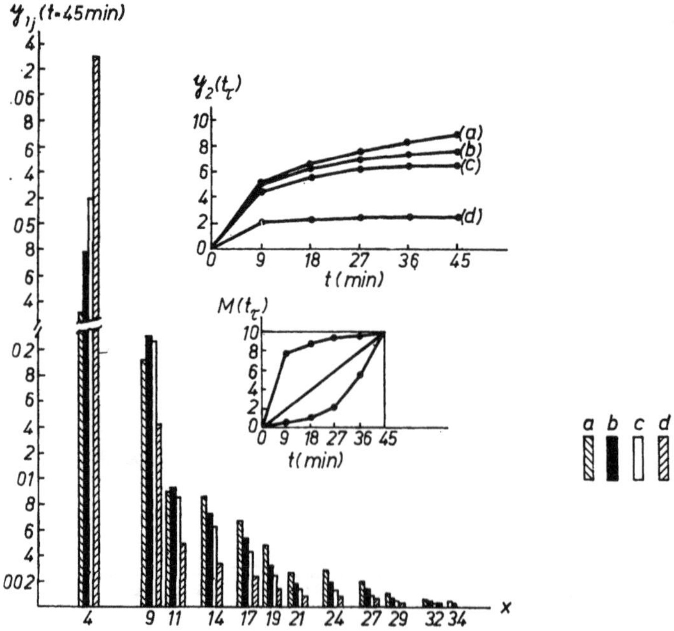

Fig, 13. Distribution y_{1_j} ($t = 45$ min) of RNA chain lengths in a simulation experiment with EUTRANSCRIPT 1 depending on 4 different polymerase mortality functions (a), (b), (c), (d); and time depending amount $y_2(t_\tau)$ of RNA.

* (I4): The shape of $F1_0(t_\tau)$ proved to be of an optimum type with $F1_0(t_0^{\cdot}) = F1_0$ $(t = 18 \text{ min}) = 0$ and $F1_0(t = 9 \text{ min}) = 0.9$, indicating impuls onset. During the interval (0′, 9′) an increasing amount of the sites B_0 become decondensed so that the polymerases waiting in front of B_0 are allowed to start. During the interval (9′, 18′) the nbr. of polymerases moving into the transcription system decreases more and more until $D_0(t = 18 \text{ min}) = 0$. —

Con. question 2b:

In the case of *AS* deficiency ($e < 1$) we had to conclude — forced by different scenarios — that the shapes of $F1_j(t_\tau)$ and $F2_j(t_\tau)$, $j = 1, \ldots, b$, had not to be changed qualitatively, but the maxima of these functions must be settled less than in the case of no deficiency ($e = 1$).

Con. question 3:

Question 3 is concerned with polymerase mortality that proved to suppress the RNA synthesis process (fig. 13). Mortality and *AS* lack lead to equivalent effects on polymerase movement (fig. 14). Generally, the nbr. of polymerases waiting at blocker B_1 decreases. This decrease is inhibited by both *AS* lack and/or polymerase mortality. Generally, the nbr. of polymerases at blocker B_{10} increases. This increase is also inhibited by both *AS* lack and/or polymerase mortality.

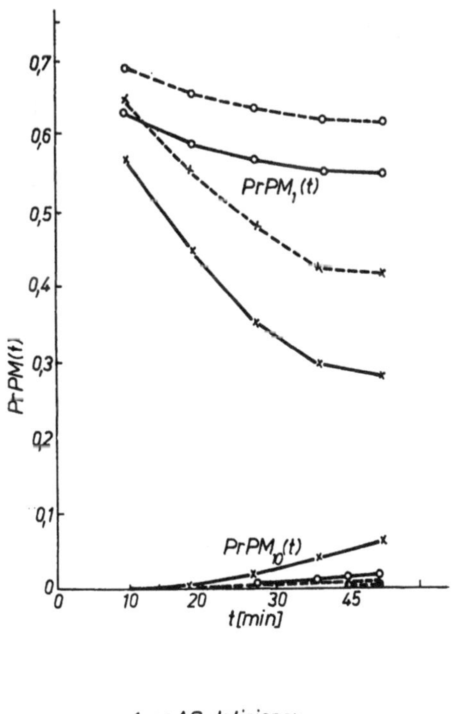

Fig. 14. Probabilities $PrPM_j(t_\tau)$ that at instant t_τ a polymerase molecule is located at blocker site B_j, $j = 1{,}10$, depending on ammonium sulfate deficiency and polymerace mortality.

By simulation studies it is not possible to decide whether polymerase mortality was to be excluded in the real system. On principle, 2 cases are possible:

A) Polymerase mortality does not happen really and is not simulated too. Consequently, $I(1)$ leads to the assumption that different deblocking intensities at different blocker sites occur, but no experimental evidence is known to support this assumption.

B) The hypothesis that polymerase decay happens really and $F1_1(t_\tau) = F1_2(t_\tau) = \ldots = F1_j(t_\tau) = \ldots = F1_b(t_\tau)$ holds seems to be more probable. Consequently, in simulation runs without mortality of polymerase molecules relation $I(2)$ is needed to find correspondence between $y_1(\cdot), y_2(\cdot)$ and $\hat{y}_1(\cdot), \hat{y}_2(\cdot)$ and is to be interpreted as being due to polymerase mortality in the real object, (not due to increasing "tenacity" of the blockers of far from B_0); it seems reasonable to us to assume a homogeneous biophysical/biochemical nature of all blocker sites $U_i \in I_B$.

Con. question 4:

As a further result of simulation studies, the answer to question 4 is: correspondence between $y_1(\cdot), y_2(\cdot)$ and $\hat{y}_1(\cdot), \hat{y}_2(\cdot)$ resp. may be observed only if $PrPM_j(t_0) = 0$ $j = 1, 2, \ldots, b$. Fig. 15 shows, that all chain lengths are possible in the case that at each blocker site B_j a polymerase may be localized which is ready at instant t_0 to move on, i.e. in the case

$$PtPM_j(t_0) > 0, \quad j = 0, 1, \ldots, b$$

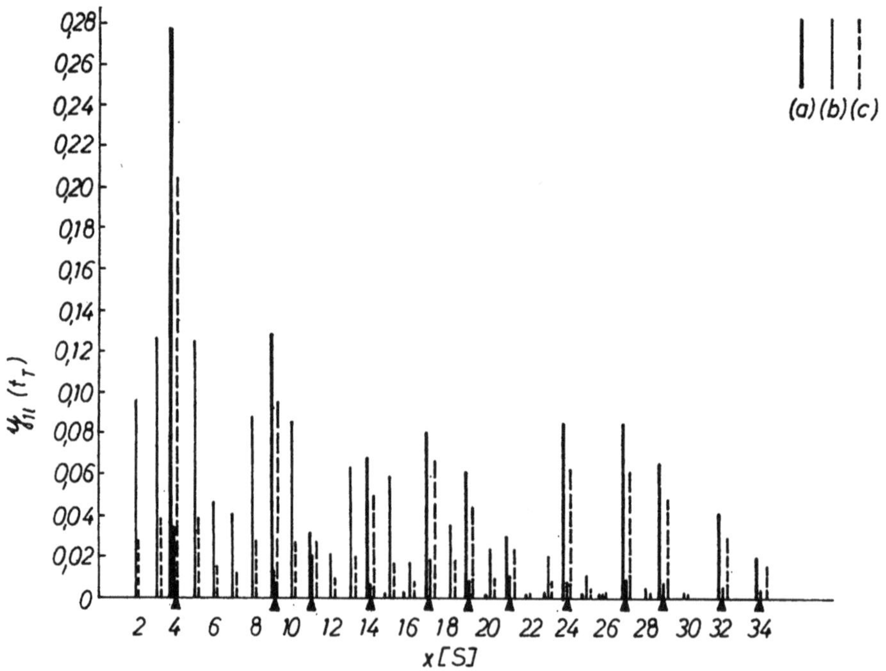

Fig. 15. Distribution $y_{1_j}(t_T)$, $T = 45$ min, of lengths of de novo synthesized RNA chains in a simulation experiment with EUTRANSCRIPT 1 depending on the polymerase start distribution on the templates: (a) all polymerases at the onset site B_0; $PrPM_0(t_0) = 1$; (b) 66% of polymerases at B_0, 34% of them equally distributet at $B_1 \ldots B_{12}$; $PrPM_0(t_0) = 0{,}66$; $PrPM_j(t_0) = 0{,}029$, $j = 1, \ldots, 12$; (c) all polymerases equally distributed at all blockers; $PrPM_j(t_0) = 0{,}077$, $j = 0, 1, \ldots, 12$. In each case no polymerase mortality.

no distinct chain lengths are observable. Since from the results of real experiments it was concluded that the de novo RNA synthesis was performed by polymerases initiated anew, in the following simulation experiments all polymerases are considered to be localized at B_0 at instant t_0, i.e. $PrPM_0(t_0) = 1$, as mentioned above. –

Summarizing these results with EUTRANSCRIPT 1 the following is notable. Correspondence of $\boldsymbol{y}_1(\cdot)$, $y_2(\cdot)$ and $\hat{y}_1(\cdot)$, $\hat{y}_2(\cdot)$ resp. is observed if

(i) $I_B \neq 0$, i.e. existence of blockers
(ii) experimentally determined $\hat{\boldsymbol{A}}$ is used
(iii) $PrPM_0(t_0) = 1$
(iv) shapes of $F1_j(t_\tau)_j = 1a) \ldots, b$, are used, leading to $F2_j(t_\tau)$ of an optimum type with strong early increase up to the optimum and after that slowly decreasing, with $F2_j^{max}$ ($t = 4$ min.) independent on AS shortage and with $F2_1^{max}, \ldots, F2_7^{max} > F2_8^{max} > \cdots > F2_b^{max}$ in the case of $M(t_\tau) = 0$, $\tau = 0, 1, \ldots, T$ or $F2_1^{max} = F2_2^{max} = \cdots = F2_j^{max} = \cdots = F2_b^{max}$ in the case of $M(t_\tau) > 0$, $\tau = 1, \ldots, T$.
(v) a shape of the polymerase onset function $F1_0(t_\tau)$ is utilized corresponding to an optimum type with $F1_0(t_0) = F1(18 \text{ min.}) = 0$ and $F1^{max}(9 \text{ min.})$.

To get sufficient certainty in statements about the shapes of the functions $F2_j(t_\tau)$ (see fig. 15), especially since they proved essential in governing the movement of the polymerases in the whole system, these shapes were investigated extensively by means of an adaptation method for solving extremal problems in large scale systems, especially in simulation systems. This work is broadly outlined in the paper BORN and BELLMANN in this volume. The adaptation problem was solved utilizing a very quick reduced algorithm based mainly on the functions $F2_j(t_\tau)$, $j = 1, \ldots, b$, and $F1_0(t_\tau)$, called EUTRANSCRIPT 2 (see its short description in the following paragraph). The final result was that always such shapes proved to be insufficient which were not of the optimum type, i.e. only shapes of the $F2$-optimum type guaranteed a correspondence between the results $\boldsymbol{y}_1(\cdot)$, $y_2(\cdot)$ of the algorithm and the results $\hat{y}_1(\cdot)$, $\hat{y}_2(\cdot)$ of the real eucaryotic transcription system under experimental investigation (see fig. 16).

In coincidence with the argumentation outlined in paragraph 4 (question 2) fig. 16 demonstrates, that only the first 6 blockers are useful for conclusions about the deblocking process. They become deblocked with maximal probability of approx. 0.5 to 0.6. This maximum is reached after 10–15 min. The shapes shown in fig. 16, being sequentielly selected by the parameter search algorithm, confirm the empirically realized shape types shown in fig. 11.

Con. question 5:

Finally, we try to give an answer in form of some hypotheses about template decondensation at blocker sites, supported by computer simulation experiments. Attempting to fit together the results of all simulation studies in order to get a logical interpretation the following hypotheses are proposed:

1. Each blocker $B_{j\varrho}$ ($j = 1, \ldots, b$; $\varrho = 1, \ldots, r$) at a template site $U_{i\varrho}(i \in I_B$; $\varrho = 1, \ldots, r)$ be bound by m_ϱ biochemical links; $m_\varrho = 1, 2, \ldots, M_\varrho$ (M_ϱ unknown and not needed to be known).
2. A decondensation of a blocker site B_ϱ means changing its non-permeable state $D_\varrho = 0$ to the permeable state $D_\varrho = 1$. The decondensation event happens only if w_ϱ bonds of the m_ϱ links are dissolved ($w_\varrho \leqq m_\varrho$).

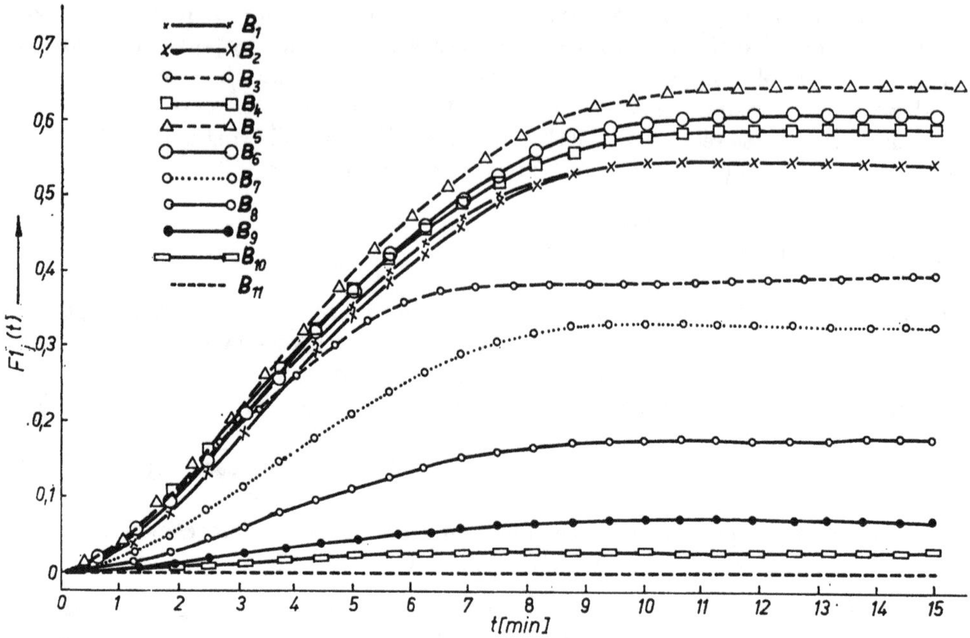

a)

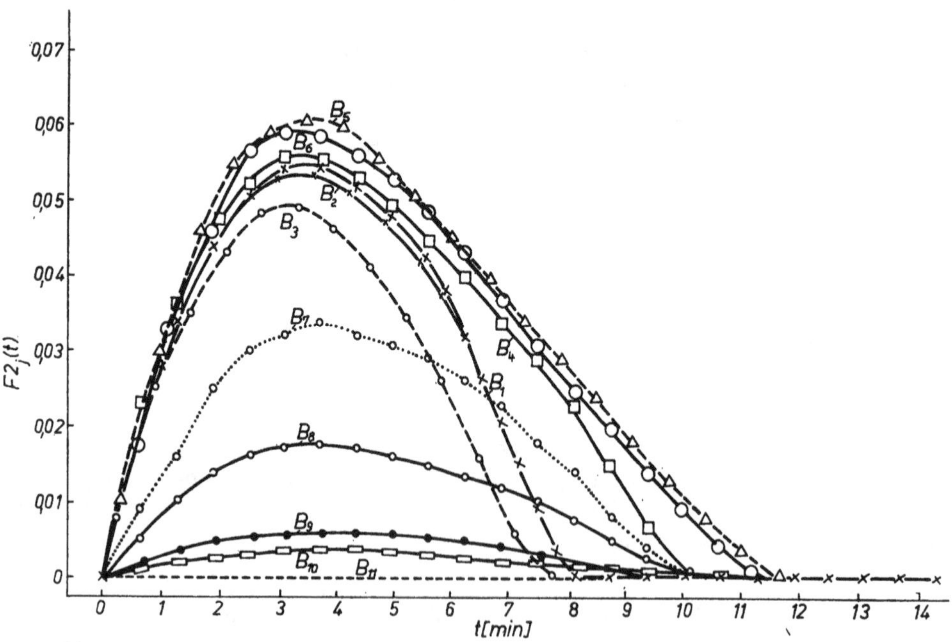

b)

Fig. 16. (a) locker removal functions $F1_j(t_\tau)$, $j = 1, \ldots, 12$, and (b) the corresponding blocker removal rates $F2_j(t_\tau)$ as recognized by solving a 52 dimensional parameter adaption problem utilizing a special stochastic search algorithm.

3. The dissolving process of such a single links needs a constant time interval δt, not depending on j and ϱ, and the w_ϱ dissolving processes, needed for the event "decondensation of blocker B_ϱ", does not happen simultaneously but sequentelly.
4. Consequently, the time consumed in realizing the decondensation event of blocker $B_{j\varrho}$ on a template U_ϱ, may depend linearly on w_ϱ:

$$DT_{j\varrho} = w_\varrho \cdot \delta t$$

5. The value w_ϱ ($\varrho = 1, \ldots, r$) be a random variable with the statistical expected value $E(w) = \overline{w} = \lambda_w$ and the variance $\sigma_w^2 = \lambda_w$, assuming a Poisson distribution. Consequently, the $DT_{j\varrho}$ become random variables too with mean and variance λ_{DT}, i.e. the probability functions $F2_j(t_\tau)$ that a blocker B_j changes from state $D_j = 0$ to $D_j = 1$ is expected to possess a Poisson-like shape (independent of j), as it was observed in simulation studies (see fig. 11) and confirmed by application of a search algorithm (see fig. 16).

5. The reduced model EUTRANSCRIPT 2 (Reservoir Floodgate model, RFM)

In order to solve the adaptation task mentioned in the preceding paragraph, and thereby to make possible a big amount of literations (5000 to 30000), the computing time of EUTRANSCRIPT 1 was to be reduced drastically, i.e. a very much quicker version of the model was needed. In the following this model (EUTRANSCRIPT 2) is described.

As outlined in the preceding paragraph, the process of RNA synthesis performed by the polymerase moving from left to right depends essentially on how the polymerase platoons, waiting at the blockers B_j become able to move on. This is governed by the functions $F2j(t_\tau)$, $j = 0, 1, \ldots, b$, $\tau = 1, 2, \ldots, T$, which determine the ratio of such waiting polymerases that are now allowed to continue RNA synthesis by overcoming their obstacles $B_{\varrho j}$. Therefore only these functions $F2_j(t_\tau)$ were included into EUTRANSCRIPT 2.

Furtheron, since simulation experiments showed that the de novo RNA synthesis starts with polymerase onset at $U_0 = B_0$, in EUTRANSCRIPT 2 it was fixed: $PrPM_0(t_0) = 1$. Polymerase mortality needed not to be considered (see the argumentation in the preceding paragraph): $M(t_\tau) = 0$, $\tau = 0, 1, \ldots, T$. $R1$, $R2$, $R3$, and DT are held constant and were equal to the values fixed in EUTRANSCRIPT 1. No lack in ionic strength was assumed.

5.1. The reservoir floodgate system

The concept of the reduced model is based on the idea that the transcription system is equivalent to a linear sequence of reservoirs R_j, $j = 1, \ldots, b$, with inflow and outflow, the latter controlled by floodgate blockes B_j, $j = 1, \ldots, b$ (Hollatz, Born, Bellmann, 1977). The reservoir-floodgate system is filled by the inflow into the first reservoir R_1, described by the inflow rate (or source function) $F2_0(t_\tau)$, $\tau = 1, \ldots, T$, indicating the amount of initiated polymerase per time interval $(t_\tau, t_{\tau+1})$; i.e. $F2_0(t_\tau)$ describes the polymerase onset process.

In EUTRANSCRIPT 2 the content of the reservoir R_j corresponds to the waiting polymerase platoon at B_j in EUTRANSCRIPT 1.

The relative outflow rate of reservoir R_j, $j = 1, \ldots, b$, is described by the outflow rates (moving-on-functions) $F2_j(t_\tau)$, $j = 1, \ldots, b$, $\tau = 1, \ldots, T$, indicating the rel. amount of the polymerases, being in the reservoir R_j at instant t_τ, that are now at the instant t_τ) allowed to overcome the floodgate R_j for further non-inhibited movement towards the (next) reservoir R_{j+1}. In both models EUTRANSCRIPT 1 and 2 the functions $F2_j(t_\tau)$ indicate the rel. amount of the polymerases — being waiting up to t_τ — which overcome the obstacle B_j at instant t_τ, corresponding to the set PR of scheme 2.

The whole system is assumed to be empty at t_0 (corresponding to the assumption that all polymerases start at B_0. At the discrete moments t_τ, $\tau = 1, \ldots, T$ a medium (polymerases) streams at (source-) site B_0 into the system, at first unrestricted until flowing into the first basin. As in EUTRANSCRIPT 1, the unresricted flow occurs with constant velocity = 34 nucleotides per sec; a single time step DT corresponds to 7.5 sec. real time. DT is the time needed by a polymerase to move along a single S-unit U_i under unrestricted conditions.

5.2. Parameters and functions

1. Position of the reservoirs R_j with their floodgates (blockers) B_j, $j = 1, 2, \ldots, b$, expressed in terms of time intervals DT beginning at B_0:

$$\hat{A} = (\hat{A}_1, \hat{A}_2, \ldots, \hat{A}_j, \ldots, \hat{A}_b), \quad A_j = A(B_j),$$

and the known and fixed values $R1$, $R2$, $R3$, all as used in EUTRANSCRIPT 1.

2. Inflow rate (source function) $F2_0(t_\tau) = a(t_\tau)$ of the medium (polymerases) streaming from the onset site B_0 into the transcription system at each moment t_τ, $\tau = 1, \ldots, T$, $j = 0$. The source function was assumed to be unknown too, although from simulation experiments its shape was derived (see fig. 11), the shape of $F2_0(t_\tau)$ was now to be the subject of the adaptation problem. The parameters x of this function had now to be adaptated (variable parameters, "soft" parameters of the model).

3. Moving-on-functions $F2_j(t_\tau)$ of the floodgates (blockers) B_j, $j = 1, \ldots, b$, corresponding to the blocker removal rates of EUTRANSCRIPT 1 (see paragraph 2.1.3). They indicate the relative amount of such polymerases waiting within the reservoirs R_j at moment t_τ, that are now allowed to continue their movement beyond B_j at that moment, corresponding to set PS of scheme 2.

   Both for $F2_0(t_\tau) = a(t_\tau)$ and $F2_j(t_\tau)$, $j = 1, \ldots, b$, $\tau = 1, \ldots, T$, polynomials of third degree with a certain maximal (= 1) and minimal level (x_4) were utilized with the coefficients (parameters for adaptation)

$$x = (x_1, x_2, x_3, x_4) \quad \text{for} \quad F2_0(t_\tau)$$

   and the coefficients

$$x_j = (x_{j_1}, x_{j_2}, x_{j_3}, x_{j_4}),$$
$$j = 1, \ldots, b, \text{ for } F2_j(t_\tau):$$

(1°) $F2_0(t_\tau) = a(t_\tau) = \max\,(x_1\tau + x_2\tau^2 + x_3\tau^3, x_4)$, with $x_4 \geqq 0$, $\tau = 1, \ldots, T$

and

(2°) $F2_j(t_\tau) = \min\,\{\max\,(x_{j_1}\tau + x_{j_2}\tau^2 + \tau_{j_3}{}^3, x_{j_4}), 1\}$ with $x_{j4} \geqq 0$, $\tau = 1, \ldots, T$, $j = 1, \ldots, b$.

(4) The state of the *RFM* is discribed by the nbr. $y_j(t_\tau)$ of polymerases (waiting) within the reservoirs R_j at moment t_τ, corresponding to set *PU* of scheme **2**,

$$\boldsymbol{y}(t_\tau) = (y_1(t_\tau), \ldots, y_j(t_\tau), \ldots, y_b(t_\tau)), \tau = 1, \ldots, T.$$

The output is identical with the state $\boldsymbol{y}(t_\tau)$.

5.3. Output computation

For each moment t_τ the state is calculated recursively by the *REM*-algorithm

$$\boldsymbol{y}(t_{\tau+1}) = REM\big(\boldsymbol{y}(t_\tau), \tilde{\boldsymbol{y}}(t)_\tau, F2_0(t_\tau), F2_j(t_\tau)\big)$$

with $\tilde{\boldsymbol{y}}(t_\tau) = \big(\tilde{y}_1(t_\tau), \ldots, \tilde{y}_i(t_\tau), \ldots, \tilde{y}_u(t_\tau)\big)$ and $\tilde{y}_i(t_\tau)$ – nbr. of polymerases at site U_i at moment t_τ as follows.

Until the moment $t_\tau = A_j$ the reservoir R_j is empty, $j = 1, \ldots, b$:

A) $y_j(t_\tau) = 0; t_\tau = 1, \ldots, A_j$.

The medium (polymerases) needs at least A_j time intervals to flow from the source (onset site) B_0 to the reservoir B_j.

- The polymerases which starts at B_0 at moment t_1 arrive after $A_1 + 1$ time intervals at the reservoir R_1 (unrestricted movement between B_0 and B_1 with velocity v), i.e.

B1) $y_1(A_1 + 1) = a(t_1)$

- The polymerases, which overcame floodgate (blocker) $B_{j-1}(A_j - A_{j-1})$ time steps ago, arrive at the reservoir of $B_j (j = 2, \ldots, b)$ at moment $A_j + 1$, i.e.

B2) $y_j(A_j + 1) = F2_{j-1}(A_{j-1} + 1) \cdot y_{j-1}(A_{j-1} + 1); \quad j = 2, \ldots, b.$

- The amount of polymerases within the reservoir of $B_j (j = 1, \ldots, b)$ at moment $(A_j + \tau + 1), \tau = 1, \ldots, T$ is the amount of them at moment $(A_j + \tau)$ **plus** the amount which was allowed to continue its morement beyond floodgate (blocker) $B_{j-1}(A_j - A_{j-1})$ time steps ago and which is now arriving at B_j **minus** the polymerases which are released out of the reservoir B_j at moment $(A_j + \tau + 1)$ in order to move on, i.e., it is

C1) $y_j = (A_j + t_{\tau+1}) = y_j(A_j + \tau)$

C2) $\qquad + F2_{j-1}(A_{j-1} + \tau + 1) \cdot y_{j-1}(A_{j-1} + \tau + 1)$

C3) $\qquad - F2_j(A_j + \tau) \cdot y_j(A_j + \tau)$

$\tau = 1, \ldots, T - A_j - 1; \quad j = 1, \ldots, b.$

5.4. The adaptation problem

The task is to find the coefficients

$$\boldsymbol{x} = (x, x_1, \ldots, x_j, \ldots, x_b),$$

with

$$x = (x_1, x_2, x_3, x_4)$$

$$x_j = (x_{j_1}, x_{j_2}, x_{j_3}, x_{j_4})$$

of the $b + 1$ polynominals, describing the shapes of the functions $F2_j(t_\tau)$ and $F2_0(t_\tau)$ mentioned above in (1°) and (2°) resp., in such a way that the a generalized distance between the output $\hat{y}(t_{\tau*})$ of the real experiment and the output $y(t_{\tau*}, x)$ of the model becomes minimal. The following (quality) functional was to be minimized,

$$\min \left\{ Q = \sum_{\substack{\tau^*=1,2,3 \\ j=1,\dots,12}} g_j(t_{\tau*}) \, |y_j(t_{\tau*}, x) - \hat{y}_j(t_{\tau*})| \; x \in X \right\}$$

with

$\tau^* = 1$ is equivalent to $\tau = 40$; 5 min. real time

$\tau^* = 2$ is equivalent to $\tau = 160$; 20 min. real time

$\tau^* = 3$ is equivalent to $\tau = 240$; 30 min. real time

and

$$x \in R^{52} \quad \text{with} \quad 4 \cdot (b + 1) = 52, \quad b = 12$$

and the $g_j(t_{\tau*})$ being weights for better control of the adaptation process.

5.5. Result of the adaptation

The adaptation problem was solved by means of a special search algorithm developed on the basis of the main principles of biological evolution by BORN and BELLMANN (BORN, 1978; BELLMANN, BORN, 1978) and described in detail in the paper BORN, BELLMANN in this volume. During the adaptation process, a set of qualitatively very different shapes were offered permanently to the search algorithm. Out of this set, the search algorithm rejected always such shapes which were not of a special type, called optimum function, with minimal level (OFM-type). This minimal level (x_4) is reached after approx. 20 min. experimental time and the maximum after 7—8 min., i.e. for all moving-on-functions and for the source function by the search algorithm such shapes were prefered which increase monotonuously up to their maximum and decrease after this softly to a minimal level (x_4) being maintained until the end of the experiment (see fig. 15). The OFM type corresponds very clearly to the type of shapes selected empirically by simulation experiments with EUTRANSCRIPT 1, as demonstrated in fig. 11.

6. Synopsis

In order to get more insight into the mechanism of the stimulation of the in vitro transcription process of eukaryotic chromatin by high salt concentrations, a simulation model with special structural features of the template was elaborated, based on the results of in vitro RNA synthesis experiments in rat liver nuclei. In applying ammonium sulfate (AS) in the experiment distinct RNA fractions were observed. The distribution of these distinct RNA chain lengths proved to be depending on time and AS concentration. On the basis of these results special structures of the chromatin were assumed hypothetically, called transcription blockers, which are assumed to be able to stop the movement of a polymerase until this blocker site becomes "inactivated" so that the polymerase moves on. The distance between the blockers appeared to be nearly constant and equal to approximately $2S$-units (540 nucleotides).

The simulation algorithm EUTRANSCRIPT 1 is a discrete simulation model and describes the polymerase movement on the template. The essential increments are (i) a set of templates of equal length, (ii) variable number of blocker sites, each with variable time functions which describe the probability that at a given moment a "closed" blocker site becomes open (deblocking event), an open blocker becomes closed (reblocking event), and a given blocker state is maintained, (iii) a "initiation" site with a variable time function indicating the "initiation" rate of polymerases, (iv) a variable polymerase mortality function, (v) a variable start distribution of polymerases on the template, (vi) variable polymerase velocity. The state of the transcription system is characterized by the amount of polymerases at different template sites, its output by the RNA chain length distribution.

The results of simulation experiments support the existence of "transcription blockers". Key processes proved to be polymerase "initiation" and deblocking processes at the first 7 blockers. Shape analyses of the "initiation" rate $(F_0(t))$ and the deblocking functions $F2_j(t)$ led to the conclusion that in the real system the derepression of polymerase happens within 18—20 min (with a maximum after approx. 9—10 min) and that the deblocking process develops very strongly during the first 4—7 min and loses its intensity for the following 10 min more and more. Consequently, the probability of polymerase waiting at a blocker site strongly decreases; the RNA synthesis process is damped with increasing time. Hypothetical conclusions about the biochemical nature of the blocker sites are drawn.

In order to increase the certainty of knowledge about the shapes of the deblocking functions $F2_j(t)$ and of the "initiation" function $F1_0(t)$ which were realized as being the key processes of RNA synthesis in these experiments a reduced model (called EUTRANSCRIPT 2) with strongly reduced computing time (some tenth sec.) was elaborated. By application of a special parameter adaptation algorithm for simulation models with ill conditioned mathematical structures the arising 52 dimensional adaptation problem was solved. The results support the optimum type of the functions. The algorithm is based on the principles of evolution strategy, it is outlined in the contribution Born, Bellmann in this volume.

References

Auer, G., G. P. M. Moore, N. R. Ringertz and A. Zelterberg (1973): DNA-dependent RNA synthesis in nuclear chromatin of fixed cells. Relationship between dye-binding properties, nuclear protein content and RNA polymerase activity. Exptl. Cell Res. 76, 229—233.

Bellmann, K., and F. Collatz (1975): Stochastisches, dynamisches Simulationsmodell zur Analyse des Transcriptionsprozesses bei Eucaryonten. Arb.-Bericht, unveröffentl.

Bellmann, K., and J. Born (1979): Numerical solution of adaption problems by means of an evolution strategy. In: Modelling and Optimization of Complex Systems. Proceedings of the IFIP-TC 7 Working Confernce. Novosibirsk 1978. Springer-Verlag, Berlin, Heidelberg, New York, 157—167.

Born, J. (1978): Evolutionsstrategien zur numerischen Lösung von Adaptationsaufgaben, Dissertation (A). Humboldt-Univ., Berlin.

BORN, J., and K. BELLMANN (1981): Numerical adaptation of parameters in simulation models using evolution strategies. In: Bellmann, K. (Ed.), Modelling and Simulation of Molecular Genetic Information Systems. Akad.-Verlag, Berlin, 291–319.

BLACKBURN, K. J., and H. G. KLEMPERE (1967): Biochem. J. 102, 118–176.

BRADBURY, E. M., T. MOSS, H. HAYASHI, R. P. HJELM, P. SUAU, R. M. STEPHENS, J. P. BALDWIN and C. CRANE-ROBINSON (1978): Nucleosomes, Histone Interactions, and the Role of Histones H3 and H4. Cold Spring Harbor Symp. XLII, 277–286.

CEDAR, H., and G. FELSENFELD (1973): Transcription of Chromatin Assembled in Vitro. J. Mol. Biol. 77, 237–254.

CHAMBON, P., M. RAMUZ, P. MANDEL, J. DELY (1968): The influence of ionic strength and a polyanion on transcription in vitro. I. Stimulation of the aggregate RNA polymerase from rat liver nuclei. Biochim. Biophys. Acta 157, 504–519.

DANEHOLT, B., S. T. CASE, M. M. LAMB, L. NELSON and L. WIESLANDER (1978): The 75 S RNA transcription unit in Balbiani rings and its relation to chromosome structure. Phil. Trans. Roy. Soc. B 283, 383–389.

FELSENFELD, G. (1978): Chromatin. Nature 271, 115–122.

GISSINGER, F., C. KEDINGER and P. CHAMBON (1974): Biochemie (Praxis) 56, 319–333.

HAMEISTER, H. (1977): RNA synthesis in isolated polytene nuclei from Chironomus tentans. Chromosoma 62, 217–242.

HOLLATZ, H., J. BORN and K. BELLMANN (1977): A reservoir floodgate system. Working paper, unpublished.

KORNBERG, R. D. (1977): Structure of Chromatin. Ann. Rev. Biochem. 46, 931–954.

LINDIGKEIT, R., K. BELLMANN, H. FENSKE, M. BÖTTGER, M. HOLTZHAUER and I. EICHHORN (1974): Effect of removal of f1-histone on the conformation of nuclear chromatin and on the transcription process FEBS-Lett. 44, 146–152.

LINDIGKEIT, R., M. HOLTZHAUER and K. BELLMANN (1974): On the transcription process in the isolated rat liver nucleus. FEBS Meeting, Budapest 343–348, Publishing House of the Hungarian Academy of Sciences.

MANDEL, J. L., and P. CHAMBON (1969): Characterization of RNA synthesized at high ionic strength by rat liver aggregate RNA polymerase. Biochem. Biophys. Res. Commun. 35, 868 – 874.

MOORE, G. P. M., G. AUER and A. ZELTERBERG (1974): Reactivation of nuclear RNA polymerase activity in growth stimulated epithelial cells. Exptl. Cell Res. 88, 375–381.

OUDET, P., J. E. GERMOND, M. BELLARD, C. SPADEFORA and P. CHAMBON (1978): Nucleosome structure. Phil. Trans. R. Soc. Ldn. B 283, 241.

PEDERSON, T. (1978): Intern. Review of Cytology 55, 1–22.

PELLING, C. (1964): Ribonukleinsäure – Synthese der Riesenchromosomen. Autoradiographische Untersuchungen an Chironomus tentans. Chromosoma 15, 71–122.

PELLING, C. (1969): Synthesis of nuclei acid in giant chromosomes. Progress in Biophys. and Molecular Biology 237–270 (J. A. V. BUTLER and D. NOBLE, ed.). Oxford and New York, Pergamon Press.

POGO, B. O. (1969): Modification of RNA synthesis in isolated rat liver nuclei by low salt concentration and specific divalent cations. Biochim. Biophys. Acta 182, 57–65.

RO, T. S., and H. BUSCH (1967): Studies on RNA polymerase of nuclear fractions of regenerating liver. Biochem. Biophys. Acta 134, 184–187.

SASS, H. (1978): Thesis, Tübingen.

WASYLIYK, B., and P. CHAMBON (1979): Transcription by Eukaryotic RNA Polymerases A and B of Chromatin Assembled in vitro. Eur. J. Biochem. 98, 317–327.

WASYLYK, B., G. THEVENIN, P. OUDET, P. CHAMBON (1979): Transcription of in vitro assembled Chromatin by *Echerichia coli* RNA Polymerase. J. Mol. Biol. 128, 411–440.

WEINTRAUB, H., and M. GROUDINE (1976): Chromosomal Subunits in Active Genes have an altered Conformation. Science 93 (oder 193), 848–858.

WEISBROD, ST., M. GROUDINE and H. WEINTRAUB (1980): Interactions of HMG 14 and 17 with actively transcribed genes. Cell **19**, 289–301.

WIDNELL, C. G., and J. R. TATA (1966): Studies on the stimulation by ammonium sulfate of the DNA-dependent RNA polymerase of isolated rat-liver-nuclei Biochim. Biophys. Acta 123, 478–492.

WILLIAMSON, P., and G. FELSENFELD (1978): Transcription of Histone-Covered T7 DNA by *E. coli* RNA Polymerase. Biochemistry 17, 5695–5705.

ZYLBER, E. A., and S. PENMAN (1971): Products of RNA polymerase in HeLa cell nuclei. Proc. Natl. Acad. Sci. USA, 68, 2861–2865.

8. Discrete simulation of replication of a RNA-bacteriophage prototype system

A. Knijnenburg, U. Kreischer

Contents

1. Introduction

Biological subjects display some properties that make it difficult to infer the system's overall behaviour on the next higher level of consideration from a sum of single phenomena. They are complex, non-linear, hierarchicly-grained, and dynamic systems. In molecular-biological systems knowledge about the informational and functional integration of a lot of single control mechanisms, each of which is studied keeping constant the parameters of the others, cannot be gained unambigously. To trace inherent causal mechanisms of interaction of biochemical processes it is indispensable to separate the effects of the participating particular processes by keeping constant the influences of all processes except the specific one under consideration. Due to the practical unavailability of experiments which rigorously fulfil this requirement in the cellular regime, we have been searching for images of a biological system with a very high degree of segregation of the single processes relative to each other. This intent has been followed by generating a discrete simulation model for a RNA-bacteriophage prototype. Such a mathematical model enables us to study the variation of parameter sets, even under conditions where the structures of the modelled system elements are variable in themselves. The following features of the group of RNA-bacteriophages motivate their choice as the modelling object:

1. RNA-bacteriophages represent, during the infection process, a molecular-biological system which due to the very high specifity of all phage-coded particles is functionally segregated and informationally relatively independent of the subordinated cell system of the bacterium.
2. The specific processes of RNA-bacteriophage replication in the host cell involve four epigenetic unit operations:

   a) Replication of the system-specific memory, the phage-RNA-replication;
   b) Production of system-specific mRNA, now the complementary (—)RNA-strands acting as the genetic matrix more commonly formed of DNA in other species;
   c) Translation of system-specific messengers;
   d) Production of system-specific nucleoprotein complexes for regulation of gene expression on the level of translation.
3. The control mechanisms of the single processes for coordination of the time-dependent development of the phage-coded subcellular apparatus function informationally independently of the cellular metabolic level. They can be mapped on the molecular level without supplementary suppositions related to qualitative influences of the cellular metabolic network. Effects of single phage-coded processes on the behaviour of the complete system of all phage-specific particles (phage system) are directly measurable.
4. The phage system, while growing in the host cell, is qualitatively as well as quantitatively surveyable, sufficient for modelling on the molecular level: an ample number of experimental data is available.

The global objectives of this modelling are:

I. How is the process of growth of a self-organizing information-processing molecular-genetic system controlled by coupling different single control mechanisms on the molecular level?

II. Can the dynamic of this growth process be attributed to properties of the genetic memory of that system?

The task of modelling the RNA-phage system, which is to be derived from the objectives, is to predict the qualitative nature of the time-dependent development of the phage system in the host cell from a knowledge of the structure of the genetic memory, the phage-RNA.

2. Suppositions for the model design

The second objective requires to perform in the model the condition that the structures, numbers, and points of appearence of all phage-specific gene products are determined by the phage-RNA only. Furthermore in the model processes of RNA-replication and virion assembly shall be mapped independently of the host cell system. The host cell environment must be introduced into the phage model as a marginal condition with a modifying effect of the magnitude 1. More restrictively, the following basic assumptions regarding the object of modelling are made:

1. The development of phages will be simulated under such conditions that no inhibitional influences are exercised by the host cell. It will thus be assumed that cell substances utilized by the phages as ribosomal subunits, replication factors, t-RNAs,

amino-acids, nucleoside triphosphates etc. are present in abundance. This condition corresponds in character with **in vitro** experiments and ensures that we can attribute the behaviour of the phage system exclusively to its structure alone.

2. The translation apparatus of the bacterial cell will be used by the phage system without functional alternation. We assume that knowledge gained from protein synthesis in bacteria is transferable to bacteriophage-specific polysomes.
3. As causal mechanisms, responsible for the structure of phage development, the biochemical processes which result from the mutual relations between the system elements of the phage system will suffice. Spatial distributions of the phage-specific molecule compartments in the host cell are neglected.

Furthermore, it is necessary to separate the processes of RNA-replication from those of RNA-translation, so that the multiplication of the genetic memory and the expression of the genetic memory can be treated as independent phenomena in the model.

At last, the activities of various mechanisms for translation control must be separated from each other. That means that the model must enable us to distinguish whether the change in the frequency of translation of a cistron derives from the specific configuration of secondary RNA-structures or from protein-nucleic acid interactions. Therefore we exclude control mechanisms whose direction of action is from the phage-specific apparatus for RNA-multiplication toward the apparatus for RNA-translation, specifically we exclude:

1. The translational control of the cistron for the maturation protein by the RNA-polymerisation process;
2. The translational control of the cistron for the coat protein by polymerase (WEISSMANN et al., 1972).

The result of these theoretical simplifications is a prototype of a very simple viral system whose global behaviour is solely conditioned by the control mechanisms of genetic expression.

3. The RNA-bacteriophage prototype system

The RNA-bacteriophage prototype system has been defined retaining all common features of the different froms of this phage group (R17, M12, MS2, f2, Qβ etc.) according to the objectives of modelling and the restrictions which have been derived from them not considering all other specifities of the single forms. The prototype is given as a tupel

$$\mathfrak{P} = [M, V, \mathfrak{C}, \mathfrak{R}, \mathfrak{W}]$$

where M is the structure of one parental RNA-strand,
V is the virion structure,
$\mathfrak{C}$ is the set of equivalence classes of system elements,
$\mathfrak{R}$ is the set of relation between these classes, and
$\mathfrak{W}$ is the set of features describing the global replication process behaviour.

M: The RNA contains three cistrons with the sequence 5′-maturation-coat-polymerase-3′ in the relation of length 12 : 4 : 15. To each cistron corresponds one separate ribosome binding site with a length between 20 and 40 nucleotides (fig. 1). The total RNA-length is about 3350 ± 50 nucleotides.

V: Each virion consists of 180 coat proteins, 1 molecule RNA and 1 maturation protein. Spatial relations between the molecules within the virion are neglected.

$\mathfrak{E}$: The totality of system elements contains the following equivalence classes:

i single-stranded RNA ((+)RNA);
ii single-stranded complementary RNA ((−)RNA;
iii partially double-stranded RNA, formed by (+)RNA with active polymerases (RI);
iv double-stranded RNA ((±)RNA);
v intermediate RNA, formed by (±)RNA or (−)RNA with active polymerases synthesizing nascent (+)RNA-strands (RF);
vi phage-RNA coded ribosomes bound to (+)RNA-strands of each form;
vii coat protein;
viii maturation protein;

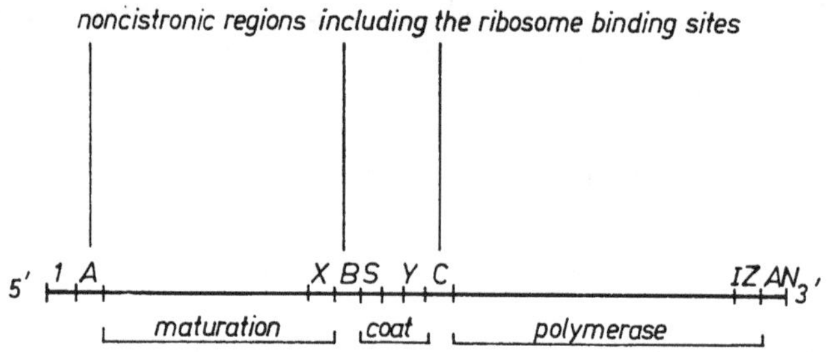

Fig. 1. Linear structure of the RNA-bacteriophage prototype (+)RNA-strand.

ix polymerase (according to the above assumptions the concentration of active replication enzymes shall only depend on the phage-coded subunits (GARWES et al., 1969)
x ribosome bound (+)RNA with coat protein repressors at the polymerase cistron ribosome binding site (STAVIS, AUGUST, 1970)
xi complexes formed by ribosome bound (+)RNA with coat protein repressors and bound maturation protein STAVIS, AUGUST, 1970; NATHANS et al., 1969)
xii virion assembly complexes formed by (+)RNA with bound maturation protein to which coat protein is bounded in succession (NATHANS et al., 1969)
xiii virions.

$\mathfrak{R}$: r_1: i results from the replication process of v;
r_2: ii results from the replication process of v;
r_3: iii results from binding ix to i;
r_4: iv results from the replication process of iii;
r_5: v results from binding ix to iv or ii;
r_6: vi results from binding ribosomal subunits to i or to nascent (+)RNA-strands of v;
r_7: vii results from total translation of the coat cistron at vi;
r_8: viii results from total translation of the maturation cistron at vi;
r_9: ix results from total translation of the polymerase cistron at vi;
r_{10}: If cistrons are not totally translated, the resulting amino acid chains have no phagespecific biochemical activity. They don't become system elements;

r_{11}: Ribosome binding to the polymerase cistron binding site depends additionally on the translation of the aminoterminal end of the coat cistron. This phenomenon is called the polarity effect of the coat cistron (see fig. 1);

r_{12}: x results from binding vii to i or to nascent (+)RNA-strands of v;

r_{13}: xi results from binding viii to x;

r_{14}: xii results from binding vii to xi;

r_{15}: xiii results from finishing the virion assembly process at xii.

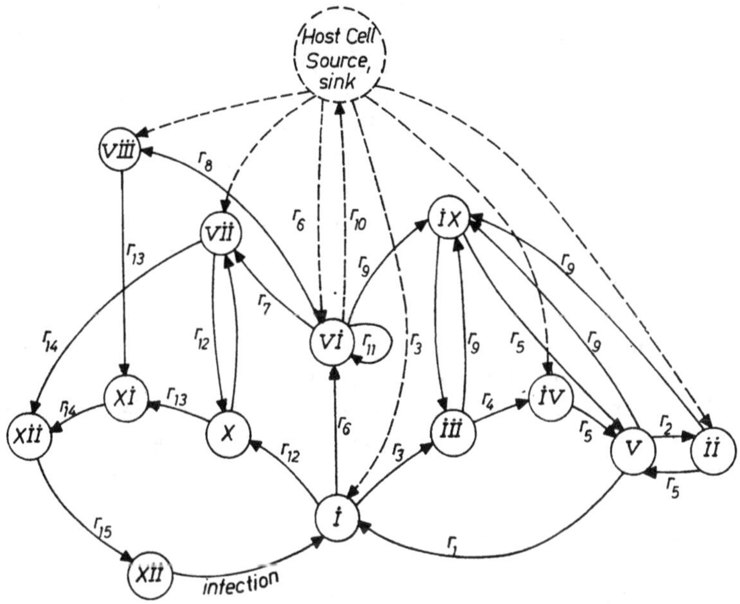

Fig. 2. The functional structure of the RNA-bacteriophage prototype system. Notations are explained in text.

The above qualitative description of 𝔈 and ℜ supplies the functional structure of the phage prototype system as it is depicted in the scheme of fig. 2. The quantitative description of 𝔈 and ℜ in a more detailed manner is the contents of the simulation of 𝔓 during the infection process. During simulation the changes of the numbers of molecules or molecule complexes in all nodes are calculated.

𝔚: w_1: The in-vivo-infection process takes about 40 minutes until lysis.

w_2: 10 minutes past infection an exponential increase of polymerase activity begins.

w_3: About 20 minutes past infection the polymerase activity achieves a maximum, and the relation of coat protein to (+)RNA is about 5···10:1, and repression of the polymerase cistron translation begins.

w_4: After 20 minutes the polymerase activity remains nearly constant.

w_5: The virion concentration increases linearly from about 20 minutes past infection on.

w_6: At the end of the infection process the relation of (+)RNA-synthesis to (±)RNA-synthesis is about 8···10:1.

These features are to be taken as the basis to adapt the dynamic model behaviour to the behaviour of the subject.

4. The model of the phage prototype system

4.1. General view

The definition of the phage prototype system includes a static description of the subject to be modelled. To study the dynamic behaviour of this system a model is needed that encloses the speed-relations between the processes of RNA-replication, RNA-translation and virion assembly, and the resulting time-consumption of all realizations of the elements of $\mathfrak{R}$. The translation is the slowest process. It fixes a time in a natural way. The mean period which is needed to incorporate one amino-acid into the growing peptide chain by an active ribosome has been defined as the system time. We assume that the time-consumption of all association and dissociation processes of molecule complexes be small against the defined system time, so that these processes have not to be regarded molecularly in detail. The remaining time-consuming processes are the elongation of phage-RNA coded ribosomes and polymerases on phage-RNA and the conformation change of the virion assembly complexes which is called virion maturation (Stavis, August, 1970; Nathans et al., 1969).

Considering these three phage-specific process types in the above described way we obtain a discrete model. The time-consumption of all processes is measured in numbers of the defined system's time-step. All RNA-positions of alongating ribosomes and polymerases and the development stages of virion assembly complexes are expressed by magnitudes of the time-steps consumed since the process initiation points.

The principal book-keeping procedure is the following: One of the above considered processes takes n_i time-steps from its initiation up to the termination. Now we define n_i different processes stages each of which distinguishes itself from the next-neighbouring by a one-time-step difference. At the process association all molecules or molecule complexes forming the processing complex are allocated to the first stage. Every time the members of all stages up to number $n - 1$ are advanced into the next stage, and new numbers of complexes start processing with stage one, and a number of complexes leave the last stage n_i because they dissociate and release the products. We assume that the (+)RNA-strand replication and the (−)RNA-strand replication have the same time-consumption, and that the affinity of the phage coded polymerase to (+)RNA is much greater than the affinity of the cellular ribosomal subunits to (+)RNA.

The number of products released by the mentioned processes depends on the number of synchronously initiated complexes and on the process control regime.

The simplest process description is:

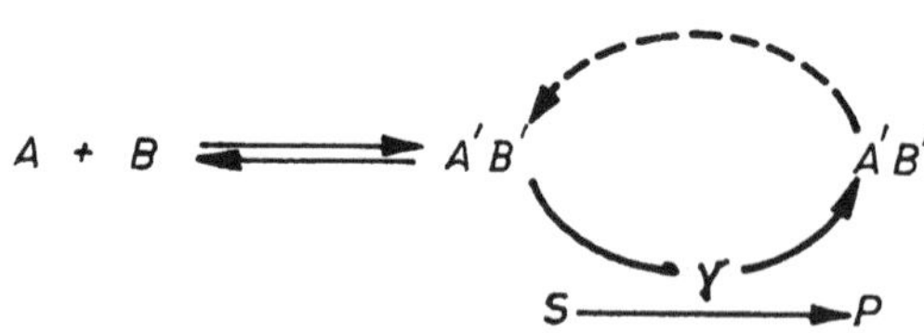

where A and B are the complex forming units, $A'B'$ is the processing complex, P is the product ($A, B, A'B', P \in \mathfrak{R}$), S is the substrate (according to the above assumptions unlimited), and γ is the process under consideration mapped by the explained book-keeping procedure.

One example. γ be the RNA-polymerisation process. Then A is the (+)RNA-strand (or (—)RNA in the other case), B is the pool of free polymerases, $A'B'$ is the matrix-enzyme-complex, S is the cellular nucleoside-triphosphate pool, and P is the (—)RNA-strand (or (+)RNA in the opposite case).

The elongation of one enzyme molecule on one RNA-strand both forming the complex $A'B'$ is called "**one** process". In general the development of **one** complex $A'B'$ is defined as "one process γ".

The number card (γ) of synchronously initiated equivalent processes is

$$\text{card}\,(\gamma_{t_0}) = \alpha\beta_{t_0} \min\,[C_{t_0}(A); C_{t_0}(B)] \tag{1}$$

where α is the binding probability of B to A resulting from the stereo-chemically fixed affinity, β_{t_0} is the reduction factor of the maximal process initiation rate caused by the relation of the concentrations $C(A)$ to $C(B)$ at time t_0, $0 \leqq \beta_{t_0} \leqq 1$.

Under the assumption that no inhibitional factors influence the initiation of the process, after a period τ that represents the process duration expressed in time-steps the increase of products P is

$$C_{t_0+\tau}(P) = C_{t_0+\tau-1}(P) + \text{card}\,(\gamma_{t_0}). \tag{2}$$

A more realistic description includes influence of initiation control factors and of the substrate concentration. Then the product concentration increase is

$$C_{t_0+\tau}(P) = C_{t_0+\tau-1}(P) + \alpha\beta_{t_0}\gamma_{t_0}(1 - \delta_{t_0})\,\varepsilon_{t_0} \min\,[C_{t_0}(A); C_{t_0}(B)] \tag{3}$$

where γ_{t_0} is the influence of a positive control, $0 \leqq \gamma_{t_0} \leqq 1$, δ_{t_0} is the influence of a negative control, $0 \leqq \delta_{t_0} \leqq 1$, ε_{t_0} is the mean reduction factor of the maximal substrate transformation rate as a function of the substrate concentration $C_{t_0}(S)$, $0 \leqq \varepsilon_{t_0} \leqq 1$. ε_{t_0} is not relevant for those processes where cellular substrates are utilized which have been defined as unlimited.

Because polymerase can bind to different RNA-matrices ((+)RNA, (±)RNA, (—)RNA, RF, RI) and ribosomal subunits can bind to different phage-RNA-messengers (free (+)RNA-strands, (+)RNA-strands of RF, nascent (+)RNA-strands of RI) the initiations of several processes γ_j must be seen in connection. A simple example shall be used for explanation. Two matrices, A_1 and A_2, be used by the same enzyme B. Then the following picture results

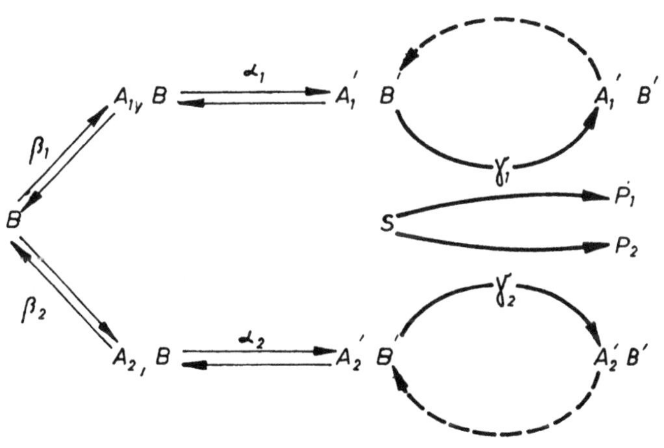

and equation (3) can be generalized to

$$C_{t_0+\tau}(P_i) = C_{t_0+\tau-1}(P_i) + \min \left[\mathfrak{R}_i C_{t_0}(A_i);\ \mathfrak{R}_i C_{t_0}(B) \cdot \frac{\mathfrak{R}_i C_{t_0}(A_i)}{\sum\limits_{j=1}^{l} \mathfrak{R}_j C_{t_0}(A_j)} \right] \tag{4}$$

where $\mathfrak{R}_i$ equals $\left(\alpha_i \beta_{i_{t_0}} \gamma_{i_{t_0}} (1 - \delta_{i_{t_0}})\, \varepsilon_{l_0}\right)$ and l is the number of competing matrices A_j, $j = 1, \ldots, i, \ldots, l$. It is to be seen that we assume the distribution of B to the different A_j to be proportional to the binding probabilities and the matrix concentrations.

Thus the distribution of B is a stochastic process, and, because such coupled process initiations take place at each time during the phage prototype development the infection process can only be described by means of a stochastic algorithm in its enormous complexity on the molecular level. This algorithm "A" is used to establish which system elements by which manner will be transformed into others.

Consequently, the numbers of the affected elements will change and thereupon the system state that has been defined as a vector of the numbers of the elements assigned to $\mathfrak{C}$, the set of element equivalence classes of the phage prototype system. Hence, algorithm "A" assigns to each system state not a determined but a probable sequential state. Due to the high complexity of the state definition of the phage prototype and the Markovian character of the state transfer function the state transformation can only be given in a generative form

$$z_t = \text{"}A\text{"}^t \circ z_0, \quad t = 0, 1, \ldots, t_{\max}$$

that requires a computer simulation.

z_t is the system state at time t. "A"$^t \circ z_0$ means that "A" was t-times applied to the initial state z_0, first to z_0, then to the resulting state and so forth, thus generating a state sequence that describes the development of the phage prototype system up to an expected point of time $t_{\max}$. The alternations of the different molecular pools are exactly discernible. The chosen cycle length of "A" corresponds to 0.3 seconds in vivo, $t = 1000$ in the model is equivalent to five minutes of phage development. According to the used book-keeping procedure the problem of quantifying the three process types to be described can be reduced to the quantification of the process initiations:

1. process initiation $C_{t_0}(A'B') = W_{t_0}$ described by the term min [...] of equation (4)
2. processing with substrate consume $C_{t_0}(A'B')$

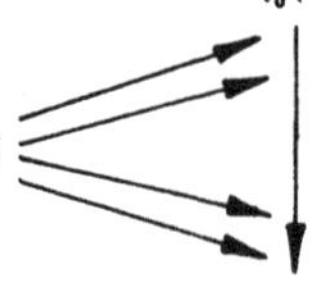

$C_{t_0+\tau}(A'B')$

3. process termination and dissociation of the complex $A'B'$ into $A + B$

$$C_{t_0+\tau+1}(P) = C_{t_0+\tau}(A'B').$$

Specific duration τ_j of the following processes have been regarded in the model:

Synthesis of maturation protein, coat protein and polymerase by the transcription process, synthesis of $(\pm)$RNA- and $(+)$RNA-strands by the RNA-replication process, and the virion assembly.

Algorithm "A" is figured in its principal structure in the following scheme:

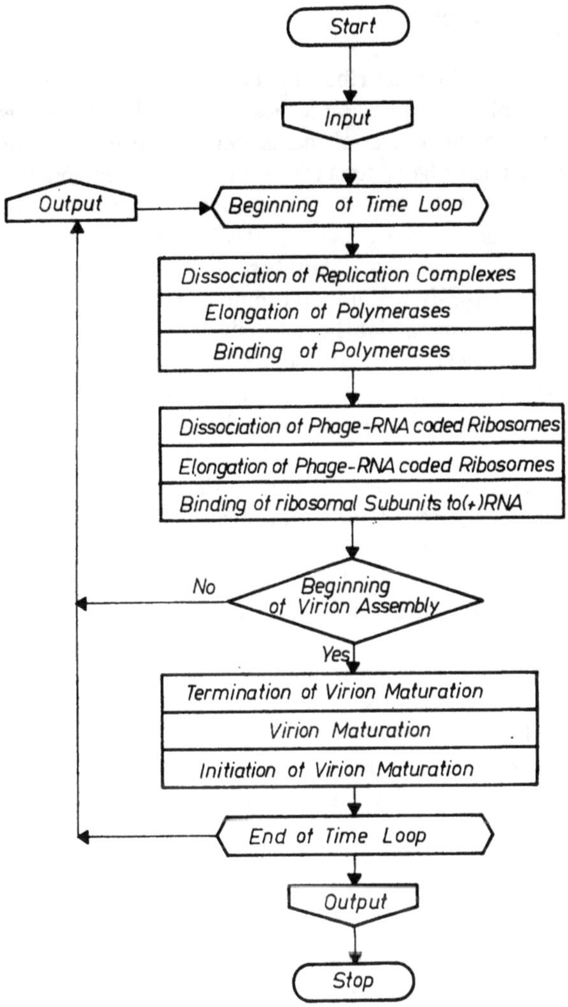

Fig. 3. Principal flow chart of the simulation model.

4.2. The phage prototype subsystems

The progeny (+)RNA-strands can introduce into three different processes:

a) They can serve as messengers for the production of phage proteins.
b) They can be used as matrices for the production of complementary (—)RNA-strands.
c) They can introduce into the virion maturation process to produce infectious progeny phage particles, the virions.

Hence, we establish three subsystems that are functionally connected:

a) The translation subsystem which supplies the replication enzyme for the replication subsystem and the viral coat proteins for the virion maturation subsystem.
b) The replication subsystem which supplies the progeny genetic material for the virion maturation subsystem.
c) The virion maturation subsystem which performs the system's goal function, the production of progeny infectious phage particles.

4.2.1. The translation subsystem

According to the defined system time the ribosome elongation of one triplet takes one time-step. One ribosome occupies 9 triplets on a messenger-RNA and a mean distance of 25 triplets between ribosomes on bacterial messengers has been observed. Thus we define a mean ribosomal distance of 34 codons for two neighbouring codons which have an equivalent position within two next-neighbouring active ribosomes. The ribosome binding sites of the three cistrons of the phage (+)RNA as well as the cistronspecific translation termination sites are expressed in numbers of this "Mean ribosomal distance" related to the 5′-end of the (+)RNA. If one ribosome has been initiated on a binding site at a definite point of time, in the model the next ribosome binding to this same site can occur after a 34-time-steps delay. Two ribosomes on equal sequences of different (+)RNA-strands which have been quantified by means of the "mean ribosomal distance" cannot be distinguished by its localisation because distances smaller than 34 codons cannot be captured.

Such ribosomes are distinguished from each other by their time of initiation. Within a loop of 34 times ribosomal subunits can bind with cistron-specific binding probabilities to the ribosome binding sites (see fig. 1). These probabilities were not known exactly from experiments but working values could be estimated from cistron specific translation frequencies **in vivo** and **in vitro**. Later on we could adjust better values in the course of model adaption.

Hence, ribosomes initiated with a time-delay of 34 and bound to the same cistron species of different (+)RNA-strands can be allocated to the same compartment. Ribosomes initiated with a one-time difference are allocated to a next compartment and so on. From this consideration result 34 different ribosome compartments for each cistron species. Each of the resulting 102 ribosome compartments has been mapped by a separate book-keeping procedure as introduced above, that means each of them contains a ribosome initiation stage, a sequence of ribosome elongation stages and a translation termination stage. In the last the ribosomes dissociate and release the phage proteins. If the relation of coat protein to (+)RNA achieves a value greater than six, coat protein bind to the polymerase cistron ribosome binding site to repress the polymerase production. This process has been separately quantified for all 34 relevant compartments in the model. The polarity effect of the coat cistron has been mapped by the same principle. In each of the 34 compartment stages which contain the region S (see Fig. 1) the number of ribosomes located on the region S is used to calculate the dependence of the ribosome binding frequency at the polymerase cistron binding site on the translation of the aminoterminal end (region S) of the coat protein cistron.

4.2.2. The replication subsystem

The polymerase distribution to different types of RNA-matrices has been described in section 4.1. The affinity of polymerase to (—)RNA-strands is greater than to (+)-strands, the related model parameters are α_1 and α_2. All polymerisation processes in the model start at the 3′-ends of the RNAs. All RNA-replications are mapped by the already known book-keeping procedure, but now the difference between two neighbouring development stages is defined by the number E of nucleotides which are incorporated per one time-step into the nascent RNA-strand. We assume this number to be constant. All RF or RI with nascent RNA-strands of an equal length are captured in related compartments. But, there is a principal difference between RF and RI. The polymerases razing on (+)RNA-matrices in the model stop the translation process and, if ribosomes are released before reaching the translation termination sites of the cistrons, peptide chains occur which have no phage specific activity. Coat protein repressors are released by polymerases too, in the model they return into the coat protein pool. Therefore in every ribosome compartments the adverse (—)RNA-replication process must be considered. On the contrary, at the nascent (+)RNA-strands of the RI ribosomal subunits can bind to the binding sites before the (+)strands are released from the complementary (—)RNA-matrix. At each time the question is to be asked for all nascent (+)RNA strands, which of the ribosome binding sites have already been synthesized to decide whether these (+)RNA-strands can act as messengers for certain proteins to be introduced into the corresponding compartments of the translation subsystem.

Additionally each RI-compartment must be divided into a lot of subcompartments because more than one polymerase can work on one (—)RNA-matrix. All considerations about the book-keeping of the growing (+)RNA and its connection with the translation subsystem have to be transferred to these RI-subcompartments. We assume that the binding probability of another polymerase to a (—)RNA-matrix under consideration depends on the distance of the polymerase bound before to the 3′-end of the (—)RNA. We establish for this dependence the function

$$\mathrm{PROB(V)} = (\mathrm{V/NC})^{H} \tag{5}$$

where V is the distance of the last bound polymerase to the 3′-end of the (—)RNA expressed in units of E, NC is the RNA-length in units of E and PROB(V) ist the probability of the binding of another polymerase in dependence on the distance of the polymerase bound before, H is a free parameter. At each time some relations of the stochastic process of polymerase binding to RI can take place thus requiring a number of V — 1 subcompartments of each RI-compartment in the stage V.

4.2.3. The virion maturation subsystem

The virion maturation process is initiated by binding one molecule maturation protein to (+)RNA-strands. This induces a RNA-conformation change that represses the maturation cistron translation. The polymerase cistron translation is already repressed at this time so that at the virion initiation complexes only the production of coat proteins carries on. The maturation protein binding depends on the threshold relation of maturation protein to free (+)RNA and nascent (+)RNA. Because in this late phase of the phage system development practically no free (+)RNA exists the maturation proteins bind to the nascent (+)strands of the RI-complexes. If a nascent (+)RNA has been used by a maturation protein to form a virion initiation complex, this RNA is introduced into the virion maturation subsystem, but, the coat cistron translation is further considered in the translation subsystem. Because at this time of process development the number of coat proteins is not large enough for the formation of complete coats for all progeny (+)RNAs we assume that the maturation process of virions — that means in the model the incorporation of coat proteins into the maturing coats — depends additonally on a further threshold-relation of coat proteins to virion initiation complexes. In **in-vivo** experiments at such complexes a high production rate of coat proteins occurs. The coat protein production will be repressed too with the starting incorporation of coat proteins into the maturing virions. A definite number of coat proteins is incorporated into the coat per time-step. We use the known book-keeping procedure with a start stage, a lot of maturing stages and an end stage where the maturation process is finished and complete virions leave the virion maturation subsystem.

4.3. The simulation system

The implementation of the model on a computer cannot be explained here in detail, this would fill a separate booklet. The organisation of the algorithm "A" in the central processor of the Russian computer BESM6 has been realized in FORTRAN. The whole mainstorage was needed for the program, and one simulation run took about five hours[1] — this shall be mentioned only to have not kept back two disadvantages of a "portrait-modelling" on the molecular level. The molecular processes are mapped in subroutines which are controlled by a main program. The organisation of the simulation system is depicted in fig. 4.

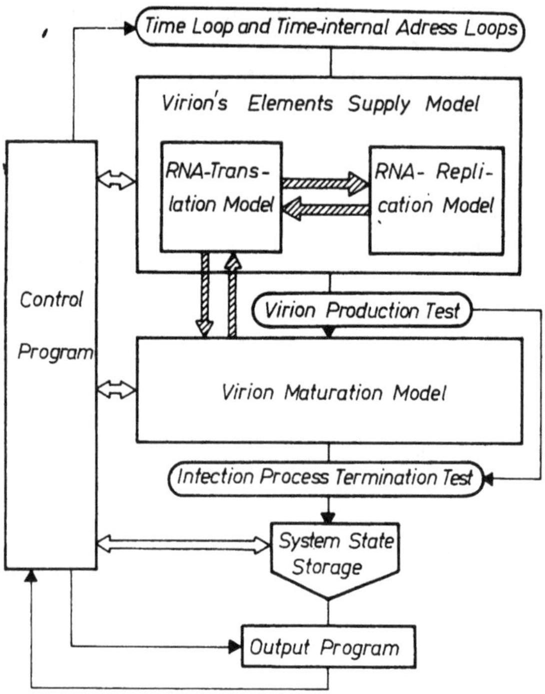

Fig. 4. Simulation system of the RNA-bacteriophage prototype.

5. Model validation

In this paper model validation is ment to reproduce the set of features of the defined phage prototype system describing the global replication process behaviour by means of simulating the functional structure of the phage prototype system as mentioned in the section before.

---

[1] 10^6 operations per second are performed on an average.

As to the first feature, the 40 minutes duration of the **in vivo**-infection process (W_1): The modelled process does not come to a natural end, it must be stopped at a given point. Under the conditions simulated in the model the available cellular materials are unlimited and the phage replication process-internal dynamics which results from the structural information of one parental (+)RNA-strand alone does not include an abbreviation within itself. The phage infection process termination seems to be caused indeed by the host cell lysis. More about this topic in the discussion.

The second feature, the exponential polymerase activity increase in the early phase of infection (W_2) is perfomed by the model, see fig. 5.

The third feature, the repression of the polymerase cistron translation at a concentration relation of coat protein to (+)RNA greater than six (W_3) can be observed from fig. 6.

The required quasi-constance of the polymerase activity (W_4) and the linear increase of maturing virions (W_5) after 20 minutes past infection are depicted in fig. 7 and fig. 8.

The last defined feature of the phage prototype behaviour (W_6) is the eight to ten times higher (+)-synthesis in relation to the ∓))RNA-synthesis near the lysis. This

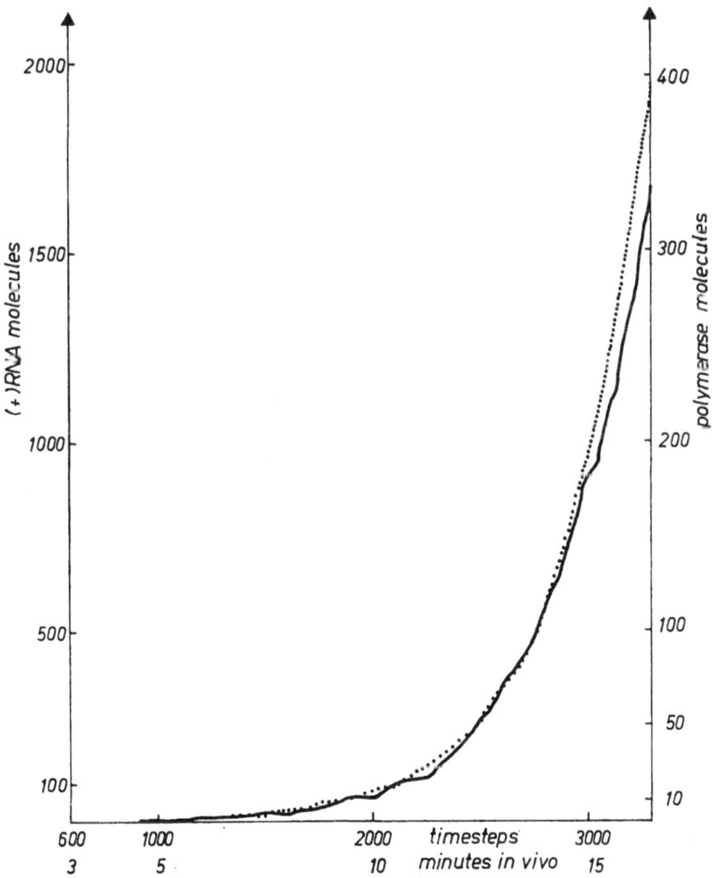

Fig. 5. Polymerase concentration rise (full line) and (+)RNA-concentration rise (dotted line).

finding could not be reproduced with the model in the described form. The achieved (+)RNA — to (±)RNA relation was about 3 to 1. This problem will be pointed out in the discussion.

Two statements can be made:

1. The model is valid regarding the qualitative process course, that means, the entity of the phage infection process dynamic is reproducible.
2. Two quantitative features, the topic of W_6 and the time consumption of the virion maturation cannot be reproduced with the phage prototype simulation model. The other quantitative attributes of the phage infection process are performed by the model.

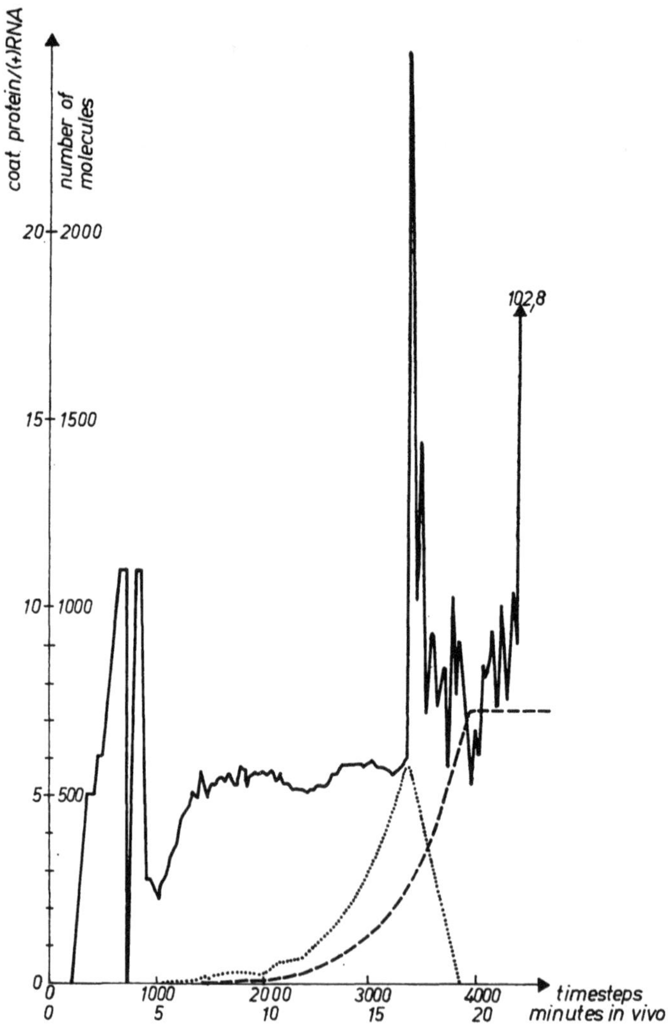

Fig. 6. Repression of the polymerase cistron translation.
Full line: coat protein in relation to (+)RNA,
Dashed line: polymerase concentration,
Dotted line: number of ribosomes elongating on polymerase cistrons.

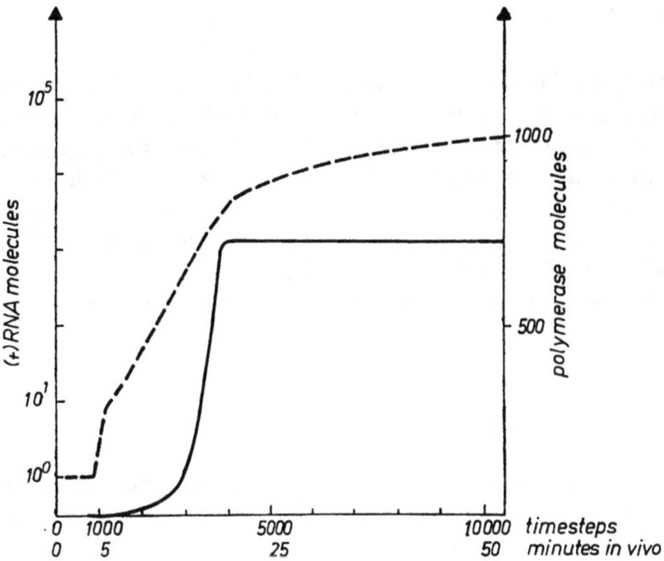

Fig. 7. (+)RNA-concentration rise after repression of polymerase cistron translation (dashed line) and constant polymerase concentration (full line).

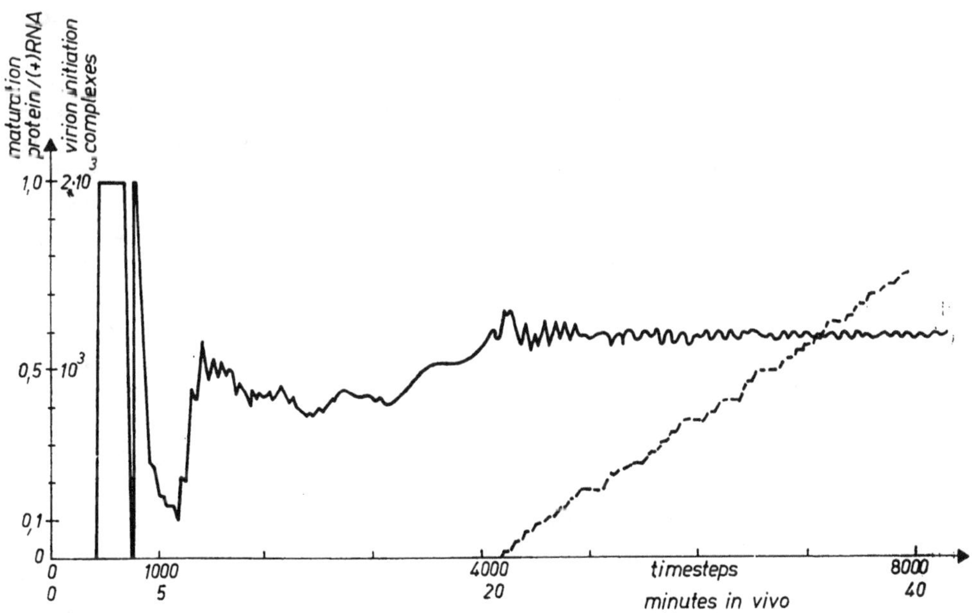

Fig. 8. Initiation of virion production. Linear increase of concentration of virion initiation complexes (dashed line) and maturation protein in relation to (+)RNA (full line).

6. Results and discussion

In this section we have to reply to the questions established in the introduction, how is the phage system growth controlled by the coupling of different single mechanisms, and how can the entity of this process be attributed to "properties" of the parental (+)RNA. From a model one can never give an absolute answer to questions regarding the **in vivo**-reality but one can reduce the set of possible interpretations of one phenomenon. Moreover, the discrepancies between some qualitative features of the modelled phage system dynamic and the corresponding in vivo-results are to be discussed. Such quantitative descrepancies must be expected because of the restrictions of the phage prototype definition (compare with section 4.).

6.1. The integrated phage replication control system

As mentioned above the **simulated** phage prototype replication process is not terminated by an internal abbreviation condition. Since the time where coat protein is going to suppress the polymerase production by binding to the ribosome binding sites of the polymerase cistron of the (+)RNAs the phage system grows linearly. Thus we can handle the topic of discussion in a two-step sequence: first, the exponential phase of the phage system growth and second the linear phase. There remains to study how the three elements (+)RNA, polymerase and coat protein interact to switch over the growth phases because no virion production occurs within the exponential phase of growth. The role of maturation protein must be elucidated within the second phase of growth. What properties should curves have that reflect a signal for the intiation of a new process? The first process with respect to this question is the repression of the polymerase production the meaning of which for the total phage system development has been explained above, and the second process of this type is the initiation of the virion meturation by binding maturation porteins to nascent (+)RNA-strands. The curves to be searched for must show an impulse and after the controlled process has been initiated the magnitudes have to remain above a characteristic threshold, if the considered process is not stabilised by another mechanism. No curve figuring the growth of the molecule number of one of the system elements fullfills these requirements. During the exponential system growth all molecule species concentrations grow exponentially and during the linear phase all concentrations grow linearly that represent the strong system's growth dependence on the polymerase concentration rise. But, the calculated curves of relations of molecule species concentrations show the required properties: the relation of coat protein to (+)RNA in the exponential growth phase (fig. 6 and fig. 8). The simulated phage prototype development starts with a single strand of parental (+)RNA (fig. 9a).

First ribosomal subunits of the host cell bind to the binding site of the coat cistron with the probability Q. If the aminoterminal region S of the coat cistron is translated, the ribosome binding site of the polymerase cistron becomes available which had been bound by partial hybridisation with the region S. Now ribosomal subunits bind to the polymerase cistron binding site with the probability R, thus the ribosome frequency on the polymerase cistron depends on $(Q \cdot R)$ (fig. 9b). If the polymerase cistron translation is terminated, the nascent polymerase is just available at its special binding site that is formed by a secondary RNA-structure in which a region at the 3′-end of the (+)RNA and a region including the ribosome binding site of the coat cistron by partial hybridisation are participating. This mechanism ensures a relative preference of the

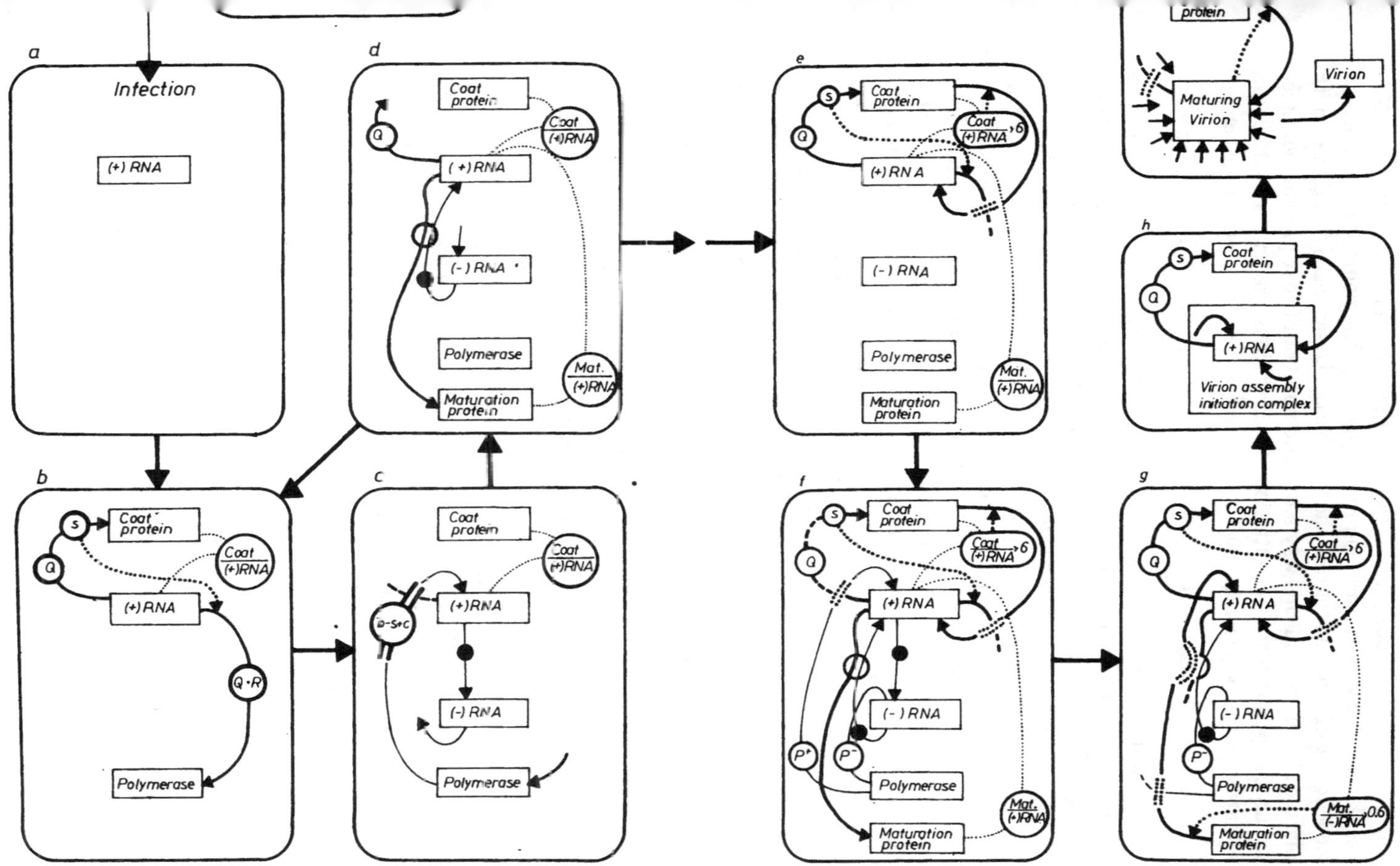

Fig. 9. The integrated phage replication control system.

▭ phage system elements, ○ control variables, —●— polymerase bound to RNA,

—¦¦—, —||— repression of a process, coupled processes.

(+)RNA against (−)RNA-strands with respect to the polymerase affinity to RNA, and it represses the coat cistron translation process (fig. 9c). The point of this repression depends on the length b of the coat cistron, the length s of the sequence S and the length c of the polymerase cistron, and in this way a definite number of coat protein molecules is produced within the period from the initiation of a first ribosome on the coat cistron up to the dissociation of the polymerase cistron coded ribosome. At this time the control term $\frac{\text{coat protein}}{(+)\text{RNA}}$ achieves a magnitude of 10 or 11 but, the (+)RNA-pool contains the parental (+)-strand only and no repression of the polymerase cistron translation occurs by binding one coat protein-hexamer to the ribosome binding site of the polymerase cistron. Progeny (+)RNA-strands seem to be required for this coat protein binding. After the first (−)RNA-strand is synthesized at the parental RNA-matrix this (−)-strand is accepted as matrix by the same polymerase forming an intermediate replication complex. The nascenting progeny (+)strand gets near its 5'-end the ribosome binding site of the maturation cistron where the translation can be initiated now by a ribosome. The speed of the polymerase elongation is much higher than that of ribosomes. The (+)strand synthetisation is finished while the maturation cistron translation is still going on (fig. 9d), the coat cistron translation can be initiated again and the further processes run down as it has been explained before. Back to the maturation cistron. After the initiated ribosome left the binding site this free sequence binds to a complementary sequence at the end of the maturation protein cistron now to suppress a further maturation cistron translation. Only one, or two in few cases, molecules maturiation protein are supplied per one progeny (+)RNA-strand by this mechanism, and, remember that one molecule maturation protein per one (+)RNA is needed for the virion production.

Because of the autocatalytic effect of the growth of the (+)-RNA-pool realised by the single mechanisms of the gene expression and RNA-replication the phage system grows exponentially — or even more rapidly but, this has not been analysed. If the characteristic threshold coat protein to (+)RNA greater than six is exceeded, the repression of the polymerase production begins by binding coat protein hexamers to the polymerase cistron specific ribosome binding site (fig. 9e). The key parameters for the activation of this control mechanism to the right time have been found to be the relative preference frequency of (+)RNA-strands by nascent polymerase against (−)RNA-strands and the relation of Q to R (Knijnenburg, Kreischer, 1975). These parameters have been adapted as about one third for the first one and 0.7 for the second one. Thus for the further system development in the model the foundation was laid for the two discrepancis between the model and the subject of the (+)RNA- to (−)-RNA-synthesis in the late phase and the inadequate supply of coat proteins for the virion assembly. The model contains the constraint that the control mechanism for the coat cistron translation by the polymerase was not captured by a detailed book-keeping procedure. The parameter found out for the relative preference frequency of (+)RNA by polymerase is correct but it could not be refered to this special mechanism in the model, so we didn't change this parameter with the overswitching of the system's growth phases that led to a to small (+)RNA-production during the late phase in the model. After the repression of the polymerase production is realized molecules of the free enzyme pool bind to (+)RNA-strands or (−)RNA-strands, respectively with the binding probabilities P^+ and P^-. Now the (−)strands are absolutely preferred to a much higher extent, $P^- \gg P^+$ (fig. 9f) and this leads to a relatively low repression of the coat protein production at the (+)-

RNA-strands because most polymerase work at (−)strands. The coat protein production at one (+)RNA is not interrupted by the binding of the nascenting polymerase thus supplying much more than 11 coat proteins between the production and the replication of one (+)RNA-strands. So our second problem has been solved too. This trend to rise coat protein supply is supported by another process, the binding of one maturation protein each to a (+)RNA-strand (fig. 9g). It depends on the control term $\frac{\text{maturation protein}}{(+)\text{RNA}}$. The threshold to be exceeded was adjusted to 0.6 in the model. Such (+)RNA-strands are introduced into the virion maturation process. Polymerase cannot bind to it, which automatically leads to a repression of the maturation protein production. This can be interpreted **in vivo** as an efficiency increase in using the host cell translation apparatus for the production of coat proteins because in this stage of infection the availability of host cell factors and materials gets already stressed more and more. At these complexes there is no more break of the coat protein production by the RNA-replication (compare fig. 9g with 9f). The bound maturation protein induces a conformation change of the RNA-secondary-structure and the virion formation begins by the self-assembly of coat proteins (fig. 9h). In an early phase of this process called "virion maturation" the coat protein production goes on but later on it will be stopped by the growing coat or by the structural change of the RNA in the progressing process of "enwrapping" the RNA into the core, respectively (fig. 9i). If the coat consists of 180 protein molecules, the virion, the progeny phage particle, gets ready forming an icosaeder [3]. The virions accumulate in the bacterial cell until the lysis. Now the life cycle of the RNA-bacteriophage-system is closed.

This system's analysis by means of a simulation model is a good illustration for the role of mathematical models as a tool in the process of cognition.

Much detailed knowledge about these single processes was put into the model but only little was known about the effects of single mechanisms within the integrated control system of the phage with respect to the quantitative process behaviour. Especially the impossibility to adapt the model quantitatively with fixed parameters gave us an insight into the central role of the coupling of the coat protein production and the (+)RNA-replication by the discussed control mechanism which has not been mapped in the model. This mechanism has the function within the whole control system to switch the key process parameter of the relation of the (+)RNA- to (−)RNA-synthesis and the coat protein- to (+)RNA-synthesis. In the early phase, during the exponential growth of the phage system, the (+)RNAs are prefered by the nascenting polymerases and the coat protein production rate per one (+)RNA is small. In the late phase, when the polymerase production is repressed the (−)RNAs are preferred by the free polymerases and the coat protein production rate per one (+)RNA rise immediately to supply enough coat material.

Moreover, we can make a general remark. This phage system shows us where the information for the self-organisation process of biological systems is stored in the genetic material and how this information is expressed during the self-organisation of the system. All control mechanisms concerning the genetic storage itself on the molecular level are bond to specific nucleic-acid secondary-structures, and this seems to be a direct parallel as to how proteins function. This is not a new idea but it makes possible a better understanding, especially by a phage system, by which fundamental principle life is organised.

6.2. The role of gene expression in relation to the replication process dynamics

In this section the development of the phage system shall be compared with the ontogenesis of higher organised biological organisms from a more abstract viewpoint. What features could serve as a basis for a comparison of the ontogenesis of non-cellular with cellular systems? There are two obvious common features concerning the process of self-organisation of biological system: 1. They pass through a well-defined species-specific sequence of development stages during their ontogenesis starting with a determined initial stage. 2. The ontogenetic self-organisation process achieves a natural end that means the system remains in an end stage. First, some terms have to be clarified regarding the phage ontogenesis, and then we are going to discuss the integration of the control mechanisms depicted in figure 9 with respect to the resulting differential gene expression. The modelled phage system supplies an opportunity to study whether differential gene expression principally causes something like a "division of labour" between different "parts" of the developing phage system which would have an equivalent meaning like the cells of a multicellular system during the cellular differentiation process.

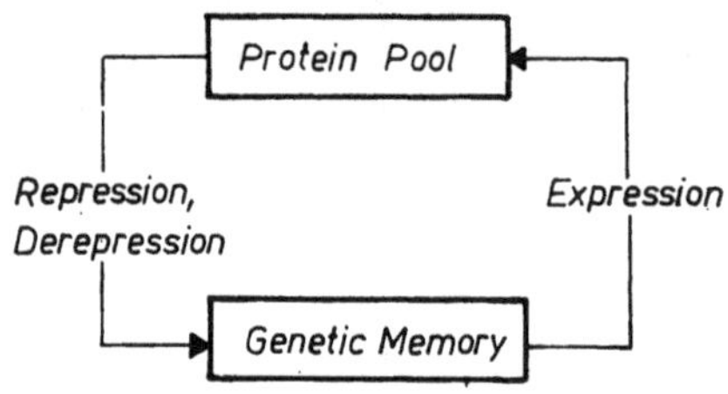

Fig. 10. Macroscopic consideration of interaction of genetic memory and protein pool.

The phage-RNA replication in the host cell is multiplication of the genetic memory of the phage system. The equivalent process to that is multiplication of the DNA during the ontogenesis of a multicellular system by mitotic cell division. In both system types specific sequences of patterns of the expression of the genetic memories can be observed, and in multicellular systems the DNAs of different cells show in parallel different patterns of gene expression at one convenient point of consideration. At the end of this process there is a set of patterns of gene expression, each performed by a group of equal cells, where a further development does not seem to be possible under natural conditions.

The above mentioned abstract viewpoint we intend to apply to the discussed topic is that we emphasize two subsystems interacting together within a cell, the "genetic memory" and the "protein pool" (fig. 10). The protein pool of our modelled phage prototype system contains the three gene products maturation protein, coat protein and polymerase. Dependent on the concentration regime, each protein can exercise several functions which have been explained in the section before and in fig. 9. A definite configuration of functions of the single proteins leads to a specific state if this partial system.

The genetic memory of the phage system contains three functional units each of which consists of one cistron species with the attached translation control sequences summarized about all (+)RNA-strands. One functional unit of the genetic memory is denoted as a "cistron compartment". At different RNAs a different number of phage proteins which act as translation repressors or ribosomes can be bound within the considered section at different times. The **state of one cistron compartment** has to be defined as the relative amount of equal RNA-sections which are just being expressed by the translation process. This value is called the "translatability" of the cistron compartment. **The state of the total genetic memory**, therefore, is given as the translatability pattern of the three cistron compartments. The translatability of one cistron compartment is defined by:

$$\xi_i = \frac{C_{\mathrm{Rib}_i}}{C_{\mathrm{RNA}} \cdot L_i \cdot p_i} \tag{6}$$

where ξ_i is the translatability of the cistron compartment i, C_{RNA} is the total number of (+)RNAs, C_{Rib_i} is the number of phage coded ribosomes on cistron i, L_i is the length of cistron i, scaled in units of the mean ribosome distance (see section 4.2.1.), and p_i

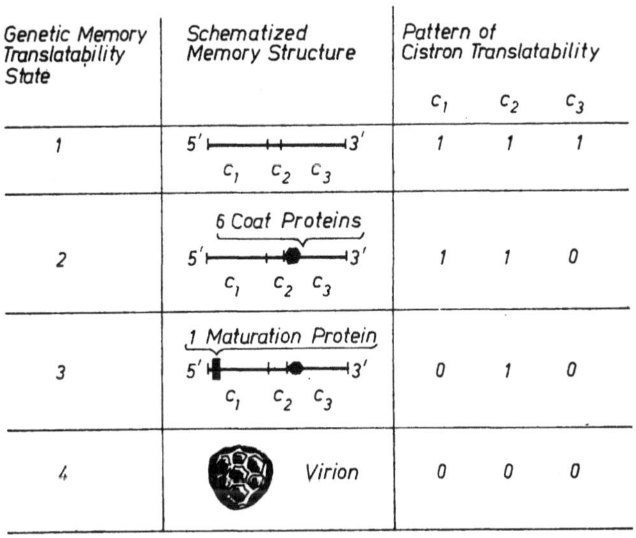

fig. 11. Translatability states of a single (+)RNA-strand.

is the ribosome binding probability to cistron i. The cistron indices i are 1 for the maturation cistron, 2 for the coat cistron, and 3 for the polymerase cistron. The adjusted ribosome binding probabilities are $p_1 = 1$, $p_2 = 0.7$, $p_3 = 0.1$.

In just one state, that of the RNA-bacteriophage at the start of the infection process, the genetic memory is composed solely of one nucleic acid (fig. 9a). If the genetic expression of this strand wouldn't be coupled with the RNA-replication the states of the genetic memory taking into account this single RNA only could be as represented in figure 11. In this case where $C_{\mathrm{RNA}} = 1$ the ξ_i can be only 1 or 0 because the number of ribosomes bound to a cistron results from a stochastic realisation of $L_i \cdot p_i$, that means $C_{\mathrm{Rib}_i} = L_i \cdot p_i$ if no repression by a protein is exercised, and $C_{\mathrm{Rib}_i} = 0$, if the translation is repressed.

The following questions are of interest for the growing genetic memory of the phage system: Do the translatability patterns change the defined memory states in the course of the phage replication process continuously for quasi-discretely? What sequence of the memory states occurs and or which periods do states exist? Is the state sequence of the total memory comparable with the state sequence that the single RNAs run through?

Figure 12 shows the time-dependent translatability for the three cistron compartments. The curve depicting the growing number of (+)RNA-strand in fig. 12 indicates that the translatability of the cistron compartments is independent of the growth of the genetic memory of the phage system. All translatabilities distinguish themselves by characteristic qualitative levels whose terminal states remain stable ad infinitum in the model. Within a short time-interval the curves change their level erratically. A discre-

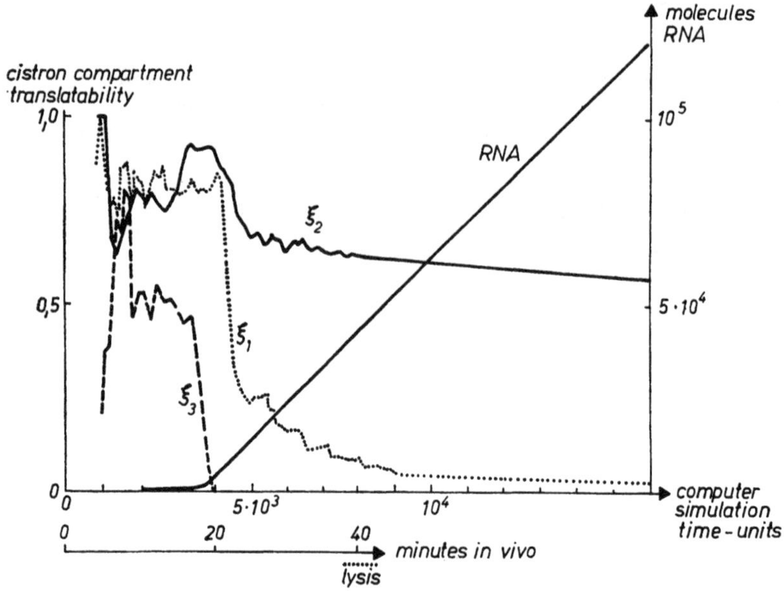

Fig. 12. Time-dependent development of translatability of the three phage specific cistron compartments.

ξ_1 — maturation cistron compartment,
ξ_2 — coat cistron compartment,
ξ_3 — polymeres cistron compartment.

tisation of the single translatability curves seems possible using the following notation: If a cistron compartment is mainly translatable in relation to the whole process, we write $\xi_i \to 1$. If not, we write $\xi_i \to 0$. When we apply this view to the curves in Fig. 12 the obtained results are:

1. The cistron compartment of the coat protein is mainly translatable during the whole process, $\xi_2 \to 1$.
2. a) The cistron compartment of polymerase is in relation to the whole process mainly translatable up to about 3350[1] time-steps, $\xi_3 \to 1$.
   b) Between 3350 and 3900 time-steps this compartment becomes untranslatable $\xi_3 \to 1 \cdots \xi_3 \to 0$.
   c) Beyond 3900 time-steps this compartment is untranslatable, $\xi_3 = 0$.
3. a) The cistron compartment of the maturation protein is mainly translatable up to about 4000 time-steps, $\xi_1 \to 1$.

[1] 1000 time-steps in the simulation model correspond to 5 minutes in vivo.

b) Between 4000 and 4400 time-steps this compartment mainly becomes untranslatable, $\xi_1 \to 1 \cdots \xi_1 \to 0$.

c) Beyond 4400 time-steps the same compartment is mainly untranslatable, $\xi_1 \to 0$.

Using the introduced notation for the genetic memory of the total system, as defined above, patterns of translatability are discernible. The sequence of these patterns is given in Fig. 13 including the periods in which the single patterns are expressed. A comparison of fig. 13 with fig. 11 shows that the state sequence of the total genetic memory is **qualitatively** almost identical with the possible state sequence for a single RNA. The dynamic state changes of the memory are reflected in states 2 and 4. The states 1, 3, and 5 are

Genetic Memory Translatability State	Pattern of Cistron Compartment Translatability			Timing	
	$\mathcal{L}_1$	$\mathcal{L}_2$	$\mathcal{L}_3$	Timesteps	Minutes in vivo
1	$\to 1$	$\to 1$	$\to 1$	0 – 3350	0.00 – 16.75
2	$\to 1$	$\to 1$	$\to 1 \cdots\cdots \to 0$	3350 – 3900	16.75 – 19.50
3	$\to 1$	$\to 1$	$\to 0$	3900 – 4000	19.50 – 20.00
4	$\to 1 \cdots\cdots \to 0$	$\to 1$	$\to 0$	4000 – 4400	20.00 – 22.00
5	$\to 0$	$\to 1$	$\to 0$	4400 – ∞	22.00 – ∞

Fig. 13. Genetic memory translatability states of the total phage system and the periods are explained in text.

equivalent to the possible translatability states of one (+)RNA strand 1, 2, and 3. The translatability state of one virion, state 4 in fig. 11, cannot be taken into this comparison because the functional subsystem "genetic memory" of the phage system during the infection process comes to an end by the lysis of the host cell **in vivo**. In the model where no lysis is mapped the genetic memory of the phage system comes to a stable end state, where the translatability of the maturation cistron compartment is low, and the translatability of the polymerase cistron compartment is zero. The action of the total memory is based on the relative synchrony of the processes in all single (+)RNA-strands. There is insufficient time between the formation and the replication of one (+)RNA-strand to realize consequently resulting asynchronies between different RNA-strands. These asynchronies evidently have no influence on the dynamics of the total system. The divergence of the function of each cistron compartment appearing in the temporal sequence of the process of phage development does not lead to a simultaneous appearence of different sub-states of any single cistron compartment and consequently not to "differentiated parts" of the phage system. A "division of labour" of single RNA-strands by the summed partial gene expressions of RNA-submemories remains impossible, since all RNA-strands are submitted to the influence of the same protein pattern. Hence, this phage-specific partial realization of information by the RNA-memory is not equivalent to the partial realization of information of the genetic memory of a multicellular system.

In multicellular systems a partial reading of the DNA-information in the course of one cell cycle is the timed splitting of existing states within the epigenetic network of a number of relatively disjuncted sub-states, each realized in cells separated by space. Each cell is labelled by a specific protein pattern and in its turn able to activate specific DNA-sections.

Thus, the causes of a differential gene expression during phage ontogenesis become transparent; they are exclusively of regulative nature and serve for tuning the temporal appearence and concentration of proteins in the process of the system's self-organization.

In the phage development the process of self-organization ends if the state 5 of the genetic memory has been reached. Then the system-internal control mechanisms don't change any more and allow a continuous production of progeny phage particles. The only limitation is the damaging of the host cell. The translatability of the polymerase cistron compartment is zero, a constant number of polymerases synthesize progeny (+)RNAs with a constant rate at the (−)RNAs. The maturation cistron compartment of these (+)RNAs has a low translatability to allow the production of maturation proteins, which by binding to the (+)RNAs introduce the (+)strands into the virion maturation process. The high translatability of the coat cistron compartment ensures a constant supply of proteins for the self-assembling virion coats. Whether this supply is sufficient depends on the relation of the process speed of the coat cistron translation to that of the virion coat self-assembly.

References

Garwes, D., A. Sillero and S. Ochoa (1969): Virus specific proteins in Echerichia coli infected with phage Q_β. Biochim. Biophys. Acta **186**, 166–172.

Knijnenburg, A., and U. Kreischer (1975): A new aspect of the RNA-bacteriophages translation control mechanism. BioSystems **7**, 245–249.

Nathans, D., P. M. Oeschger, S. K. Polmar and K. Eggen (1969): Regulation of protein synthesis directed by coli phage MS 2 RNA. I. Phage protein and RNA synthesis in cell infected with suppressible mutants. Lourn. Mol. Biol. **39**, 279–292.

Stavis, R. L., and I. T. August (1970): The biochemistry of RNA bacteriophage replication. Ann. Rev. Biochem. **39**, 527–560.

Weissmann et al., 1972: FEBS Proceed. 8-th Meeting Amsterdam, **27**, 1–18.

9. Numerical adaptation of parameters in simulation models using evolution strategies

J. Born, K. Bellmann

Contents

1. Introduction

Biological systems with their typical features of complexity and non-linearity are described very often on a phenomenological level by relative simple models. If insights into biochemical or biophysical mechanisms underlying the phenomenon considered are envisaged more details should be included in the model in order to map the structures necessary for causal understanding. This leads to an exponential increase in system elements and interactions being considered and consequently to algorithmic simulation models. Such simulation models are used to investigate the dynamical behavior of the system in dependence on its basic structure, i.e. depending on the basic structural assumptions (hypotheses) introduced in the model and on its parameters as well. In general, these parameters may be grouped in two categories:

- parameters relatively easily determinable by means of experiments (fixed parameters)
- parameters which are unknown or nearly unknown (soft parameters, variable parameters).

The first one are used in the model by means of good estimates; about the other ones hypotheses are to be made. Thus the problem arises of testing the suitability of such hypotheses by means of an appropriate procedure.

Based on the structural assumptions introduced into the model the task is to find parameter values that, when used in the model, lead to a good correspondence between simulation system and real system dynamical behaviour, i.e. the variable parameters are to be chosen in such a way that the model shows a special behaviour. More exactly, the following adaptation problem is to be solved:

$$\inf \{f(\boldsymbol{y}(\boldsymbol{x}), \boldsymbol{y}^0) \mid \boldsymbol{x} \in X\} \tag{A}$$

with
a set of variable parameters $X \subseteqq R^n$, $X \neq \emptyset$;
a vector function $\boldsymbol{y}$, $\boldsymbol{y}: R^n \to R^q$, depending on $\boldsymbol{x}$ and describing the dynamical behaviour, i.e. the system output, determining for each parameter vector $\boldsymbol{x} \in X$ an output vector $\boldsymbol{y}(\boldsymbol{x})$;
a special output vector $\boldsymbol{y}^0$, $\boldsymbol{y}^0 \in R^q$, describing the output behaviour of the real system to which $\boldsymbol{y}$ is to be adapted;
a weighting functional $f: R^{2q} \to R^1$, which evaluates a parameter vector $\boldsymbol{x}$ on the basis of a distance between $\boldsymbol{y}(\boldsymbol{x})$ and $\boldsymbol{y}^0$, e.g.

$$f(\boldsymbol{y})(\boldsymbol{x}), \boldsymbol{y}^0) = \sum_{i=1}^{q} c_i(y_i(\boldsymbol{x}) - y_i{}^0)^2, \quad c_i \geqq 0, \quad i = 1, \ldots, q$$

In general, two controversial aspects are to be considered:

- complex simulation models necessarily include many details and consequently a vast number of parameters may be "soft parameters",
- there are technical and economic restrictions (mainly storage location and computing time resp.) in computer simulation, especially in computing of the output vector $\boldsymbol{y}(\boldsymbol{x})$.

A compromise must be found. In very many cases for extended computer simulations and in adaptation of variable parameters a reduced (aggregated) simple model is used, capturing the essential behavioural features of the system considered. To get an appropriate reduced model the key processes and parameters must be detected by means of various simulation runs and sensitivity analysis. An example is the reduction of the complex detailed simulation model EUTRANSCRIPT to its hard core, i.e. to its reduced form, called reservoir floodgate system (see Bellmann, Born, Lindigkeit, Schulz — in this volume; Hollatz, Born, Bellmann, 1977). By means of this model compression the computing time of $f(\boldsymbol{y}(\boldsymbol{x}), \boldsymbol{y}^0)$ was reduced from 1.2 min to 0.15 sec., an economically suitable computing time for solving the adaptation problem.

Returning to the basic idea, the problem (A) must be considered as a high dimensional extremum problem

$$\inf \{g(x) \mid x \in X\} \ (= g^*).^1 \tag{E}$$

Here the quality functional $g: R^n \to R^1$ is defined by $g(x) = f(y(x), y^0)$ and in the following we will assume that always $g^* > -\infty$.

The problem (E) is a general extremum problem in the sense that in practical cases no restrictive assumptions are to be made for both the parameter set X and the quality functional g, such as convexity, differentiability, continuity, and unimodality.

[1] In the following we will omit the characterization of a vector by underlining.

2. Numerical methods for general extremum problems

As mentioned above, only approximative solutions of general extremum problems are to be envisaged, i.e. one has to expect a solution obtained by an iterative process of the type

$$x^{k+1} = x^k + d^k, \tag{1}$$

with $x^{k+1} \in X$ and $g(x^{k+1}) \leqq g(x^k)$.
Quite a large number of such iterative solution methods have been developed. They can be distinguished by the different calculation of the descent d^k, which is the main problem in elaborating such methods. In tuning the techniques for calculation of d^k assumptions about the local shape of the function g are used. A special group of methods uses the explicit knowledge of the gradient of g in x^k:

$$g'_{x^k} = \left(\frac{\partial g}{\partial x_1{}^k}, \ldots, \frac{\partial g}{\partial x_n{}^k}\right)$$

for the calculation of d^k

$$d^k = d^k(g'_{x^k}). \tag{2}$$

Under restricting assumptions (e.g. assuming that g is repeatedly differentiable), one can show the convergence to a (local) optimum and it is possible to determine the rates of convergence of the iteration method.

Solving a general extremum problem such gradient methods are useful in a qualified sense only. It is possible to replace the gradient g'_{x^k} by an approximation, for instance by the difference quotient:

$$g'_{x^k} \cong \frac{g(x^k + h) - g(x^k)}{h}, \quad h > 0.$$

The explicit use of the gradient is evaded by certain methods calculating d^k directly by means of a construction using only the values of the function g.

Such direct search methods often are heuristic ones; generally, convergence can not be proved mathematically. These methods were shown to be successful in application to practical problems. A comparative evaluation of direct methods by means of solving different test problems was illustrated by SCHWEFEL (1977).

In both groups, the descent d^k is calculated deterministically on the basis of a fixed rule. d^k may also be calculated stochastically using the random process:

$$d^k = d^k(\Theta) \qquad \Theta - \text{a random variable.} \tag{3}$$

The main advantage of iterative, stochastic search methods is their high chance of finding the global solution. With only some weak restrictions the convergence with probability one of the process to a global extremum can be shown, as demonstrated in the following.

3. Structure of evolution strategies

3.1. On the elaboration of evolution strategies

Evolution strategies are iterative, stochastic methods for numerical solution of extremum problems, using the basic principles of biological evolution by algorithmic simulation. In elaborating of adaptation methods based on evolution strategies, i.e. in simulating evolution of natural populations in order to improve adaptation methods, experience shows that the mechanism of biological evolution has proved to be an excellent strategy of adapting a given genetic population to a given environment as well as possible, by "evaluating" each of its elements (fitness). On the other hand, the characteristic features of an extremum problem (A), as described above, may be considered as abstract concepts corresponding to the main characterics of the natural evolution process.

We imagine that a population of individuals is given with the following properties:

— the genotype of each individual is characterized by a parameter vector $x \in X$, i.e. each of the components of a parameter vector $x \in X$ can be interpreted as a single gene and

— the parameter set X corresponds to the set of all genotypes possible in a given genetic population.

— The phenotype of an individual corresponds to the output vector $y(x)$, $x \in X$.

— The selection value of a genotype (fitness) coincides with the qualities $g(x) = f(y(x), y^0)$, where by the special output vector y^0 can be interpreted as the phenotype best adapted to a given environment, i.e. as the fittest phenotype.

As in the natural selection process only the fittest are allowed to reproduce, discarding the remainder of the population (truncation of a distribution). In an evolution strategy all elements x with insufficient quality $g(x)$ are not taken into the next iteration step population. Consequently, a successive improvement of the mean quality of the population is to be expected. In elaborating a numerically efficient evolution strategy only the essential processes of evolution are to be included:

(i) variation by mutation, inversion and recombination,
(ii) reproduction and
(iii) selection.

How important in addition to the main principles (i)—(iii) a simulation of the phenomenon of the

(iv) genetic load or gene bank may be, will be illustrated in the following.

As one of the earliest workers in this field, BREMERMANN (1968, 1970) developed evolution strategies. The evolution strategies stated by RECHENBERG (1973) for the optimization of technical systems can also be utilized in the solution of extremum problems. Recent papers on evolution strategies have been published by HOLLAND (1976) and SCHWEFEL (1977).

The most evident characteristic of an evolution strategy is that it operates on a sequence of point sets — a sequence of populations — in contrast to the formation of a sequence of iteration points in the case of "usual" methods. The evolution strategies of the above mentioned authors are variants (partially with modifications) of an iterative pattern including:

A. initial step ($k = 0$).

Formation of a starting population P_1 of m ($m \geqq 1$, arbitrary, but fixed) parameter vectors in X:

$$P_1 = (x_1^1, \dots, x_1^m); \quad x_1^i \in X, \quad i = 1, \dots, m.$$

B. k-th iteration step ($k \geqq 1$).

1. Variation

Starting from the k-th population P_k, l ($l \geqq 1$, arbitrary, but fixed) new variants $x^{v_1}, \dots, x^{v_l}$ are generated.

Each of the new variants is determined by:

1.1. A random choice of r ($r \geqq 1$, arbitrary, but fixed) parents $x^{l_1}, \dots, x^{l_r}$ from the parameter vectors $x_k^i \in P_k$, $i = 1, \dots, m$, where every vector of parameters is selected with a certain probability (which may be determined by its quality).

1.2. Realization of mutation (i.e. stochastic modification) in each component of the parents x^{l_i}, $i = 1, \dots, r$ with a certain probability and a certain spectrum of variability.

1.3. Realization of inversion (i.e. inversion of the order of parameters within a stochastically determined component section) for the mutated parents.

1.4. Realization of crossing-over (i.e. an interchange of parameters within stochastically determined, equal component section) for the mutated and inverted parents, with a certain probability.

1.5. Realization of interchromosomal recombination (i.e. stochastic choice of one of the thus modified parents as a new variant).

2. Selection

Formation of the population P_{k+1} by selecting the m best (by their qualities) new variants $x^{v_1}, \dots, x^{v_l}$, $x^{v_i} \in X$, $i = 1, \dots, l$, and (or else without considering) the parameter vectors $x_k^i \in P_k$, $i = 1, \dots, m$.

3.2. On the efficiency of evolution strategies

In solving practical adaptation problems, the computing time required — being rather long in most cases — for calculating the quality $y(x)$ of a parameter vector x determines an economic limit on the number of possible quality computations. For such a limited number of iterations, the object is to achieve as "close" an approximation as possible to the solution set $X^* \left(X^* = \{x \in X \mid g(x) = g^*\}, X^* \neq \emptyset\right)$, or at least to improve the quality as much as possible.

An algorithm devised to fit this concept will always be a compromise between the object of achieving a high efficiency and the requirement to maintain security of global convergence and a satisfactory global efficiency, respectively. For an evolution strategy, the term "efficiency" shall be understood to be a measure of the variations (improvement) of the quality of the best parameter vector existing in the population after a fixed number of iterations, whereas the term "global efficiency" means a measure of the variations (improvement) of the distance of the best parameter vector from the solution set X^*.

In particular the efficiency and the global efficiency, respectively, are influenced by:

i. the formation of the starting population;
ii. the potency of the population and
iii. the determination of the strategy parameters.

As regards iii, the essential point is a suitable determination of the variability spectrum for mutations.

Among the evolution strategies of the authors mentioned in Section 2, SCHWEFEL's multi-component evolution strategy (abbreviated as MCE strategy) is suited best to the object mentioned above. In the MCE strategy, during the iteration the strategy parameters (only the variability spectra for mutations) are varied by applying the evolution strategy to the strategy parameters themselves in order to achieve a high efficiency.

SCHWEFEL's tests certify a relatively high security of global convergence for the MCE strategy. Our own numerical tests show that the global efficiency of the MCE strategy is unsatisfactory (cf. Section 3.3.4.). One cause of this deficiency is the neglect of the reproduction principle. In the MCE strategy, the elements of the population P_k are not taken into account in selection, which may result in a reduction of quality from one iteration to another, especially for large variability spectra and a small number l of descendants.

Although a population is involved, the resulting possibilities of determining new variants are utilized only inadequately and after a few iterations the variability in the population is determined by the realization of mutation alone. This phenomenon is observed in all evolution strategies of the authors mentioned in section 3.1. The conception of the following evolution strategy remedies this deficiency.

3.3. Developing an extented evolution strategy which realizes the principle of genetic load — the EGL method

3.3.1. Basic principles

The EGL method can be considered as a variant of the general scheme mentioned in Section 2.1., where $l = 1$, $r = 2$, the population P_k of the k-th iteration step being taken nto account in selection. The extensions consist in

- operating on strategy populations, and hence a variation of the strategy parameters during the iterations by applying the evolution strategy to the strategy parameters themselves;
- realizing the principle of genetic load. Every population contains parameter vectors which are introduced at the initial step and never subjected to selection, irrespective of their qualities.

Possible natural mechanisms which could cause the phenomenon of genetic load (diploid heredity with dominance relations and recessivity) will not be taken into account. Only the effect of these mechanisms (permanent retention of alleles within the gene pool independent of their haploid selection value) will be considered in order to avoid unnecessary complexity of the algorithm.

In the following the EGL method is described in detail.

3.3.2. Algorithmic structure of the EGL method

A. Initial step ($k = 0$)

1. $m_0 (m_0 \geqq 2)$ parameter vectors

$$(x_1^1, \ldots, x_1^{l_0+1}, \ldots, x_1^{l_0}, x_1^{m_0}) = 0_1^A \cup 0^L = 0_1;$$

with $x_1^l \in X, l = 1, \ldots, m_0$;
are composed to a starting object population O_1, whereby their qualities are calculated. The parameter vectors

$$(x_1^{l_0+1}, \ldots, x_1^{m_0}) = 0^L \quad 1 \leqq l_0 \leqq m_0$$

are the genetic load 0^L of the object population.

2. $m_s (m_s \geqq 2)$ strategy vectors

$$({}^Rs_1^1, \ldots, {}^Rs_1^{l_s}, {}^Rs_1^{l_s+1}, \ldots, {}^Rs_1^{m_s}) = {}^RS_1^A \cup {}^RS^L = {}^RS_1;$$

$$\text{with } {}^Rs_1^l \in R_{0.1}^{n+2}, \quad l = 1, \ldots, m_s{}^1;$$

are composed for a starting strategy population ${}^RS_1$ for probabilities of realization of inversion, crossing-over and mutations in the object population and m_s strategy vectors

$$({}^Vs_1^1, \ldots, {}^Vs_1^{l_s}, {}^Vs_1^{l_s+1}, \ldots, {}^Vs_1^{m_s}) = {}^VS_1^A \cup {}^VS^L = {}^VS_1;$$

$$\text{with } {}^Vs_1^l \in R_+{}^n, \quad l = 1, \ldots, m_s{}^2;$$

are composed to a starting strategy population ${}^VS_1$ for the variances in realization of mutations in the object population.

The qualities of the stategy vectors are set equal to zero. The strategy vectors

$$({}^Rs_1^{l_s+1}, \ldots, {}^Rs_1^{m_s}) = {}^RS^L$$

and

$$({}^Vs_1^{l_s+1}, \ldots, {}^Vs_1^{m_s}) = {}^VS^L, \quad 1 \leqq l_s \leqq m_s,$$

are the genetic loads ${}^RS^L$ and ${}^VS^L$ resp. of the corresponding strategy populations.

3. (Constant) strategy vectors are fixed

$${}^Rc \in R_{0,1}^{n+4}$$

and

$${}^Vc \in R_+^{n+2}$$

for the probabilities of the realization of inversion, crossing-over and mutations as the variances utilized in realizing mutations in the strategy populations.

4. A (constant) parameter $g_s^{\min}$ for a minimal strategy quality is fixed.

B. k-th iteration step ($k \geqq 1$)

1. Variation step

1.1. Stochastic determination of strategy variants ${}^Rs^A$ and ${}^Vs^A$ out of the strategy populations ${}^RS_k$ and ${}^VS_k$ utilizing the strategy vectors ${}^Rc$ and ${}^Vc$ resp.

---

[1] Let $R_{0,1}^n = \{x \in R^n \mid 0 \leqq x_i \leqq 1, \quad i = 1, \ldots, n\}$.
[2] Let $R_+{}^n = \{x \in R^n \mid x_i \geqq 0, \quad i = 1, \ldots, n\}$.

1.2. Stochastical determination of a variant x^A from the object population O_k utilizing the strategy vectors ${}^Rs^A$ and ${}^Vs^A$ resp.

2. Selection step

2.1. Investigation of the feasibility of x^A.
For $x^A \notin X$, we put

$$0_{k+1} = 0_k$$

and

$${}^RS_{k+1} = {}^RS_k$$

and

$${}^VS_{k+1} = {}^VS_k$$

and one has to go to *3.*, otherwise the quality of x^A and the quality of the strategy variants ${}^Rs^A$ and ${}^Vs^A$ are calculated.

2.2. Selection within the actual object population $0_k{}^A$ according to the quality of the variant x^A, with this generation of the next object population 0_{k+1}.

2.3. Selection within the actual strategy population ${}^RS_k{}^A$ and ${}^VS_k{}^A$ according to the quality of the strategy variants ${}^Rs^A$ and ${}^Vs^A$, resp., with this generation of the next strategy populations ${}^RS_{k+1}$ and ${}^VS_{k+1}$, resp.

3. Final step (without operation).

Stochastic determination of variants within a population

$$P = (p^1, \ldots, p^m); \quad p^i \in R^n, \quad i = 1, \ldots, m \cdot$$

utilizing strategy vectors ${}^Rs \in R_{0,1}^{n+2}$ and ${}^Vs \in R_+{}^n$ means:

i. Random choice of two parents p^{E_1} and p^{E_2} from the population P with equal probability $1/m$;
ii. Realization of mutation on each component of p^{E_1} and p^{E_2} with the probabilities ${}^Rs_{i+2}$ and the variances ${}^Vs_i$ for $p_i{}^{E_1}$ and $p_i{}^{E_2}$, $i = 1, \ldots, n$;
iii. Realization of inversion within the mutated parents p^{E_1} and p^{E_2} with the probability ${}^Rs_1$;
iv. Realization of crossing over between the mutated and inverted parents p^{E_1} and p^{E_2} with the probability ${}^Rs_2$;
v. Realization of interchromosomal recombination by means of random choice of one of parents modified in this way p^{E_1} and p^{E_2} with the same probability 1/2 as variant p^A.

More exactly: let p^E be a parent chosen for mutation in (ii) and let p^V be the resulting (mutated) element. In every case mutations are realized normally distributed and in such a way that holds

$$p_i{}^V = \delta({}^Rs_{i+2}) \cdot \min\{|p_i{}^E + p_i{}^E \cdot \vartheta({}^Vs_i)|, 1\},^1 \quad i = 1, \ldots, n,$$

if P is a strategy population for probabilities ${}^RS_k$;

$$p_i{}^V = \delta({}^Rs_{i+2}) \cdot |p_i{}^E + p_i{}^E \cdot \vartheta({}^Vs_i)|, \quad i = 1, \ldots, n,$$

---

[1] Let be δ: $\langle 0,1\rangle \to \{0,1\}$ with $\delta_\xi(s) = \begin{cases} 1 \text{ if } \xi \leq s \\ 0 \text{ else} \end{cases}$, where ξ is a random variable uniformly distributed on $\langle 0,1\rangle$. $\vartheta(s)$ characterizes a random variable $(0, s)$-normally distributed.

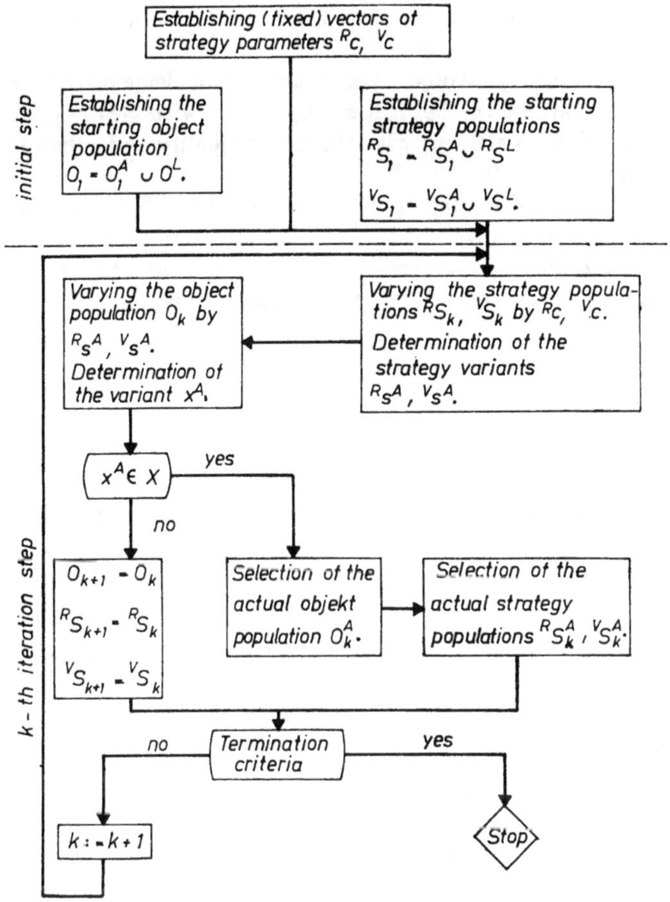

Fig. 1. Block diagram of the algorithmic realization of the EGL method.

if P is a strategy population for variances ${}^VS_k$ and

$$p_i^V = p_i^E + \delta({}^Rs_{i+2}) \cdot \vartheta({}^Vs_i), \quad i = 1, \ldots, n,$$

if P is an object population O_k.

In realizing inversion and crossing over the sites of the component sections (for crossing over always the same in p^{E_1} as in p^{E_2}) are fixed stochastically with equal probability $1/n$.

The quality of the strategy variants ${}^RS^A$ and ${}^VS^A$

$$g_s^A = g_s({}^Rs^A) = g_s({}^Vs^A)$$

is calculated at selection step 2.1. by the relative difference between the quality of the variant x^A and the quality of the parent x^E, where x^E is that parent x^{E_1} or x^{E_2} whose modification had been chosen by a variation in 1.2. as x^A:

$$g_s^A = \frac{g(x^E) - g(x^A)}{g(x^E)}\,^{1}$$

[1] Without loss of generality here let $g(x) > 0$ for all $x \in X$.

Selection within an actual object population $0_k{}^A$ according to the variant x^A (step 2.2.) means:

If there exists within the actual object population $0_k{}^A$ an element with a lower quality than the determined variant x^A, i.e. if a $\tilde{x} \in 0_k{}^A$ exists with $g(\tilde{x}) > g(x^A)$, then x^A is included in $0_k{}^A$ instead of an element with the lowest quality and consequently we get 0_{k+1} otherwise it holds $0_{k+1} = 0_k$.

Selection within the actual strategy populations ${}^RS_k{}^A$ and ${}^VS_k{}^A$ according to the quality $g_s{}^A$ of the strategy variants ${}^Rs^A$ and ${}^Vs^A$, resp., (see selection step 2.3.) means:

If there exists within the actual strategy populations ${}^RS_k{}^A$ and ${}^VS_k{}^A$ a pair of strategy vectors with a lower quality than $g_s{}^A$, i.e. if there exists a pair

$$ {}^R\tilde{s},\ {}^V\tilde{s}, \qquad {}^R\tilde{s} \in {}^RS_k{}^A, \qquad {}^V\tilde{s} \in {}^VS_k{}^A $$

with

$$ g_s({}^R\tilde{s}) = g_s({}^V\tilde{s}) < g_s{}^A, $$

${}^Rs^A$ and ${}^Vs^A$ are included in ${}^RS_k{}^A$ and ${}^VS_k{}^A$, resp., instead of an element with the lowest quality; consequently we get ${}^RS_{k+1}$ and ${}^VS_{k+1}$; otherwise a constrained selection is realized by putting the qualities of all elements within ${}^RS_k{}^A$ and ${}^VS_k{}^A$ equal to zero before the selection is made.

3.3.3. On the convergence with probability one of the EGL method

Firstly, some prerequisites for further investigations are introduced:

The general extremum problem

$$ \inf \{g(x) \mid x \in X\} \ (= g^*) \tag{E} $$

with $X \subseteqq R^n$, $X \neq \emptyset$ and $g: R^n \to R^1$, whereby $g > -\infty$ is considered.

$h(0)$ designates the quality of an object population 0 (here without considering the genetic load):

$$ h: R^n \times R^{m_0} \to R^1 \quad \text{with} $$

$$ h(0) = \frac{1}{l_0} \sum_{i=1}^{l_0} g(x^i), \quad \text{whereby} $$

$$ 0 = 0^A \cup 0^L \qquad \text{and} \qquad 0^A = (x^1, \ldots, x^{l_0}) $$

$$ 0^L = (x^{l_0+1}, \ldots, x^{m_0}); $$

with $x^i \in X$, $i = 1, \ldots, m_0$ and $1 \leqq l_0 \leqq m_0$.

Furthermore let

$$ X_\varepsilon{}^* = \{x \in X \mid g(x) - g^* \leqq \varepsilon\}, \quad \varepsilon > 0. $$

The problem (E) is called regular if the following conditions are satisfied:

(R_1) g is continuous.

(R_2) X is closed.

(R_3) For every $\bar{x} \in X$, $\bar{X} = \{x \in X \mid g(x) \leqq g(\bar{x})\}$ is bounded.

(R_4) For any $\varepsilon > 0$, int $X_\varepsilon{}^* \neq \emptyset$[1].

[1] For a set $M \in R^n$, let int M denote the set of all interior points.

The EGL method is called regular if the following conditions are satisfied:

(R_5) For the vector of strategy parameters ${}^Rc$: $\max_{i=1,\dots,n+4} {}^Rc_i < 1$.

(R_6) In the genetic load ${}^RS^L$, there exists an element ${}^Rs$ such that $\min_{i=1,\dots,n} {}^Rs_i > 0$[1] and ${}^Rs_{n+1} < 1$[2], ${}^Rs_{n+2} < 1$[3].

(R_7) In the genetic load ${}^VS^L$, there exists an element ${}^Vs$ such that ${}^Vs_i = s > 0$, $i = 1, \dots, n$.

In proving the convergence with probability one of the EGL method the following lemma is used.

Let $f_{z,s}: R^n \to R^1$ be the density function of an n-dimensional random vector x with (z_i, s)-normally distributed components x_i:

$$f_{z,s}(x) = \left(\frac{1}{s\sqrt{2\pi}}\right)^n \exp\left(-\frac{1}{2\cdot s^2}\sum_{i=1}^{n}(x_i - z_i)^2\right).$$

and $K_{y,r}$ an n-dimensional sphere with centre y and radius $r > 0$:

$$K_{y,r} = \left\{x \in R^n \,\middle|\, \sum_{i=1}^{n}(x_i - z_i)^2 \leqq r^2\right\}.$$

Furthermore let

$$v(z, y) = \overbrace{\int \cdots \int_{K_{y,r}}}^{n} f_{z,s}(x)\, dx$$

(i.e.: $v(z, y)$ is the probability that $x \in K_{y,r}$), than the following lemma holds:

LEMMA:

If $\|z_1 - y\| \leqq \|z_2 - y\|$, then $v(z_1, y) \geqq v(z_2, y)$.[4]

With the regularity conditions the following theorem holds:

THEOREM:

Let both the problem (E) and the EGL method be regular. Then, for a sequence of object populations $\{0_k\}$ generated by the EGL method,

$$P\{h(0_k) \xrightarrow[k\to\infty]{} y^*\} = 1.$$[5]

Proof:

According to the construction of the EGL method for the sequence of the qualities the following holds

$$h(0_{k+1}) \leqq h(0_k)$$

and

$$h(0_k) \geqq g^* \quad \text{for all} \quad k \geqq 1.$$

---

1 Probabilities of the mutations on the individual components.

2 Probability of inversion.

3 Probability of crossing over.

4 For $x \in R^n$ let $\|x\|$ be the Euclidian norm of x:

$$\|x\| = \sqrt{\sum_{i=1}^{n} x_i^2}.$$

5 In this paragraph the probability of an event A is denoted by $P\{A\}$.

Consequently a limit $\bar{g}$ exists with $\bar{g} \geqq g^*$ and

$$h(0_k) \xrightarrow[k\to\infty]{} \bar{g}.$$

We assume $\bar{g} > g^*$.

By the regularity conditions (R_2) and (R_3) an accumulation population of $\{0_k\}$ exists. The accumulation population of $\{0_k\}$ let 0^*. And there is a subsequence $\{0_{k_j}\} \subseteqq \{0_k\}$ for which

$$0_{k_j} \xrightarrow[j\to\infty]{} 0^*$$

holds.

By the condition (R_1) it holds in particular that:

$$h(0_{k_j}) \xrightarrow[i\to\infty]{} h(0^*) = \bar{g}.$$

Assuming $\varepsilon = \frac{1}{2} (\bar{g} - g^*)$.

A $\qquad \bar{j} \geqq 1$

exists such that

$$0_{k_j}^A \nsubseteqq X_\varepsilon^* \quad \text{for all} \quad j \geqq \bar{j}. \tag{1}$$

Without loss of generality let $\bar{j} = 1$.

By condition (R_4) it is int $X_\varepsilon^* \neq \emptyset$; moreover a n-dimensional sphere K exists, e.g.

$$K_{y,r} = \{x \in R^n \mid \|x - y\|^2 \leqq r^2\}, \quad y \in \text{int } X_\varepsilon^*, \quad r > 0 \quad \text{with}$$

$$K_{y,r} \subseteqq X_\varepsilon^*. \tag{2}$$

Furthermore by (R_2) and (R_3) a compact set U exists with:

$$x_k^i \in U \quad \text{for all} \quad x_k^i \in 0_k^A, \quad i = 1, \ldots, l_0, \quad k \geqq 1.$$

Based on the construction of the EGL method and (R_5), (R_6), and (R_7) the following assertions hold:

i. The probability p_1 that in generating a variant x^A for each component of a parent only mutations are realized is at least $\overline{p_1}$, whereby

$$\overline{p_1} = \frac{1}{2} \cdot \frac{1}{m_s} \cdot (1 - {}^Rc_1) \cdot \ldots \cdot (1 - {}^Rc_{n+4}) \cdot (1 - {}^Rs_1) \cdot$$
$$\cdot (1 - {}^Rs_2) \cdot {}^Rs_3 \cdot \ldots \cdot {}^Rs_{n+2} > 0. \tag{3}$$

ii. The probability p_2 that in generating the mutations a minimal variance s is realized is at least $\overline{p_2}$, whereby

$$\overline{p_2} = \frac{1}{2} \cdot \frac{1}{m_s} \cdot (1 - {}^Rc_1) \cdot \ldots \cdot (1 - {}^Rc_{n+4}) > 0. \tag{4}$$

Obviously, by (2) and the Lemma mentioned above the following holds:

For all $x \in U$ the probability p_3 that in realizing mutation at each component of vector x with the minimal variance s, the element, modified in this way can be found in

$X_\varepsilon{}^*$ is at least $\overline{p_3}$, whereby

$$\overline{p_3} = \left(\frac{1}{s \cdot \sqrt{2\pi}}\right)^n \overbrace{\int \cdots \int}^{n}_{K_{y,r}} \exp\left(-\frac{1}{2s^2} \sum_{i=1}^{n} (x_i - x_i{}^M)^2\right) dx > 0, \tag{5}$$

with x^M such that $\|x^M - y\| = \max_{x \in U} \|x - y\|$.
(3), (4), and (5) imply

$$\begin{aligned} P\{x^i_{k_j} \in X_\varepsilon{}^*\} &\geqq \frac{1}{2} \cdot \overline{p_1} \cdot \overline{p_2} \cdot \overline{p_3} \\ &= p_4 > 0, \quad i = 1, \ldots, l_0, \quad j \geqq 2 \end{aligned} \tag{6}$$

Now, without loss of generality assume

$$k_j \geqq l_0 \cdot j, \quad j \geqq 2.$$

Corresponding to (6) it holds that

$$P\left\{0^A_{k_j} \subseteqq X_\varepsilon{}^* \,\middle|\, \bigcap_{l=1}^{j-1} (0^A_{k_l} \nsubseteq X_\varepsilon{}^*)\right\} \geqq (p_4)^{l_0} = p_5 > 0, \quad j \geqq 2 \tag{7}$$

consequently

$$\begin{aligned} P\left\{\bigcap_{j=1}^{l} 0^A_{k_j} \nsubseteq X_\varepsilon{}^*)\right\} &= P\left\{0^A_{k_1} \nsubseteq X_\varepsilon{}^*\right\} \cdot \prod_{j=2}^{l} P\left\{0^A_{k_j} \nsubseteq X_\varepsilon{}^* \,\middle|\, \bigcap_{i=1}^{i-1} (0^A_{k_i} \nsubseteq X_\varepsilon{}^*)\right\} \\ &\leqq (1 - p_5)^l, \quad l \geqq 2. \end{aligned}$$

Since (7) holds

$$(1 - p_5)^l \xrightarrow[l \to \infty]{} 0$$

and because of monotony

$$\{0^A_{k_1} \nsubseteq X_\varepsilon{}^*\} \supseteqq \left\{\bigcap_{i=1}^{2} (0^A_{k_j} \nsubseteq X_\varepsilon{}^*)\right\} \supseteqq \cdots \supseteqq \left\{\bigcap_{j=1}^{l} (0^A_{k_j} \nsubseteq X_\varepsilon{}^*)\right\}$$

it holds that

$$P\left\{\bigcap_{j=1}^{\infty} (0^A_{k_j} \nsubseteq X_\varepsilon{}^*)\right\} = \lim_{l \to \infty} P\left\{\bigcap_{j=1}^{l} (0^A_{k_j} \nsubseteq X_\varepsilon{}^*)\right\},$$

and consequently

$$P\left\{\bigcap_{j=1}^{\infty} (0^A_{k_j} \nsubseteq X_\varepsilon{}^*)\right\} = 0$$

i.e.

$$P\left\{\bigcup_{j=1}^{\infty} (0^A_{k_j} \subseteqq X_\varepsilon{}^*)\right\} = 1.$$

But this is in contradiction to (1), therefore the assumption of the theorem follows. *

The regularity conditions may be satisfied. This holds obviously for the conditions of the EGL method.

But the regularity conditions $(R_1)-(R_4)$ of problem (E) are also satisfied for example under the assumptions:

i. g is continous

ii. X is convex, compact and int $X \neq \emptyset$.

Conditions i and ii are given in many practical problems. Frequently the parameter set X is given in form of a parallelepiped:

$$X = \{x \in R^n \mid a_i \leqq x_i \leqq b_i, i = 1, \ldots, n\}, \quad \text{whereby} \quad a_i < b_i,$$
$$i = 1, \ldots, n$$

and it may be assumed that g is continous.

In the proof it was assumed that normally distributed mutations are realized in each case with positive probability (and minimal variance). The pure mutation-selection mechanism guarantees the theoretical convergence with probability one. The advantage of taking higher evolution principles into account is the increased practical efficiency, which will be demonstrated in the following paragraph.

3.3.4. On the efficiency of the EGL method

For practical application the EGL method was implemented in FORTRAN-IV, using numerical basic modules, available in BESM and ESER assembler (see Hollatz, Richter, 1978). All tests were computed on the computer BESM-6. For the programme system 662 words in the main storage of this machine are needed, the computing time needed to solve a problem of dimension $n = 10$, $m_0 = m_s = 10$ is about 0.015 sec. per iteration. The EGL method was tested in several variants and with different problems (comp. Born, 1978). The numerical tests were designed in such manner that the pure and combined effects of the different evolution factors could be investigated and their efficiency and global efficiency evaluated. For comparisons the (10, 100)-standard version of the MCE strategy (i.e. following the pattern outlined in 3.1. $m = 10$, $l = 100$, $r = 10$) in form of the subroutine GRUP (represented in Schwefel (1977) was used.

In the following we will discuss the results of the multimodal "oscillation problem"

$$X = \{x \in R^n \mid -500 \leqq x_i \leqq +500, \quad i = 1, \ldots, n\}$$
$$g(x) = -\sum_{i=1}^{n} x_i \cdot \sin\sqrt{|x_i|}, \tag{T}$$

serving as an example for a practical problem. A close look at the problem (T) shows:

Within X there exist 6^n local minima. For the components of the minimum points $\bar{x}$ it holds that

$$\bar{x}_i \cong (-1)^k \left(\pi \cdot (0{,}5 + k)\right)^2, \quad i = 1, \ldots, n,$$

whereby $k = 1, \ldots, 6$.
For the global minimum point $\bar{x}^0$ within x

$$\bar{x}_i^0 \cong 420.96876, \quad i = 1, \ldots, n$$

holds and

$$g(\bar{x}^0) = -418{,}98288 \cdot n.$$

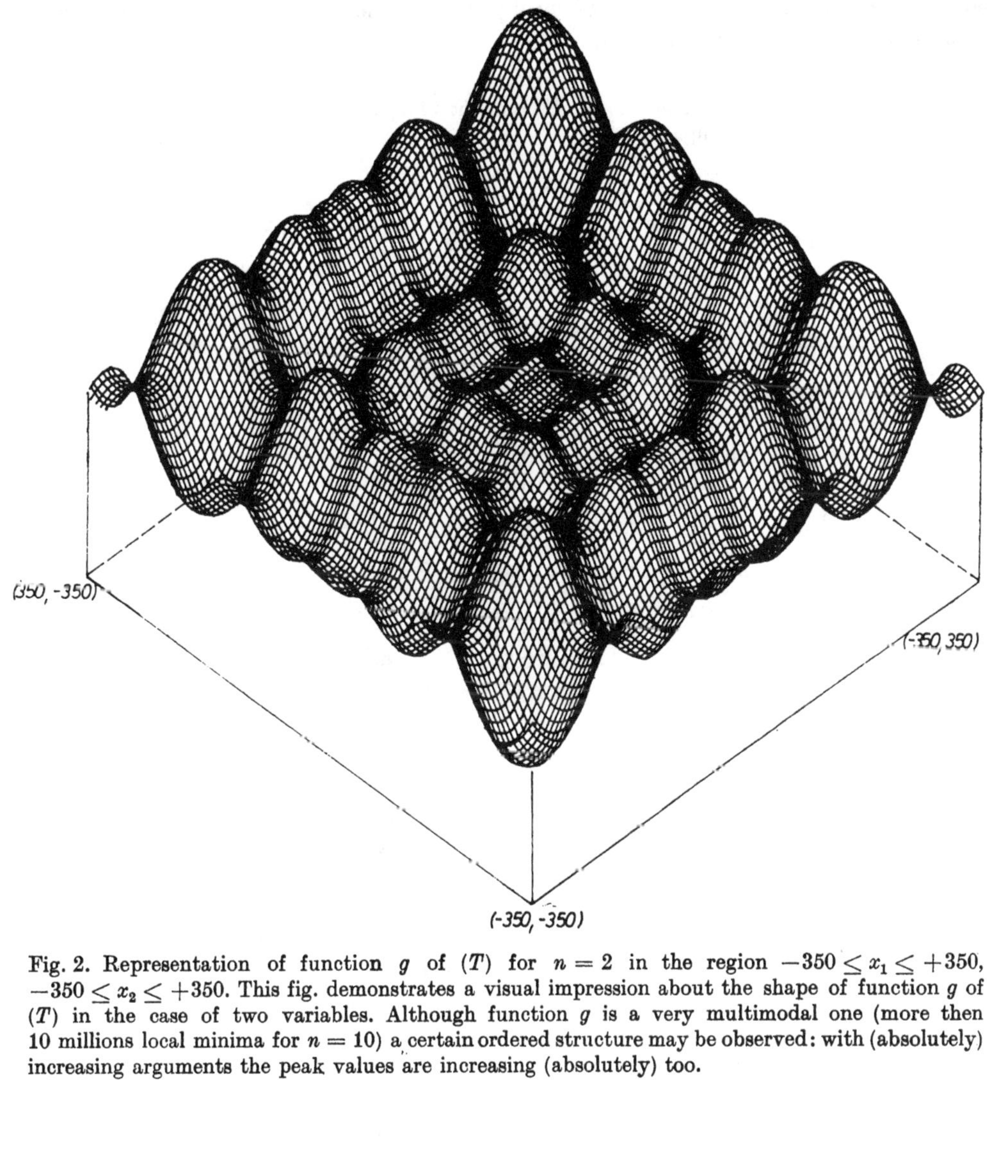

Fig. 2. Representation of function g of (T) for $n = 2$ in the region $-350 \leq x_1 \leq +350$, $-350 \leq x_2 \leq +350$. This fig. demonstrates a visual impression about the shape of function g of (T) in the case of two variables. Although function g is a very multimodal one (more then 10 millions local minima for $n = 10$) a certain ordered structure may be observed: with (absolutely) increasing arguments the peak values are increasing (absolutely) too.

For economic reasons in every test calculation the dimension $n = 10$ and a maximum of 5000 iterations were fixed (in the MCE strategy 50 iterations are equivalent to 5000 calculations of the quality functional). Within the restricted iteration nbr. of 5000 the problem (T) could be solved by the best variants of the EGL method with sufficient precision.

Holding the other factors constant, the effect of the factor population (population potency) is demonstrated by means of the object population in table 1. Two effects could be observed with increasing potency:

Table 1. Test results with the EGL method for the factor population potency (of the object population).
with:

$l_0 = m_0$ $\quad {}^{R}c_i = 0, \quad i = 1, \ldots, n + 4$
$l_s = m_s = 1$ $\quad {}^{V}c_i = 0, \quad i = 1, \ldots, n + 2$
$g_s^{\min} = 10^{18}$
${}^{R}s^{1}_{1,i} = 0.75, \quad i = 1, \ldots, n + 2$
${}^{V}s^{1}_{1,i} = 50, \quad i = 1, \ldots, n$
$x^{l}_{1,i} = \vartheta(0,50), \quad i = 1, \ldots, n, \; l = 1, \ldots, m_0$

m_0	k	$g_k^{\min}$	ϱ_k
1	0	−17.91	$1.64 \cdot 10^6$
	2000	−810.90	$2.21 \cdot 10^6$
	5000	−816.40	$2.25 \cdot 10^6$
50	0	−291.38	$2.12 \cdot 10^6$
	2000	−1916.61	$4.87 \cdot 10^6$
	5000	−2412.36	$4.82 \cdot 10^6$
100	0	−311.90	$1.53 \cdot 10^6$
	2000	−1032.13	$1.68 \cdot 10^6$
	5000	−1470.52	$2.62 \cdot 10^6$

With increasing potency the "inertia" of the population sequence is increased; a low efficiency follows. This is always the case in tests with unimodal quality functionals. Other authors also have found this effect and WEBER (1971) concludes that only a simulation of the principles of mutation and selection with only a single element makes sense.

This fact can demonstrated also with problem (T). For the relatively high potency $m_0 = 100$ a lower efficiency results than for $m_0 = 50$.

But another effect of the simulation of the factor population also becomes obvious in table 1 — the potentially greater variability.

In runs using a population potency of $m_0 = 50$ and $m_0 = 100$ significantly better qualities were observed than in the case with a single element ($m_0 = 1$).

This is an important effect of using a population rather than single elements even in the case of multimodal problems.

Here and in the following tables we use the notation

k — number of iterations
$g_k^{\min}$ — quality of a best element $x_k^{\min}$ of the k-th object population 0_k:

$$g_k^{\min} = g(x_k^{\min}) = \min \{g(x_k^i) \mid x_k^i \in 0_k, \quad i = 1, \ldots, m_0\}$$

ϱ_k — square of the Euclidian distance of $x_k^{\min}$ to the global minimum point $\bar{x}^0$:

$$\varrho_k = \sum_{i=1}^{n} (x_{k,i}^{\min} - \bar{x}_i^0)^2$$

Table 2. Test results with the EGL method for the factor spectra of variability. with:

$l_0 \quad = m_0 = 50$ $\qquad$ ${}^R c_i = 0, \quad i = 1, \ldots, n+4$
$l_s \quad = m_s = 1$ $\qquad$ ${}^V c_i = 0, \quad i = 1, \ldots, n+2$
$g_s^{\min} = 10^{18}$
${}^R s_{1,i}^1 = 0.75, \quad i = 1, \ldots, n+2$
${}^V s_{1,i}^1 = d, \quad i = 1, \ldots, n$
$x_{1,i}^l = \vartheta(0, d), \quad i = 1, \ldots, n, l = 1, \ldots, m_0$

d	k	$g_k^{\min}$	ϱ_k
0.05	0	−0.09	$1.77 \cdot 10^6$
	2000	−13.00	$1.76 \cdot 10^6$
	5000	−31.26	$1.76 \cdot 10^6$
5	0	−18.40	$1.75 \cdot 10^6$
	2000	−233.68	$1.99 \cdot 10^6$
	5000	−236.09	$2.00 \cdot 10^6$
50	0	−291.38	$2.12 \cdot 10^6$
	2000	−2073.67	$2.38 \cdot 10^6$
	5000	−2433.28	$2.15 \cdot 10^6$
500	0	−1149.09	$3.62 \cdot 10^6$
	2000	−2201.82	$4.48 \cdot 10^6$
	5000	−2341.05	$5.93 \cdot 10^6$

Table 3. Test results with the EGL method for the factors inversion and crossing over.
with:

$l_0 \quad = m_0 = 50$ $\qquad$ ${}^R c_i = 0, \quad i = 1, \ldots, n+4$
$l_s \quad = m_s = 1$ $\qquad$ ${}^V c_i = 0, \quad i = 1, \ldots, n+2$
$g_s^{\min} = 10^{18}$
${}^R s_{1,1}^1 = p_I, \quad {}^R s_{1,2}^1 = p_C$
${}^R s_{1,i}^1 = 0.75, \quad i = 3, \ldots, n+2$
${}^V s_{1,i}^1 = 50, \quad i = 1, \ldots, n$
$x_{1,i}^l = \vartheta(0{,}200), \quad i = 1, \ldots, n, l = 1, \ldots, m_0$

p_I	p_C	k	$g_k^{\min}$	ϱ_k
0	0	0	−907.62	$3.35 \cdot 10^6$
		2000	−1883.55	$2.22 \cdot 10^6$
		5000	−1885.19	$2.15 \cdot 10^6$
0	1	0	−907.62	$3.35 \cdot 10^6$
		2000	−2446.12	$1.10 \cdot 10^6$
		5000	−2446.97	$1.10 \cdot 10^6$
1	1	0	−907.62	$3.35 \cdot 10^6$
		2000	−4167.92	$1.73 \cdot 10^2$
		5000	−4174.30	$1.23 \cdot 10^2$

The results, displayed in table 1, are not satisfying, mainly due to the choice of the spectra of variability for the realization of the mutations (here maintained constant).

How important the "problem adapted" choice of the spectra of variances is, is shown in table 2.

It is evident too, that even in the case of suitable start variances (e.g. $d = 500$), in order to guarantee an effective first adaption period, the variances should be changed during the further run.

The effect of the inversion and crossing-over factors is demonstrated in table 3.

For generating the starting population a rel. large variance was chosen. An increased efficiency could be observed due to good exploitation of the given variability within the population through crossing-over. A combined realization of inversion and crossing-over brought the object population near to the global minimum point and consumed only 2000 iterations.

By means of the algorithm it is possible for each parameter to realize its own mutation probability. When in a practical problem a variation of one (or more) parameters is disadvantageous, the strategy population is able to react very quickly. Generally, the mutation probability should be chosen appr. equal to one. If the values of the parameters are not approximately equal, as in most cases, the probability for realizations of inversions must converge to zero in the final stage of the adaptation run. The probabilities for the realizations of crossing-over initially should be chosen appr. equal to one.

Tables 4 and 5 demonstrate the possibilities of utilization of the principle of genetic load.

Table 4. Test results with the EGL method for the factor genetic load.
with:

$l_s = 45$ $\qquad {}^Rc_1 = 0.5$
$m_s = 50$ $\qquad {}^Rc_i = 0.9, \quad i = 2, \dots, n+4$
$g_s^{\min} = 0$ $\qquad {}^Vc_i = 0.1, \quad i = 1, \dots, n+2$
${}^RS_1$-formation with ${}^Ra = (1; 0.8; 0.8; 0.5; 0.1)$*
${}^VS_1$-formation with ${}^Va = (500; 50; 5; 0.5; 0.05)$
0_1-formation with ${}^\circ a$

test	l_0	m_0	${}^\circ a$	k	$g_k^{\min}$	ϱ_k
4.1	45	50	(0; 50; −50;	0	−1805.89	$8.48 \cdot 10^6$
			500; −500)	2000	−4175.65	$1.12 \cdot 10^2$
				5000	−4189.82	$2.31 \cdot 10^{-2}$
4.2	45	50	(0; 5; −5;	0	−354.43	$1.37 \cdot 10^6$
			50; −50)	2000	−4186.35	$2.75 \cdot 10^1$
				5000	−4189.828	$4.91 \cdot 10^{-3}$
4.3	45	50	(0; −25;	0	0.0	$1.77 \cdot 10^6$
			−50; −250;	2000	−4070.13	$9.52 \cdot 10^2$
			−500)	5000	−4189.81	$9.11 \cdot 10^{-1}$
4.4	1	1	(0)	0	0.0	$1.77 \cdot 10^6$
				2000	−2864.21	$2.61 \cdot 10^6$
				5000	−3663.89	$1.60 \cdot 10^6$

* Let be $P \in R^n \times R^{q \cdot m}$ and $e \in R^m$. Then denotes P-formation with e:

$$P_{i,j \cdot k} = e_j, \quad i = 1, \dots, n; \quad j = 1, \dots, m; \quad k = 1, \dots, q$$

Table 5. Test results with the EGL method for factor genetic load. with:

$m_0 = m_s = 50$ $\quad$ ${}^{R}c_1 = 0.5$

$g_s^{\min} = 0$ $\quad$ ${}^{R}c_i = 0.9, \quad i = 2, \ldots, n + 4$

${}^{V}c_i = 0.1, \quad i = 1, \ldots, n + 2$

${}^{R}S_1$-formation with ${}^{R}a = (1; 0.8; 0.8; 0.5; 0.1; 1; 0.8; 0.8; 0.5; 0.1)$

${}^{V}S_1$-formation with ${}^{V}a = (5 \cdot 10^2; 5 \cdot 10^1; 5 \cdot 10^0; 5 \cdot 10^{-1}; 5 \cdot 10^{-2}; 5 \cdot 10^{-3}; 5 \cdot 10^{-4}; 5 \cdot 10^{-5}; 5 \cdot 10^{-6}; 5 \cdot 10^{-7})$

0_1-formation with ${}^{o}a = (0; 50; -50; 500; -500)$

test	l_0	l_s	k	$g_k^{\min}$	ϱ_k
5.1	45	40	0	−1805.89	$8.48 \cdot 10^6$
			2000	−4142.92	$3.73 \cdot 10^2$
			5000	−4189.828	$7.83 \cdot 10^{-4}$
5.2	50	50	0	−1805.89	$8.48 \cdot 10^6$
			2000	−4186.43	$2.62 \cdot 10^1$
			5000	−4189.79	$2.91 \cdot 10^{-1}$

The genetic loads, the starting populations and all free parameters are fixed appropriate to a practical problem (tests 4.1 and 5.1). Generally, the genetic load in the object population should be chosen such that the available heuristic information about possible solution areas is taken into account.

The genetic load in the strategy population of the probabilities had been chosen such that an actual choice of the probabilities was possible between 0.1 and 1. The genetic load in the strategy population for variability spectra was fixed such that (i) the whole area is covered and (ii) a reservoir of extremly small variances exists, which is needed especially in the final phase of the adaptation process near the minimum and is determined by the desired precision of the solution.

In all tests for the components of the parameter vectors ${}^{V}c$ for the variability spectra used in realizing mutations in the strategy populations the factor 0.1 was chosen to guarantee that the actual strategy populations change only within the spectrum determined by the corresponding genetic loads. In this way the variability was reduced, but only in a sensible degree.

The results demonstrate the possibility of increasing the efficiency and especially the global efficiency very strongly by introducing the principle of genetic load into the evolution strategy.

The importance of the factor population (i.e. of being concerned with a set of elements and not with a single one) is demonstrated by test 4.4., which indicates that in the case of considering a single element ($m_0 = 1$) the global minimum could not be attained.

On the one hand the use of the principle of genetic load leads to a maintainance of variability within the complete populations; on the other hand in the presence of a sufficiently great population potency, a sufficient stability of the actual populations is guaranteed. See e.g. test 4.3. whereby even in the case of extremly insufficient genetic load within the object population satisfying results can be observed due to high population potency. Tests 5.1. and 5.2. demonstrate the effect of the factor of genetic load within the strategy population. In both tests the start conditions are the same. In test 5.1. a relatively large genetic load with very small variances is maintained, while in test 5.2. (without genetic load) the result after 5000 iterations was not satisfying compared

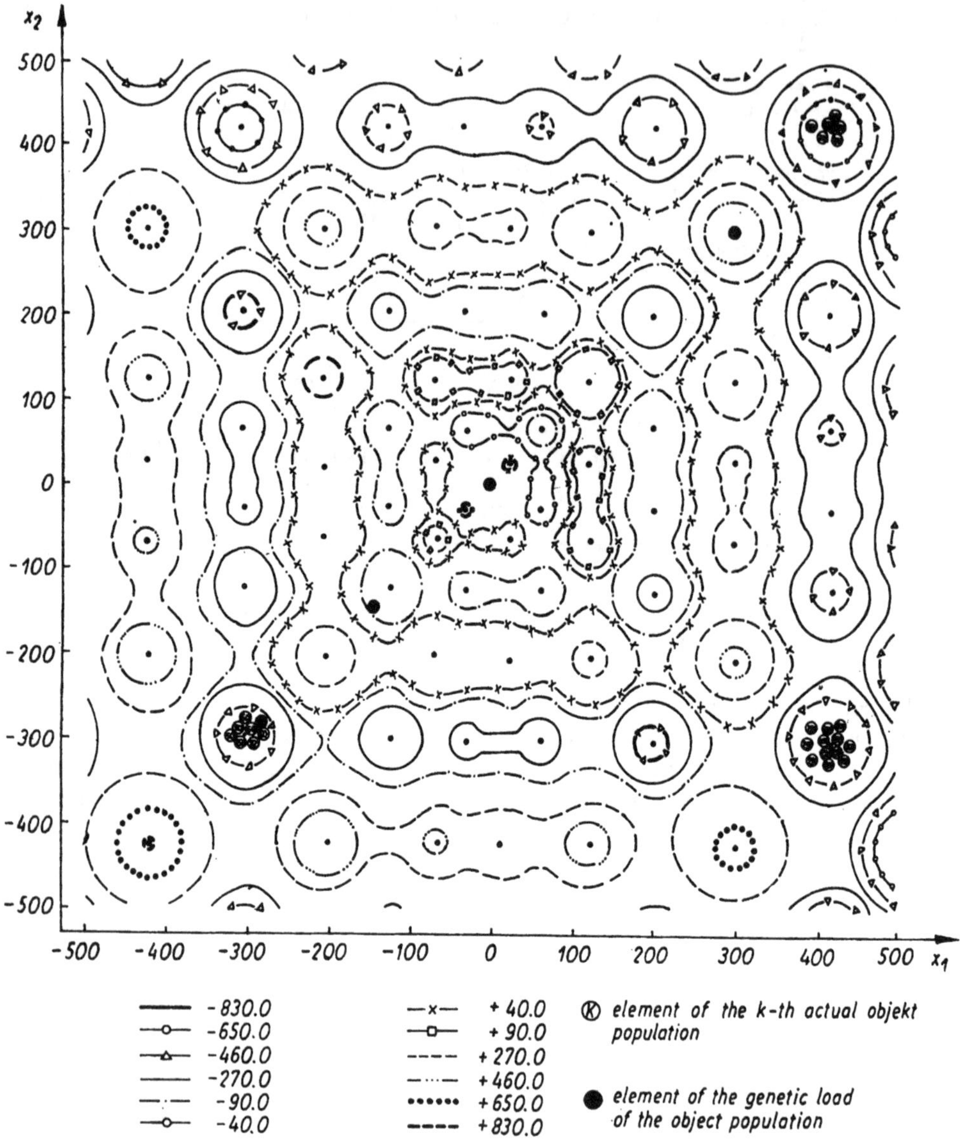

Fig. 3. Example of an adaption run for solving (*T*) realized by the EGL method with $n = 2$.

and:

$m_s = 20$ $\quad$ ${}^{R}c_1 = 0.5$

$m_0 = 20$ $\quad$ ${}^{R}c_i = 0.9, i = 2, \ldots, n + 4$

$l_s = 10$ $\quad$ ${}^{V}c_i = 0.1, \mathrm{i} = 1, \ldots, n + 2$

$l_0 = 15$

$g_s^{\min} = 0$

${}^{R}S_1$-formation with ${}^{R}a = (1; 0.8; 0.8; 0.5; 0.1; 1; 0; 8; 0; 8; 0.5; 0.1)$

${}^{V}S_1$-formation with ${}^{V}a = (5 \cdot 10^2; 2 \cdot 10^2; 5 \cdot 10^1; 2 \cdot 10^1; 5 \cdot 10^0; 2 \cdot 10^0; 5 \cdot 10^{-1}; 2 \cdot 10^{-1}; 5 \cdot 10^{-2}; 2 \cdot 10^{-2})$

0_1-formation with ${}^{\circ}a = (0; 150; -150; 300; -300)$.

This fig. demonstrates the drift of the actual object population, i. e. in which manner this population walks to global minimum. After 50 iterations the population is going to contract at a local minimum close by (−300, −300). At the 150-th iteration two elements are found at a better local minimum close by (+420, −300). Until 250 iterations the population is located completely in the valley of the global minimum close by (+420, +420).

with the result of test 5.1. Only the early phases of the run test 5.2. showed a slightly better result, due to an early discarding of small variances in the strategy population.

In test 5.1. (with genetic load) in the final phase, when the object population was near the global minimum, the increasingly smaller variances needed here were realized instantly.

Naturally the evolution strategies as special search algorithms depend on the starting values; this dependency is taken into account ingeniously by introducing the factor genetic load, as shown in a comparison with results obtained by SCHWEFEL's MCE strategy (see table 6).

Table 6. Test results with the MCE strategy.

test	x	s	k	$g_k^{\min}$	ϱ_k	$g^{\min}$	ϱ
6.1	0	$5 \cdot 10^{-2}$	50	−39.45	$1.72 \cdot 10^6$	−39.45	$1.72 \cdot 10^6$
6.2	0	$5 \cdot 10^{-1}$	50	−39.45	$1.72 \cdot 10^6$	−39.45	$1.72 \cdot 10^6$
6.3	0	$5 \cdot 10^{0}$	50	−280.38	$1.92 \cdot 10^6$	−280.38	$1.92 \cdot 10^6$
6.4	0	$5 \cdot 10^{1}$	50	−3104.11	$1.96 \cdot 10^6$	−3104.11	$1.96 \cdot 10^6$
6.5	0	$5 \cdot 10^{2}$	50	−3854.21	$5.70 \cdot 10^5$	−3854.21	$5.70 \cdot 10^5$
6.6	0	$5 \cdot 10^{2}$	100	−3854.32	$5.70 \cdot 10^5$	−3854.32	$5.70 \cdot 10^5$
6.7	420	$5 \cdot 10^{2}$	20	−4185.43	$3.54 \cdot 10^1$	−4188.61	$9.38 \cdot 10^0$
6.8	420	$5 \cdot 10^{2}$	50	−4189.82	$2.39 \cdot 10^{-4}$	−4189.82	$2.35 \cdot 10^{-4}$

The following notations are used here (see SCHWEFEL, 1977, p. 362):

x — is the formation vector for the starting object population 0_1, i.e.

$$0_1 = (x_1^1, \ldots, x_1^m)$$

is generated stochastically by

$$x_1^l = \vartheta(x, s), \quad l = 1, \ldots, m;$$

s — is the formation vector for the starting strategy population

$$S_1 = (s_1^1, \ldots, s_1^m)$$

is generated stochastically by

$$s_1^l = \vartheta(s, \bar{s})$$

whereby here the standard value

$$\bar{s} = n^{-\frac{1}{2}}$$

is used.

$g^{\min}, \varrho$ — are, resp., the quality and the distance to the global minimum point of the best element $x^{\min}$ (corresponding to $g_k^{\min}$, ϱ_k, but:) over all iterations:

$$x^{\min} = \min_k \{x \mid x \in 0_k\}.$$

Then main feature of the MCE strategy is the increasing reduction of the variances during the run, consequently in the case of overly small starting variances (tests 6.1. to

6.4.) only local minima could be found; but also in using sufficiently great starting variances (tests 6.5.—6.6.) the desired global minimum was not reached.

Already after 50 and to a greater extend after 100 iterations the variances proved too small, so that there was no chance of finding the global minimum even in the case of adding a large number of iterations. Here the lack of permanently sufficient variability in the strategy population, essential for further progress, turnes out to be disadvantageous. In test 6.7. the negative effect of ignoring the principle of reproduction in the MCE strategy is shown. Unfortunately, the very suitable starting parameter vector in the object population is already lost after the first iterations and the quality of the starting population could not be reached even after further 20 iterations.

4. Application of the EGL method to a parameter adaptation problem in the transcription simulation model EUTRANSCRIPT 2

The EGL method was used to adapt variable parameters of the transcription simulation model EUTRANSCRIPT 2 mentioned in paragraph 5 of the paper BELLMANN, BORN, LINDIGKEIT, SCHULZ, in this volume.

For the adaptation problem the reduced new algorithmic version was used, based on polynominals and consequently being a very fast simulator (0.15 sec. is needed for calculating a single quality $g(y(x), y^0)$), Here, the molecular genetic background is considered only in so far as it is concerned with special numerical features of the arising adaptation problem.

In particular, the results supporting or disconfirming different hypotheses about the real process of eucaryotic transcription are not evaluated (this was done in paragraph 5 in BELLMANN et al. in this volume).

The reduced version EUTRANSCRIPT 2 was elaborated to investigate the shapes of the blocker removal rates $F2_j$ in order to get some implications for explaining the real process, as outlined in paragraph 4 in BELLMANN et al.

The shapes of the functions $F2_j$ are desribed by third degree polynomials with minimal level, whose coefficients are to be determined. The conceptual framework of EUTRANSCRIPT 2 is characterized by a reservoir floodgate system with a medium flowing through it.

The flow depends on the permeability of the single floodgates. Consequently, the object terms polymerase molecules, blocker sites, blocker removal rates are replaced by the corresponding terms medium, floodgates, permeability of a floodgate resp. (see paragraph 5 in BELLMANN et al.).

In EUTRANSCRIPT 2 a system of blockers located sequentially is considered: $B_j, j = 1, \ldots, b$.

It is assumed that at instant t_0 the blocker system is "empty". At the time steps $t_\tau, \tau = 1, \ldots, T$ polymerase molecules flow into the system with known and constant velocity. In the model a time interval corresponds to the time needed by a polymerase molecule to move (unrestricted) a certain distance (length of a S-unit U, see paragraph 2.3.1. in BELLMANN et al.).

It is assumed

i. that the location A_j of the blocker sites B_j, $j = 1, \ldots, b$, is known, expressed in terms of time intervals, and

ii. that the flow of the polymerase through the system is controlled only by the blocker effects, wich inhibit polymerase movement.

The variable ("soft") parameters to be identified (i.e. to be adapted) are:

i. the parameters x^0 of the inflow function $F2_0(t_\tau)$ $\tau = 1, \ldots, T$, which control the amount $a(t_\tau)$ of polymerases flowing into the system at the (initiation/promotion) site B_0 are considered as the source of the system and
ii. the parameters x^j of the functions $F2_j(t_\tau)$, describing the permeability of the floodgates (i.e. blockers) B_j at instant t_τ; $\tau = 1, \ldots, T$; $j = 1, \ldots, b$. The amounts $y_j(t_\tau)$ of the polymerases "waiting" at the instants t_τ in the reservoirs of the floodgates $B_j(\tau = 1, \ldots, T; j = 1, \ldots, b)$ are calculated:

$$y_j(t_\tau) = 0 \quad \tau = 1, \ldots, A_j; \quad j = 1, \ldots, b \tag{A}$$

$$y_1(A_1 + 1) = a(t_1) \tag{B1}$$

$$y_j(A_j + 1) = F2_{j-1}(A_{j-1} + 1) \cdot y_{j-1}(A_{j-1} + 1); \quad j = 2, \ldots, b \tag{B2}$$

$$y_j(A_j + \tau + 1) = y_j(A_j + 1) \tag{C1}$$

$$+F2_{j-1}(A_{j-1} + \tau + 1) \cdot y_{j-1}(A_{j-1} + \tau + 1) \tag{C2}$$

$$-F2_j(A_j + \tau) \cdot y_j(A_j + \tau), \tag{C3}$$

$$\tau = 1, \ldots, T - A_j - 1; \quad j = 1, \ldots, b.$$

A detailed explanation of $(A)\cdots(C3)$ is given in paragraph 5 in BELLMANN et al.

The dimensions in the model EUTRANSCRIPT 2 are $b = 12$, $T = 241$ (corresponding to 30 min real time).

The inflow rate a into the system and the floodgate (blocker) permeability functions $F2_j$, $j = 1, \ldots, b$ are described by polynominals of the third degree with a certain maximal and minimal level:

$$a(t_\tau) = F2_0(t_\tau) = \max\,(x_1^0 \cdot \tau + x_2^0 \cdot \tau^2 + x_3^0 \cdot \tau^3, x_4^0), \quad \text{with } x_4^0 \geqq 0, \tau = 1, \ldots, T \tag{D1}$$

$$F2_j(t_\tau) = \min\,\{\max\,(x_1^j \cdot \tau + x_2^j \cdot \tau^2 + x_3^j \cdot \tau^3, x_4^j), 1\}, \quad \text{with } x_4^j \geqq 0, \tau = 1, \ldots, T, j = 1, \ldots, b. \tag{D2}$$

The adaptation problem with a structure described by (A), as mentioned above, arises as follows:

The parameter set X is given by

$$X = \{x \in R^{52} \mid x_{4 \cdot j} \geqq 0, j = 1, \ldots, 13\};$$

a parameter vector $x \in X$ is formed by

$$x = (x^0, x^1, \ldots, x^j, \ldots, x^{12})$$

with

$$x^0 = (x_1^0, x_2^0, x_3^0, x_4^0)$$

and

$$x^j = (x_1^j, x_2^j, x_3^j, x_4^j), \quad j = 1, \ldots, 12.$$

For each parameter vector $x \in X$ an output vector $y(x) \in R^{36}$ is calculated by the algorithm consisting of the amounts $y_j(t_{\tau_i}, x)$ within the reservoirs of the blockers B_j, $j = 1, \ldots, 12$, at 3 instants t_{τ_i}, $i = 1, 2, 3$; the 3 instants correspond to $\tau = 41, 161, 241$ and to the real times 5, 20, 30 min of the sedimentation experiment that yielded observed values at these instants: $y_j^0(t_{\tau_i})$, $j = 1, \ldots, 12$, $i = 1, 2, 3$ (see fig. 2 in BELLMANN et al.). As quality functional for the degree of correspondence between the calculated

$y_j(t_{\tau_i}, x)$ and observed $y_j^0(t_{\tau_i})$, $j = 1, \ldots, 12$, $i = 1, 2, 3$, a weighted absolute sum with weights $w_j(t_{\tau_i}) \geqq 0$, $j = 1, \ldots, 12$, $i = 1, 2, 3$ was used:

$$\inf \left\{ \sum_{j=1}^{12} \sum_{i=1}^{3} w_j(t_{\tau_i}) \, |y_j(t_{\tau_i}, x) - y_j^0(t_{\tau_i})| \;\middle|\; x \in X \right\} \qquad (A_{EU}).$$

Since one is concerned with uncertainty of y^0 and the fixation of some parameters (e.g. the assumption of constant velocity of the polymerases) and since one has chosen very simple polynomial functions to describe the blocker permeabilities, an exact solution of the problem (A_{EU}) was a priori not envisaged, but the following requirements were desired:

(R1) Determination of such parameter vectors $x \in X$, i.e. of such types of inflow functions $F2_0$ and blocker site permeability functions $F2_j$, $j = 1, \ldots, 12$, that lead to maximal 20% total deviation between the $y_j(t_{\tau_i})$ and $y_j^0(t_{\tau_i})$, $j = 1, \ldots, 12$, $i = 1, 2, 3$.

(R2) Thereby, between the calculated $y_j(t_{\tau_i})$, $j = 1, \ldots, 12$, $i = 1, 2, 3$, the same relation should hold as between the observed ones.

(R3) The result of the adaptations runs should allow an evaluation of hypotheses about the inflow (initiation process of polymerases) and about the deblocking processes, i.e. about the time dependent blocker permeability, in accordance with (R1) and (R2).

The requirements (R1), (R2), (R3) and the structure of the problem (A_{EU}) with the expection of the existence of more than one local extremum to lead to the application of the EGL method. Here the following variants and modifications were used.

(M1) The population potencies, were chosen relatively large: $m_0 = 60$ with $l_0 = 50$ and $m_s = 40$ with $l_s = 30$.

(M2) The "functional units".

$$x^0, x^j, j = 1, \ldots, 12$$

within the parameter vectors $x \in X$ are considered:

i. The parameter vectors of the genetic load of the object population 0 were chosen such that they covere a broad spectrum of hypothetical shape types of the inflow rate $F2_0$ and of the permeability functions $F2_j$, $j = 1, \ldots, 12$, of the floodgates (blockers B_j), see figures 4 and 5, taking into account especially (R3):

ii. The realization of crossing over was modified as follows:

— a crossing-over is always involved with complete functional units x^0, x^j, $j = 1, \ldots, 12$ (that means, no boundary of a crossing-over section is in the interior of a functional unit)

— a crossing-over may also occur between nonhomologous sectors of the parameter vector x, in contradiction to the genetic background (that means, a crossing-over for two parents $\bar{x}$, $\bar{\bar{x}}$ with sections $\bar{x}_{j_1}$ and $\bar{\bar{x}}_{j_2}$ with $j_1 \neq j_2$ is possible). But here this may be advantageous because of the same qualitative "meaning" of all functional units.

(M3) The genetic loads within the populations were established in such manner that the EGL method was regular.

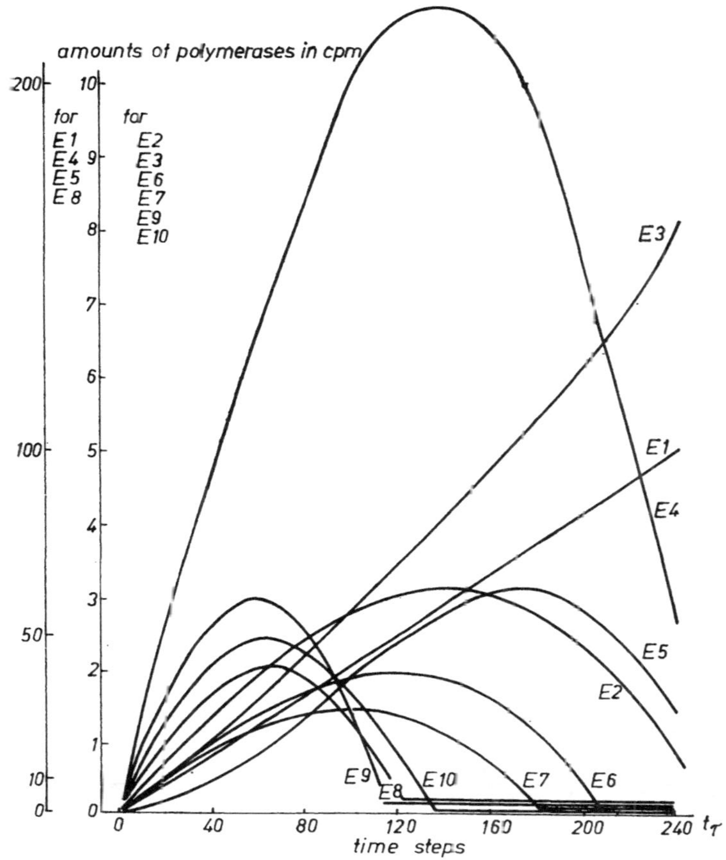

Fig. 4. Types of inflow functions.

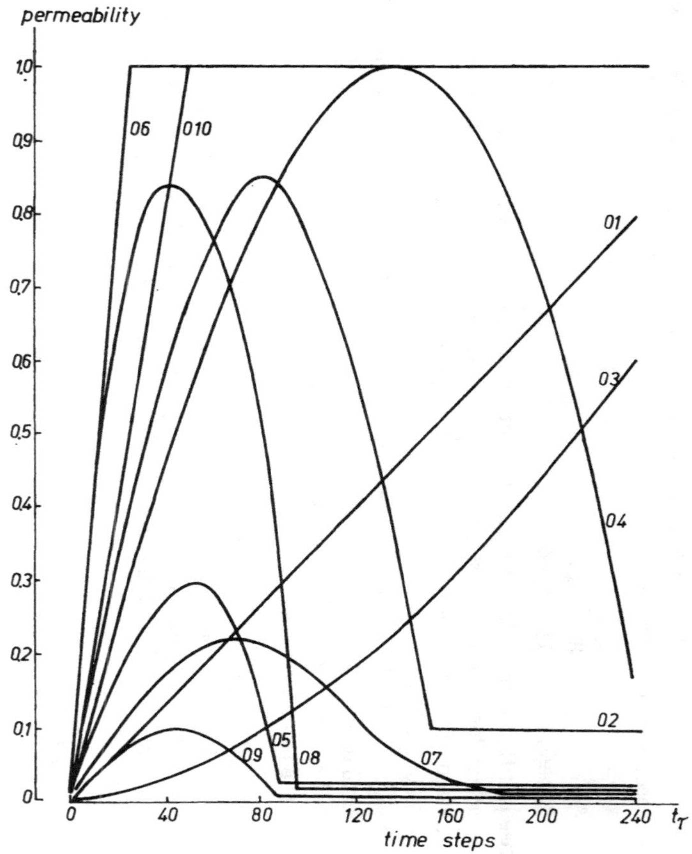

Fig. 5. Types of blocker permeability functions.

i. Furthermore, one element ${}^{R}s$ of the genetic load of the strategy population ${}^{R}S^{L}$ was chosen in such manner:

$$
\begin{aligned}
{}^{R}s_i &= 0 \quad i = 1, 2 \\
{}^{R}s_i &= 1 \quad i = 3, \ldots, 10 \\
{}^{R}s_i &= 0 \quad i = 11, \ldots, 54.
\end{aligned}
$$

Consequently, a preferential realization of mutations only within the functional units x^0 (inflow rate) and x^1 (floodgate permeability of the first blocker) was guaranteed, since the parameters x^0 and x^1 proved to be mainly governing the adaptation quality.

ii. The genetic loads within the strategy population ${}^{V}S$ were established in such manner that both the variability of the genetic load in the object population was exploited completely, and additionally a variability within the finite field of numbers of the computer was ensured.

In the fig. 4 and 5 the shape types are represented that are included in the genetic load of the object population. By means of many computations of different variants with these parameter vectors a special shape type proved to be a more and more suitable one for the inflow rate and the permeability functions of the blocker sites. It is a

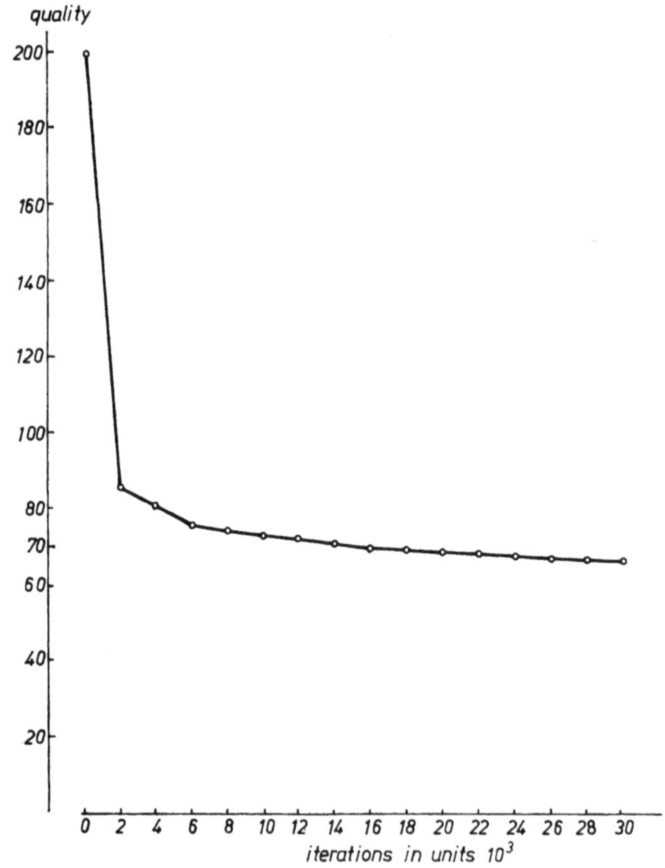

Fig. 6. Qualities of the resp. best parameter vectors of the populations.

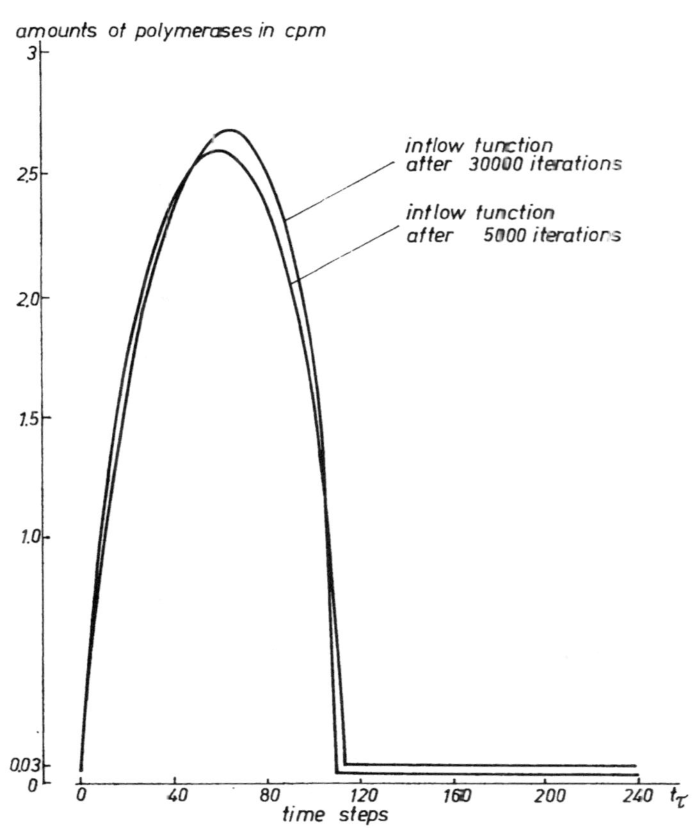

Fig. 7. Type of inflow function determined of the resp. best parameter vectors in the populations after 5000 resp. 30000 iterations.

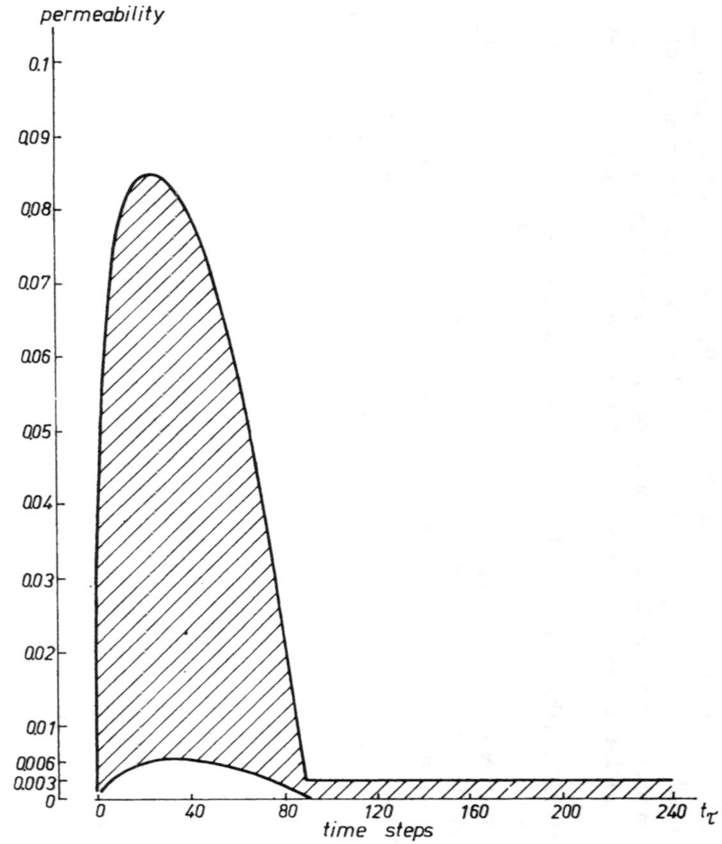

Fig. 8. Range of blocker permeabilities.

type like $E6$, $E8$, $E9$, $E10$ in fig. 4 and 02, 05, 07, 08, 09 in fig. 5, called "optimum function and minimal level" (OFM type). In every case during the adaptation runs all other function types were rejected in increasing the adaptation quality. The best qualities were obtained in the initial step in such cases where the functions $F2_0$ and $F2_1$ were of the OFM type.

In further runs the OFM type proved to be in correspondence with the real system for the other blocker sites $B_2 \cdots B_7$ and later on for $B_8 \cdots B_{12}$ too. In fig. 6 the resp. best adaptation qualities observed during the interations are shown. The requirements ($R1$) and ($R2$) could be considered satisfied after 5000 iterations.

In general, the shape of the inflow function was determined by the adaptation processes as shown in fig. 7, i.e. the initiation/promotion activity of the transcription system increases strongly within the first 7.5 min after the start of the transcription experiments; after that it decreases to a low level, maintained until the end of the experiment.

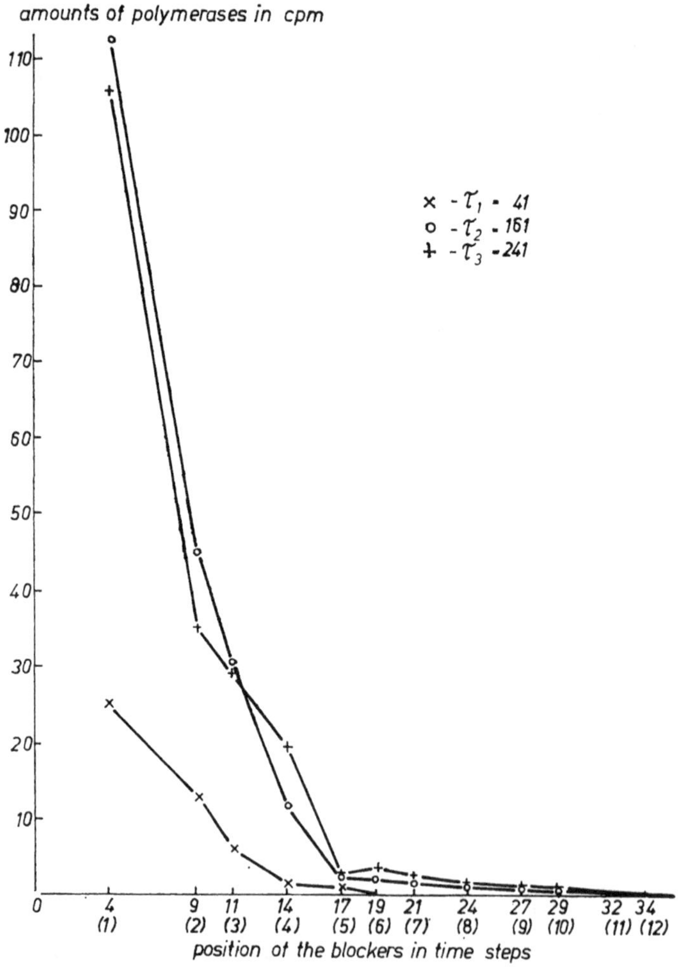

Fig. 9. Distribution of amounts of polymerases for the best parameter vector after 30000 iterations.

In fig. 8 the area of the adapted blocker permeability rates $F2_1, \ldots, F2_{12}$ is shown, which governs the further movement of polymerases waiting at the blockers $B_1, \ldots, B_{12}$, i.e. controls the outflow out of the blocker reservoirs.

Fig. 8 corresponds with the results of the unreduced model EUTRANSCRIPT 1, discussed and interpreted on the molecular genetic background in paragraph 4 in BELLMANN et al. see especially fig. 16b, p. 260.

All runs were stopped after 30000 iterations since progress was made only very slowly then.

Fig. 9 corresponds to fig. 2 in BELLMANN et al. and indicates a very satisfying correspondence between the relative amounts of polymerases at the blocker sites B_j, $j = 1, \ldots, 12$ at the different instants 5, 20, 30 min.

The results of the adaptation investigations support the hypothesis that in the real system both the inflow rate and the blocker permeability functions (responsible for the speed of polymerase movement through the whole system) are of a type with positive tendency for 7 to 8 min and after that the process declines and dies away slowly. A further improvement of the adaptation quality towards zero is expected by a correction of the restrictions given by ($D1$) and ($D2$), e.g. by means of replacing the polynomials by an appropriate exponential function, leading to an asymptotic decline of the inflow rate $F2_0$ and the blocker permeability functions $F2_j$, $j = 1, \ldots, 12$.

References

ARCHETTI, F., M. CUGIANI (1980): Numerical techniques for stochastic systems. North-Holland, Amsterdam, New York, Oxford.

BELLMANN, K., J. BORN, R. LINDIGKEIT and R. SCHULZ (1982): Modelling and simulation in analysis of eucaryotic transcription control. In this volume.

BELLMANN, K., and J. BORN (1979): Numerical solution of adaptation problems by means of an evolution stategy. In: Modelling and optimization of complex systems. Proceedings of the IFIP-TC 7 working conference, Novosibirsk, Lecture Notes in Control and Information Sciences Vol. 18, 1978. Springer, Berlin, Heidelberg, New York, 157–167.

BREMERMANN, H. J. (1968): Numerical optimization procedures derivad from biological evolution processes. In: Cybernetic problems in bionics. Gordon and Breach, New York.

BREMERMANN, H. J. (1970): A method of unconstrained global optimization. Math. Biosci. 9, 1–15.

BORN, J. (1978): Evolutionsstrategien zur numerischen Lösung von Adaptationsaufgaben. Dissertation (A), Humboldt-Universität zu Berlin.

DEMYANOV, V. F., and A. M. RUBINOV (1970): Approximate methods in optimization problems. Amer. Elsevier, New York.

DIXON, L. C. W., and G. P. SZEGÖ (Eds.) (1975, 1978): Towards global optimization vol. 1 (1975), vol. 2 (1978). North-Holland, Amsterdam, New York, Oxford.

FEISTEL, R. (1977): Zur Beschreibung von Evolutionsprozessen bei diploider Sequenzreplikation. Private communication.

HALDANE, J. B.S. (1957): The cost of natural selection. J. Genet. 55, 511–524.

HOCK, A., J. RINDERLE (1980): Zur Anwendung der Evolutionstrategie auf Schaltungen der Nachrichtentechnik. Frequenz 34, no. 7, 208–214.

HOLLAND, J. H. (1976): Adaptation. In: Progress in theoretical biology vol. 4. Academic Press, New York, San Francisco, London, 263–293.

HOLLATZ, H., J. BORN and K. BELLMANN (1977): A reservoir floodgate system. Working paper (unpublished).

HOLLATZ, H., D. RICHTER (1978): Programmoduln zur Erhöhung der Genauigkeit und Effektivität numerischer Algorithmen. ZfR — Informationen — 78.03, Berlin.

JACQUARD, A. (1974): The genetic structure of populations. Biomathematics vol. 5. Springer, Berlin, Heidelberg, New York.

KARMANOV, V. (1975): Mathematical programming (russian). Nauka, Moscow.

KIMURA, M., and T. OHTA (1974): The average number of generations until fixation of mutant gene in a finite population. Genetics **61**, 763—771.

KING, J. L., and T. H. JUKES (1969): Non-Darwian evolution: random fixation of selectively neutral mutations. Science **164**, 788—798.

PETERSOHN, U., K. VOSS, K. H. WEBER (1475): Genetische Adaptation — ein stochastisches Suchverfahren für diskrete Optimierungsprobleme. Math. Operationsforschung und Statistik 5, 7/8, 555—571.

RASTRIGIN, L. A. (Ed.) (1974): Problems of the random search vol. 3 (russian). Zinatne, Riga.

RECHENBERG, I. (1973): Evolutionsstrategie: Optimierung technischer Systeme nach Prinzipien der biologischen Evolution. Frommann-Holzboog, Stuttgart.

SCHIEMANGK, Ch. (1981): Anwendung einer Evolutionsstrategie zum Auffinden eines optimalen Subgraphen. In: 4. Fachtagung „Numerische Realisierung mathematischer Modelle, Zingst, 21.—25. 9. 1981, ZfR-Informationen-81.16, Berlin.

SCHNEIDER, B., and U. RANFT (Eds.) (1978): Simulationsmethoden in der Medizin und Biologie. Workshop Hannover 1977. Springer, Berlin, Heidelberg New York.

SCHWEFEL, H. P. (1977): Numerische Optimierung von Computer-Modellen mittels der Evolutionsstrategie. Interdisciplinary systems research vol. **26**. Birkhäuser, Basel and Stuttgart.

TRÄGER, L. (1975): Einführung in die Molekularbiologie. Fischer, Jena.

WEBER, K. H. (1971): Struktur und Optimierung eines adaptiven Klassifizierungsalgorithmus. Dissertation (A), TU Dresden.

Subject Index

U

V

W

Z

www.ingramcontent.com/pod-product-compliance
Lightning Source LLC
LaVergne TN
LVHW081330011026
849161LV00013BB/2885